高等学校土木工程专业国际化人才培养系列教材

结构力学(双语)

Structural Mechanics

黄　亮　编　著
宁永胜　刘应扬　许世展　副主编

中国建筑工业出版社

图书在版编目（CIP）数据

结构力学 = Structural Mechanics：汉、英 / 黄亮编著；宁永胜，刘应扬，许世展副主编. — 北京：中国建筑工业出版社，2024.4
高等学校土木工程专业国际化人才培养系列教材
ISBN 978-7-112-29746-7

Ⅰ. ①结… Ⅱ. ①黄… ②宁… ③刘… ④许… Ⅲ. ①结构力学－高等学校－教材－汉、英 Ⅳ. ①O342

中国国家版本馆 CIP 数据核字(2024)第 073435 号

本书依据我国高等教育国际化思路，由作者总结多年本科教学经验和教学改革成果而成，服务于结构力学教学的国际化需求。内容包括平面体系的几何构造分析、静定梁与静定刚架、静定拱、静定平面桁架、结构位移计算、力法、位移法、渐近法和影响线。本书编写过程中，注重基础性、创新性和应用性，中英对照，图文并茂，可兼顾中、外学生的教学需求。本书可作为高等学校土建、水利、力学等专业的教材，也可供有关工程技术人员参考。

本书配备教学课件，请选用此教材的教师通过发送邮件 jckj@cabp.com.cn 或拨打电话：(010) 58337285 获取。

Based on the internationalization of higher education in China, many years of undergraduate teaching experience and the teaching reform results, this book can be satisfied with the internationalization needs of structural mechanics teaching. The content of this book includes the geometric construction analysis of structures, statically determinate beams and rigid frames, statically determinate arches, statically determinate plane trusses, structural displacement calculations, force methods, displacement methods, asymptotic methods, and influence lines. Focused on foundation, innovation and application, this book can meet the teaching needs of both Chinese and foreign students because this book is written in Chinese and English with the picture accompanying essay. This book can be used as textbook of civil engineering, water conservancy, and mechanics in higher education institutions, as well as the reference of relevant engineering and technical staff.

This book is equipped with teaching powerpoints. Teachers that choose this book as textbooks can send e-mail jckj@cabp.com.cn or phone (010) 58337285 for this.

责任编辑：赵　莉　王美玲
责任校对：芦欣甜

高等学校土木工程专业国际化人才培养系列教材
结构力学（双语）
Structural Mechanics
黄　亮　编　著
宁永胜　刘应扬　许世展　副主编
*
中国建筑工业出版社出版、发行（北京海淀三里河路 9 号）
各地新华书店、建筑书店经销
北京红光制版公司制版
天津安泰印刷有限公司印刷
*
开本：787 毫米×1092 毫米　1/16　印张：21¾　字数：538 千字
2024 年 4 月第一版　2024 年 4 月第一次印刷
定价：**68.00** 元（赠教师课件）
ISBN 978-7-112-29746-7
(42666)

前　言

随着我国高等教育的不断国际化，中国学生和来华留学生对外语教材均有需求，而结构力学知识体系的难度较高，为了更好地理解结构力学概念，掌握结构力学解题思路，作者总结了多年教学教改经验，并结合来华留学生的教学实践总结，完成了本书。

本书的特色亮点包括：

1. 本书以中英双语形式呈现，旨在满足广大读者的不同语言需求，提供更便捷的学习体验。

2. 本书提供了大量的推导插图，本着“一图胜千言”的原则，将概念与思路用插图的形式呈现，使读者有直观的认识。

3. 本书吸取了近年结构力学教学经验总结，丰富了结构力学教学思路，例如位移法中关于附加刚臂和附加链杆的添加方法等，形成了较为完整的体系。

此外，本书虽然面向国际教学，但内容仍具备一定深度，可满足中国学生研究生考试的复习要求，也可作为相关专业研究生、教师的参考用书。

本书的编写也受到了全国结构力学教育工作者的启发，如单建老师的《趣味结构力学》、樊有景老师的《结构力学》等，对于提高结构力学的解题思路有很大帮助，在此对这些教材的作者表示衷心的感谢。由于编者水平有限，书中错误或不妥之处在所难免，欢迎批评指正。

本书成稿过程中，宁永胜、刘应扬两位老师做了大量编写工作，此外，侯玉洁、吴漫、张威、张春丽、徐伟、何伟、许世展几位老师也作出了很多贡献，在此表示感谢。

Foreword

With the continuous internationalization of higher education in China, both Chinese students and international students have a demand for foreign language textbooks. However, the knowledge system of structural mechanics is relatively difficult. In order to better understand the concepts of structural mechanics and master the problem-solving ideas of structural mechanics, the author has summarized years of teaching reform experience and combined it with the teaching practice of international students in China to complete this book.

The distinctive highlights of this book include:

1. This book is presented in bilingual Chinese and English, aiming to meet the diverse language needs of readers and provide a more convenient learning experience.

2. This book provides a large number of deduction illustrations, following the principle of "one picture is worth a thousand words", presenting concepts and ideas in the form of illustrations, so that readers have an intuitive understanding.

3. This book draws on the teaching experience of structural mechanics in recent years and enriches the teaching ideas of structural mechanics, such as the addition methods of additional rigid arms and additional connection rods in displacement method, forming a relatively complete system.

In addition, although this book is aimed at international teaching, its content still has a certain depth, which can meet the review requirements of Chinese graduate students and serve as a reference book for related professional graduate students and teachers.

The writing of this book has also been inspired by national structural mechanics educators, such as Professor Shan Jian's "*Fun Structural Mechanics*" and Professor Fan Youjing's "*Structural Mechanics*", which have greatly helped improve the problem-solving ideas of structural mechanics. We would like to express our sincere gratitude to the authors of these textbooks. Due to the limited level of the editor, errors or inadequacies in the book are inevitable. We welcome criticism and correction.

During the process of writing this book, teachers Ning Yongsheng and Liu Yingyang did a lot of writing work. In addition, teachers Hou Yujie, Wu Man, Zhang Wei, Zhang Chunli, Xu Wei, He Wei, and Xu Shizhan also made many contributions. We would like to express our gratitude.

目　　录

第1章 Chapter 1

绪论 Introduction

要点

- 结构力学的研究对象及研究内容
- 结构的计算简图
- 杆件结构的分类
- 荷载的分类
- 支座和节点

Keys

- The research object and content of structural mechanics
- The computing model of structure
- Classification of the structure of bar system
- Classification of load
- Support and joint

1.1 结构力学的研究对象和研究内容

1.1 The Research Object and Content of Structural Mechanics

结构力学是研究结构受力、变形规律及其分析方法的专业基础课程。

Structural mechanics is a basic professional course to study the force and deformation rules and analysis methods of structures.

1.1.1 结构

1.1.1 Structure

广义的结构：指物质"各组成部分的搭配和排列"。

Structure in the broad sense: refers to the"collocation and arrangement of the components" of a substance.

工程结构：指建筑物或工程设施中承受、传递荷载（作用）而起骨架作用的部分，通常简称结构。

Engineering structure: refers to the part of a building or engineering facility that bears and transmits load (action) and plays the role of the skeleton, usually abbreviated as structure.

由于"结构各组成部分的搭配和排列"的情况不同，按构件的几何性质又可将结构分为三大类：

Due to the different "collocation and arrangement of the structure components", the structure can be divided into three main categories according to the geometric nature of the components:

1. 杆件结构：由杆件通过不同的连接和搭配组成的结构。杆件的几何特征是其横截面尺寸要比长度小得多。梁、拱、桁架、刚架均可视为杆件结构，如图 1.1 所示。

1. Structure of bar system: Structures composed of rods that are connected and matched in different ways. The geometric characteristics of rod structures are that their cross-sectional dimensions are much smaller than their length. Beams, arches, trusses and rigid frames can be considered rod structures, as shown in Figure 1.1.

2. 板壳结构也称薄壁结构：其厚度比长度和宽度小得多。房屋建筑中的楼板和壳体屋盖、水工结构中的弧形拱坝均为板壳结构，如图 1.2 所示。

2. Plate and shell structure (thin-walled structure): The thickness is much smaller than its length and width. House construction in the floor and shell roof, arch dam in the hydraulic structures are plate and shell structures, as shown in Figure 1.2.

3. 实体结构：此类结构在长、宽、高三维尺度上大小相当，如水工结构中的重力坝（大坝），如图 1.3 所示。

3. Massive structure: These structures are comparable in size on the three-dimensional scale of length, width and height, like gravity dams (dams) in hydraulic structures, as shown in Figure 1.3.

图 1.1 杆件结构

Figure 1.1 Structure of bar system

图 1.2 板壳结构

Figure 1.2 Plate and shell structure

图 1.3 实体结构

Figure 1.3 Massive structure

1.1.2 结构力学的研究对象

1.1.2 The Research Object of Structural Mechanics

经典结构力学的研究对象为由杆件所组成的体系，即杆件结构，通常所说的结构力

The object of classical structural mechanics is the system composed of rods,

学就是指杆件结构力学。广义结构力学研究对象为可变形的结构，除可变形杆件组成的体系外，还包括可变形的连续体（板体、块体、壳体等）。

i. e. , the structure of bar system, and the commonly referred to as structural mechanics refers to the structural mechanics of rods. The object of structural mechanics in the general sense is the deformable structure, which includes the deformable continuum (plate, block, shell, etc.) in addition to the system composed of deformable rods.

结构力学的任务是根据力学原理研究在外力和其他外界因素作用下结构的内力和变形、结构的强度、刚度、稳定性和动力反应以及结构的组成规律。具体说来有以下几个方面的内容：

The task of structural mechanics is to study the internal forces and deformation, the strength, stiffness, stability, and dynamic response of the structure under the action of external forces and other external factors, and the laws of the composition of the structure according to the principles of mechanics. Specifically, there are the following aspects.

（1）研究结构的组成规律和合理形式，以及实际结构计算简图的合理选择。

(1) The study of the laws of composition and reasonable form of the structure, as well as the reasonable choice of the actual structural calculation sketch.

（2）研究结构内力和变形的计算方法，进行结构强度和刚度验算。

(2) The study of the calculation methods of structural forces and deformations, and structural strength and stiffness calculations.

（3）研究结构的稳定性以及在动力荷载作用下的结构反应。

(3) The study of the stability of the structure and the structural response under dynamic loads.

1.1.3 结构力学的研究方法

1.1.3 Research Methods in Structural Mechanics

（1）平衡条件：在外界因素的作用下，结构的整体及其中任何一部分都应满足力系的平衡条件。

(1) Equilibrium condition: Under the action of external factors, the whole structure and any part of it should satisfy the equilibrium condition of the force system.

（2）几何条件：连续的结构发生变形后，仍是连续的，材料没有重叠或缝隙。同

(2) Geometric conditions: After the continuous structure is deformed, it is still

时，结构的变形和位移满足支座和节点的约束条件。

continuous, and the material has no overlap or gap. At the same time, the deformation and displacement of the structure satisfy the constraints of supports and nodes.

(3) 物理关系：把结构的应力和变形联系起来的物理条件，即物理方程或本构方程。

(3) Physical relationship: The physical condition that links the stress and deformation of the structure, that is, the physical equation or the constitutive equation.

1.2 结构的计算简图

1.2 The Structural Calculation Sketch

实际结构复杂多变，完全按照结构的实际情况进行力学分析是不可能的，也是不必要的。因此，在进行力学分析时，总是要对实际结构做出一些假设和简化，略去某些次要因素，保留其主要的受力特征，把实际结构简化和抽象为既能反映实际受力情况又便于计算的图形。这种简化的图形就是计算时用来代替实际结构的力学模型，一般称为结构计算简图，或结构计算模型。

The actual structure is complex and changeable, and it is impossible and unnecessary to carry out mechanical analysis according to the structure' s actual situation. Therefore, in the mechanical analysis, it is always necessary to make some assumptions and simplify the actual structure, omitting certain secondary factors, retaining its main force characteristics, and simplifying and abstracting the actual structure into a graph that reflects the actual force situation and facilitates the calculation. This simplified graph replaces the mechanical model' s actual structure, generally known as structural calculation sketch or structural calculation model.

结构计算简图的选择应遵循下列两条原则：

The choice of structural calculation sketch should follow the following two principles:

(1) 结构计算简图应能正确地反映实际结构的主要受力情况和变形性能，使计算结果接近实际情况。

(1) Structural calculation sketch should correctly reflect the actual structure of the main force situation and deformation performance so that the calculation results are close to the actual situation.

(2) 保留主要因素，略去次要因素，使结构计算简图便于计算。

(2) The main factors are retained and the secondary factors are omitted, making the structural calculation model easy to calculate.

影响计算简图选取的主要因素如表 1.1 所示。

The main factors affecting the selection of the calculation sketch are shown in Table 1.1。

影响计算简图选取的主要因素 **表 1.1**

The main factors affecting the selection of calculation sketchs **Table 1.1**

影响计算简图选取的主要因素 Main factors affecting the selection of calculation sketchs		计算简图的精细程度 The fineness of the calculation sketch
结构的重要性 Importance of structure	重要结构 Important structure	精细 Exquisite
	次要结构 Secondary structure	粗略 Rough
设计阶段 Design phase	初步设计 Preliminary design	粗略 Rough
	技术设计 Technical design	精细 Exquisite
计算问题的性质 The nature of computational problems	静力计算 Static calculation	精细 Exquisite
	动力计算 Dynamic calculation	粗略 Rough
计算工具 Calculation tool	先进 Advanced	精细 Exquisite
	简陋 Shabby	粗略 Rough

1.2.1 杆件结构体系的简化

1.2.1 Simplification of the Rod

杆件的截面尺寸（宽度、厚度或直径）通常比杆件长度小得多，截面变形符合平截面假设。截面上的应力可根据截面的内力（弯矩、剪力、轴力）来确定，截面上的变形可根据轴线上的应变分量来确定。因此，在结构计算简图中，杆件用其纵轴线表示，如图 1.4 所示。

The cross-sectional dimensions (width, thickness, or diameter) of a rod are usually much smaller than the length of the rod, and the cross-sectional deformation conforms to the flat section assumption. The stress on the cross-section can be determined from the internal forces (bending moment, shear force, axial force) of the cross section, and the deformation on the cross section can be determined from the strain components on the axis. Therefore, in the structural calculation sketch, the rod is represented by its longitudinal axis, as shown in Figure 1.4.

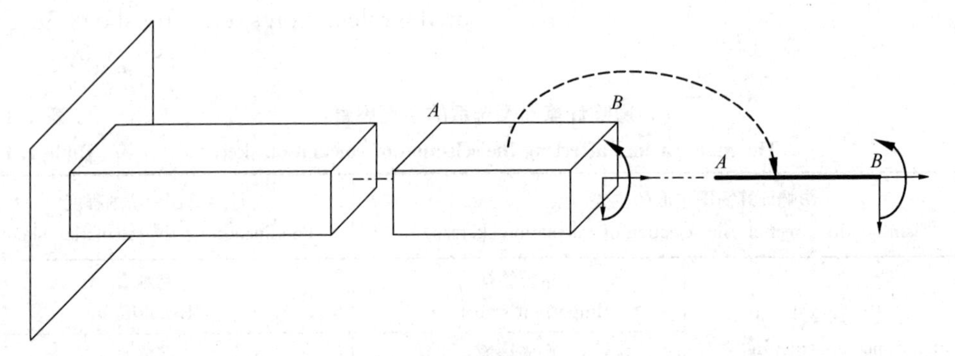

图 1.4 杆件的简化

Figure 1.4 Simplification of the rod

1.2.2 杆件间连接的简化

1.2.2 Simplification of the Linkage Between Rods

结构中杆件与杆件之间的相互连接处，简化为节点。

The interconnection between rods and rods in a structure is simplified to joint.

1.2.2.1 铰节点

1.2.2.1 Hinge Joint

铰节点：被连接的杆件在连接处不能发生相对移动，但可以发生相对转动，连接处可以承受和传递力，但不能承受和传递力矩，如图 1.5 所示。

Hinge joint: The linked bars cannot move relative to each other at the joint but can rotate relative to each other, and the joint can withstand and transmit forces but cannot withstand and transmit bending moments, as shown in Figure 1.5.

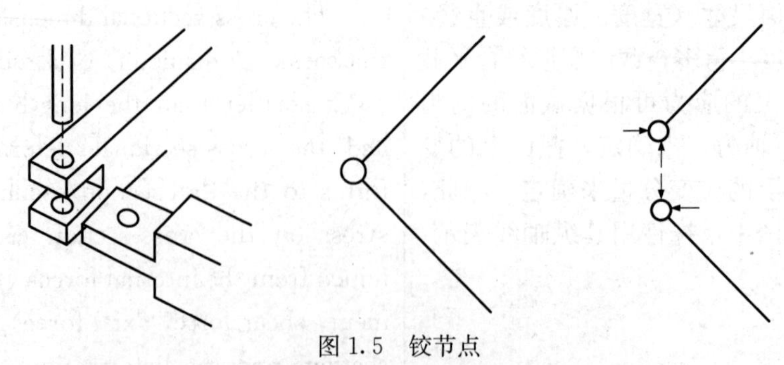

图 1.5 铰节点

Figure 1.5 Hinge joint

1.2.2.2　刚节点

刚节点： 被连接的杆件在连接处不能发生相对移动，也不能发生相对转动，连接处可以承受和传递力和力矩，如图 1.6 所示。

1.2.2.2　Rigid Joint

Rigid joint: The jointed bars cannot move relative to each other or rotate relative to each other at the joint, and the joint can bear and transmit forces and bending moments, as shown in Figure 1.6.

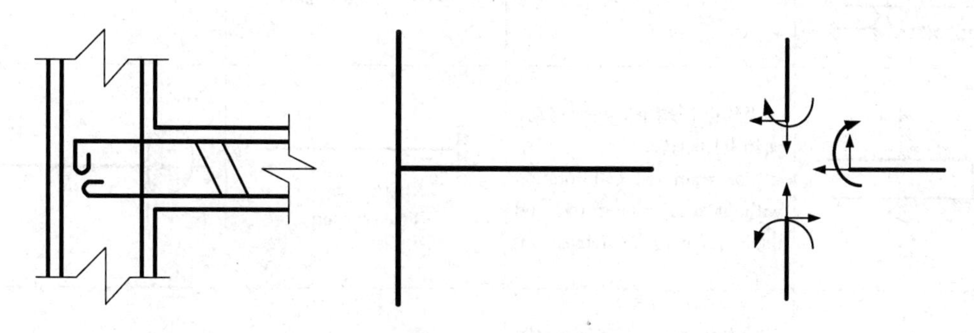

图 1.6　刚节点

Figure 1.6　Rigid joint

1.2.3　支座的简化

支座： 结构与基础或支承部分相连接的装置。支座对结构的反作用力称为**支座反力**。平面结构的支座，通常可简化为以下四种，如表 1.2 所示。

1.2.3　Simplification of the Support

Support: The device that connects the structure to the foundation or supporting part. The reaction force of the support on the structure is called **support reaction force.** The support of the plane structure can usually be simplified into the following four types, as shown in Table 1.2.

常见的支座　　**表 1.2**

Common supports　　**Table 1.2**

支座示意图 Diagram of the support	约束特点 Constraint characteristics	支座简图 Sketch of the support	支座反力 Support reaction force
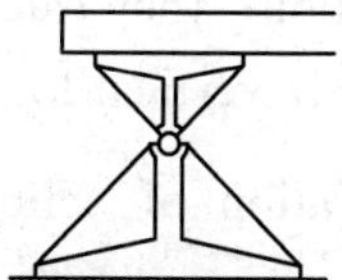	约束杆端平面移动，而不约束杆端转动 Constrain rod end plane movement without constraining rod end rotation	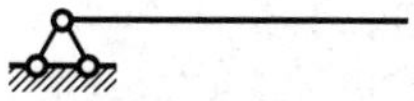 固定铰支座 Pin support	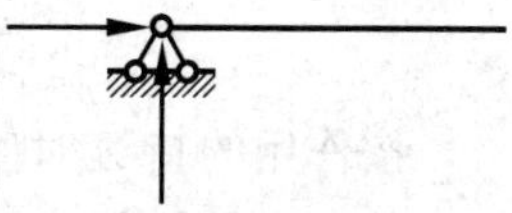

续表
Continued

支座示意图 Diagram of the support	约束特点 Constraint characteristics	支座简图 Sketch of the support	支座反力 Support reaction force
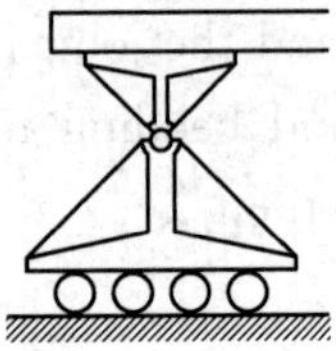	约束杆端单向平面移动 Constrain rod end unidirectional plane movement	 活动铰支座 Movable hinge support	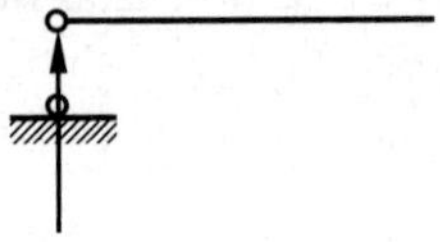
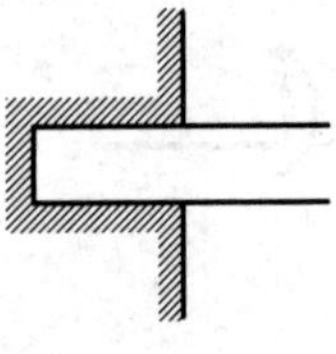	无约束杆端单向平面移动，而约束杆端转动 Constrain rod end rotation without constraining rod end unidirectional plane movement	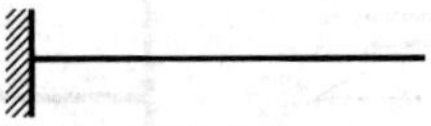 固定支座 Fixed support	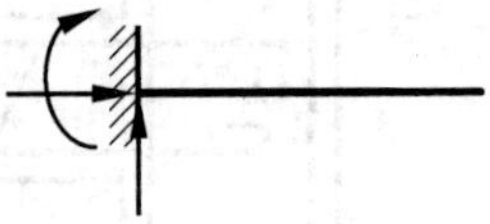
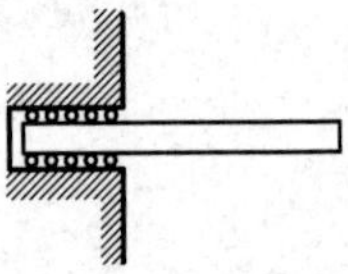	约束杆端单向平面移动和转动 Constrain rod end unidirectional plane movement and rotation	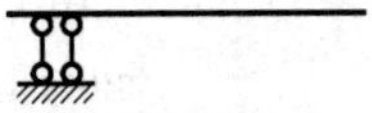 定向支座 Directional support	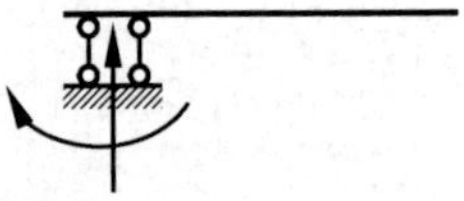

1.2.4 荷载的简化

结构受外界的力或环境温度变化等影响，会产生变形和内力，这些引起变形和内力的因素统称为作用。其中，直接或间接施加于结构的力称为**荷载**。

荷载可进行不同分类：

按方向可分为：水平荷载、竖向荷载。

按作用的持续时间可分为：永久荷载、可变荷载和偶然荷载。

1.2.4 Simplification of the Load

The structure is affected by external forces or environmental temperature changes, etc., which will produce deformation and internal forces, and these factors that cause deformation and internal forces are collectively referred to as the action. Among them, the force applied directly or indirectly to the structure is called **load.**

Loads can be classified in different ways:

According to the direction, they can be divided into: horizontal load, vertical load.

According to the duration of action, they can be divided into: permanent load, variable load and accidental load.

按时间变化性可分为：静力荷载、动力荷载。

According to the time variability, they can be divided into: static load, dynamic load.

按空间变化性可分为：固定荷载、移动荷载。

According to the spatial variability, they can be divided into: fixed load, moving load.

按作用分布情况可分为：集中荷载，分布荷载。

According to the role of the distribution, they can be divided into: concentrated load, distributed load.

在计算简图中，都需要把它们简化为作用在构件纵轴线上的分布荷载、集中荷载或力偶。

In the calculation sketch, they all need to be simplified to distributed loads, concentrated loads, or force couples acting on the member's longitudinal axis.

1.3 杆件结构的分类

1.3 Classification of the Structure of Bar System

常用的杆件结构按照其组成和受力特点分类，如表1.3所示。

The commonly used rod structures can be classified according to their composition and force characteristics, as shown in Table 1.3.

杆件结构分类 **表1.3**

Classification of the structure of bar system **Table 1.3**

名称 Designation	计算简图 Calculation sketch	组成与受力特点 Composition and force characteristics
梁 Beam	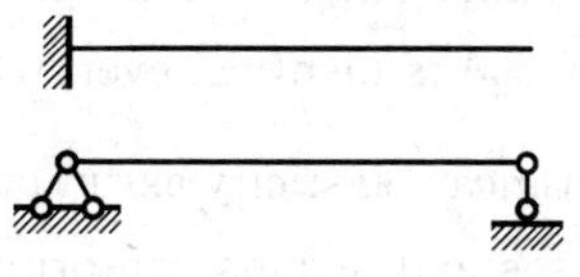	受弯构件，内力包括：轴力、剪力、弯矩 主要内力：弯矩 They are bending members, and internal forces include: axial force, shear force, bending moment Main internal forces: bending moment
刚架 Frame	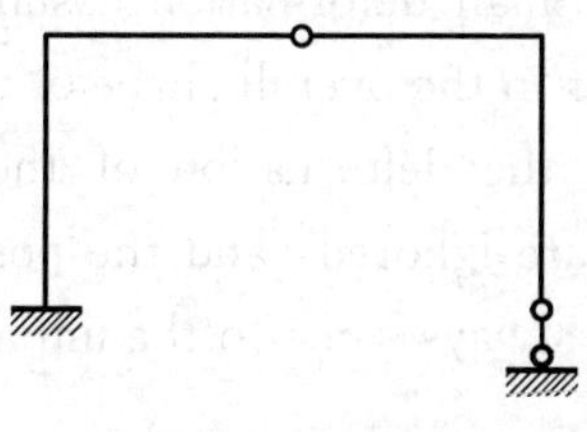	杆件间的节点多为刚节点 受弯构件，内力包括：轴力、剪力、弯矩 主要内力：弯矩 The joints of the rods are commonly rigid joints They are bending members, and internal forces include: axial force, shear force, bending moment Main internal forces: bending moment

续表
Continued

名称 Designation	计算简图 Calculation sketch	组成与受力特点 Composition and force characteristics
拱 Arch		杆轴线为曲线 在竖向荷载作用下产生水平反力 内力：轴力、剪力、弯矩 主要内力：轴力 The axis of the rods is curved line Under vertical loads，horizontal reaction force is produced Internal forces：axial forces，shear forces，bending moments Main internal forces：axial force
桁架 Truss		组成：直杆 主要内力：轴力 Components：straight rods Main internal forces：axial force
组合结构 Composite structure		组成：梁式杆、链杆 梁或杆内力：轴力、剪力、弯矩 Components：beams，truss links Internal forces of beams or rods：axial force，shear force，bending moment

1.4　结构分析的基本假定

1. 连续性假定：结构保持光滑连续，材料处处相同；

2. 线弹性假定：材料应力应变符合胡克定律；

3. 小变形假定：忽略受力后杆件变形导致的结构整体形状的变化，荷载的作用位置始终按初始位置考虑；

4. 平截面直法线假定：忽略杆截面剪

1.4　Basic Assumptions of Structural Analysis

1. Continuity assumption：The structure remains smooth and continuous，and the material is identical everywhere；

2. Linear elasticity assumption：Material stresses and strains conform to Hooke's law；

3. Small deformation assumption：The changes in the overall shape of the structure due to the deformation of the rod under forces are ignored，and the position of the load is always same to the initial position；

4. Flat section straight normal assump-

切变形产生的附加挠度，变形前后截面均保持平面。

tion：The additional deflection generated by the shear deformation of the rod section is ignored，and the section remains flat before and after the deformation.

满足上述假定的结构分析问题为线性问题，线性问题具有唯一解，在分析过程中可使用叠加原理。

The structural analysis problem satisfying the above assumptions is linear，the linear problem has the unique solution，and the superposition principle can be used in the analysis process.

思 考 题

Questions

1.1 结构力学研究的对象和任务是什么？

1.1 What are the objects and tasks of structural mechanics?

1.2 什么是结构的计算简图？它与实际结构有什么关系与区别？为什么要将实际结构简化为计算简图？

1.2 What is a calculation sketch of a structure? How is it related to and different from the actual structure? Why is the actual structure simplified to a calculation sketch?

1.3 平面杆件结构的节点通常简化为哪两种情形？它们的构造、限制结构运动和受力的特征各是什么？

1.3 Which are two cases that the joints of plane member structures usually simplified? What are their structures，features that limit structural motion and forces?

1.4 平面杆件结构的支座通常简化为哪几种情形？它们的构造、限制结构运动和受力的特征各是什么？

1.4 What kinds of situations are usually simplified for the support of the plane member structure? What are their structures，features that limit structural motion and forces?

1.5 平面杆件结构的荷载通常简化为哪几类？

1.5 What kinds of loads are usually simplified for plane member structures?

1.6 常用的杆件结构有哪几类？

1.6 What types of commonly used rod structures are?

第2章
Chapter 2

平面体系的几何构造分析
Geometrical Formation Analysis of Coplanar Structures

要点

- 几何构造分析的基本概念
- 平面几何不变体系的组成规律
- 平面杆件体系的计算自由度

Keys

- Basic concepts for geometrical formation analysis
- Laws of geometric stable system
- Computational degrees of freedom of coplanar structures

2.1 几何构造分析的几个概念

2.1 Basic Concepts for Geometrical Formation Analysis

2.1.1 刚片

刚片在几何构造分析中不计变形。

2.1.1 Rigid Body

The deformation of a **rigid body** is ignored in geometrical formation analysis.

2.1.2 自由度

自由度是物体具有独立运动方式的数目，或确定物体在任意时刻位置所需的独立坐标数目。

平面内的一个点具有2个自由度（无论在何种坐标系），如图2.1所示。

平面内的一个刚片具有3个自由度（无

2.1.2 Degree of Freedom (DoF)

Degree of freedom is the number of independent mode of movements of an object or the number of independent coordinates to determine the position of an object at any time.

A point in a plane has two degrees of freedom (at any coordinate system), as shown in Figure 2.1.

A rigid body in a plane has three de-

论在何种坐标系），如图 2.2 所示。

grees of freedom (at any coordinate system), as shown in Figure 2.2.

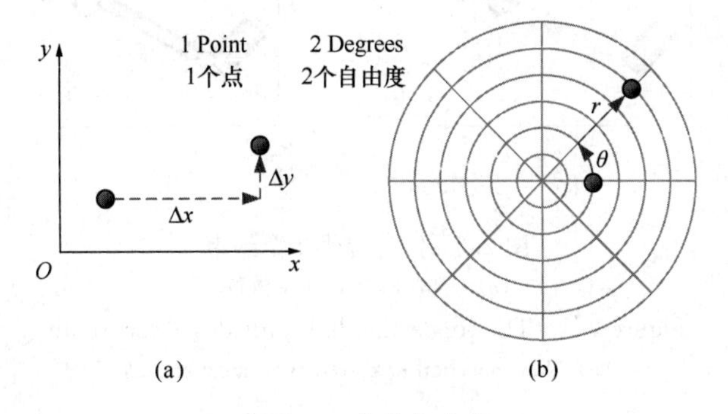

图 2.1 点的自由度

（a）直角坐标系；（b）极坐标系

Figure 2.1 Degrees of freedom of a point

(a) In Cartesian coordinate system; (b) In Polar coordinate system

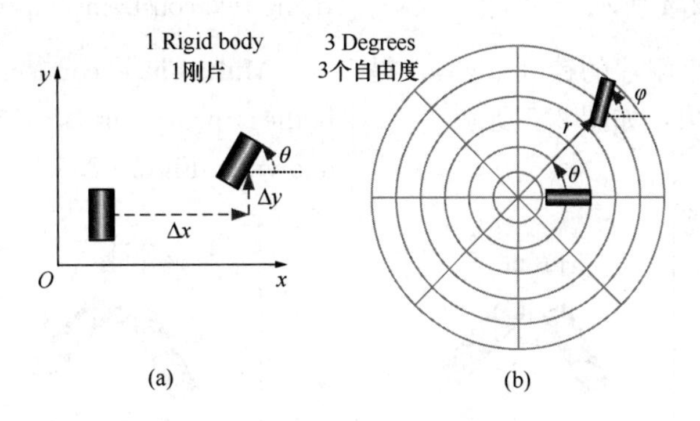

图 2.2 刚片的自由度

（a）直角坐标系；（b）极坐标系

Figure 2.2 Degrees of freedom of a rigid body

(a) In Cartesian coordinate system ; (b) In Polar coordinate system

2.1.3 约束

约束是能够限制物体或体系运动的各种装置，其作用就是减少体系的自由度。

2.1.3 Constraints

Constraints are devices that can limit the possible motions of a body or a system, the functions of which is to decrease DoF.

2.1.3.1 链杆

链杆指的是两端与其他构件铰接的杆件。1 个链杆提供 1 个约束，如图 2.3 所示。

2.1.3.1 Connection Link

Connection link is a member that connects two other components with hinges. A connection link provides one constraint, as shown in Figure 2.3.

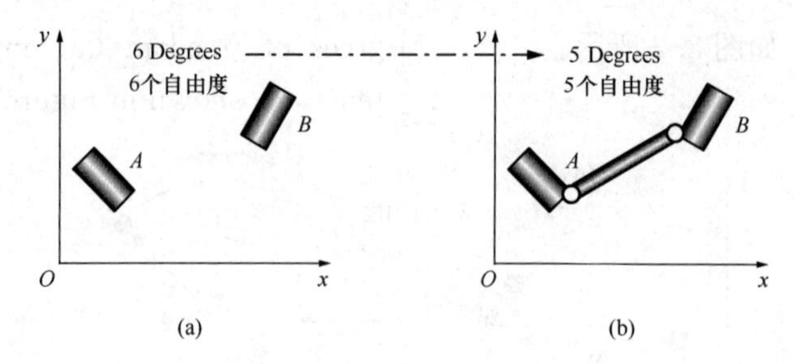

图 2.3 链杆提供 1 个约束

（a）无链杆；（b）1 个链杆

Figure 2.3 The connection link provides 1 constraint

（a）No connection link；（b）1 connection link

2.1.3.2 铰

单铰是连接 2 个刚片的铰。1 单铰提供了 2 个约束，如图 2.4 所示。

复铰是连接 n 个刚片的铰（$n \geqslant 3$），提供（$n-1$）×2 个约束，如图 2.5 所示。

2.1.3.2 Hinge

Single hinge connects two rigid bodies, providing two constraints, as shown in Figure 2.4.

Multi-hinge connects n ($n \geqslant 3$) rigid bodies, providing $(n-1) \times 2$ constraints, as shown in Figure 2.5.

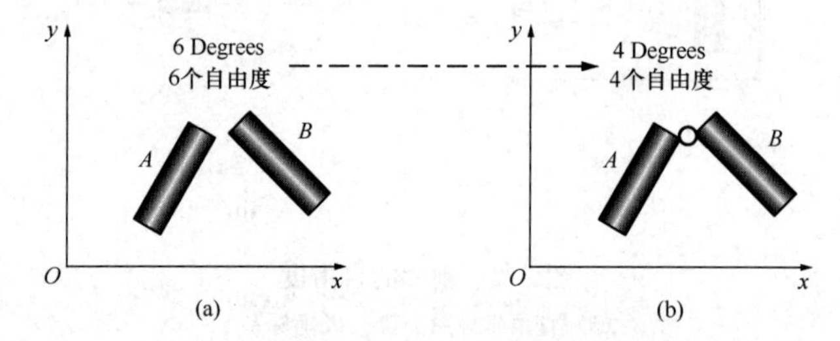

图 2.4 单铰

（a）2 个刚片无连接；（b）单铰连接的 2 个刚片

Figure 2.4 Single hinge

（a）2 rigid bodies without connection；（b）2 rigid bodies with single hinge

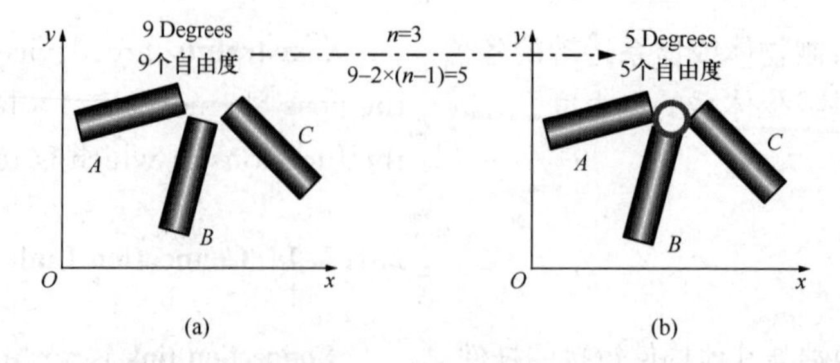

图 2.5 复铰

（a）3 个刚片无连接；（b）复铰连接的 3 个刚片

Figure 2.5 Multi-hinge

（a）3 rigid bodies without connection；（b）3 rigid bodies with multi-hinge

2.1.3.3 刚节点

单刚节点指连接两个刚片的刚节点，1个单刚节点提供3个约束，如图2.6所示。

2.1.3.3 Rigid Joint

Single rigid joint connects two rigid bodies, providing three constraints, as shown in Figure 2.6.

复刚节点指连接n个刚片的刚节点（$n\geqslant3$），提供$(n-1)\times3$个约束，如图2.7所示。

Multi-rigid joint connects n ($n\geqslant3$) rigid bodies, providing $(n-1)\times3$ constraints, as shown in Figure 2.7.

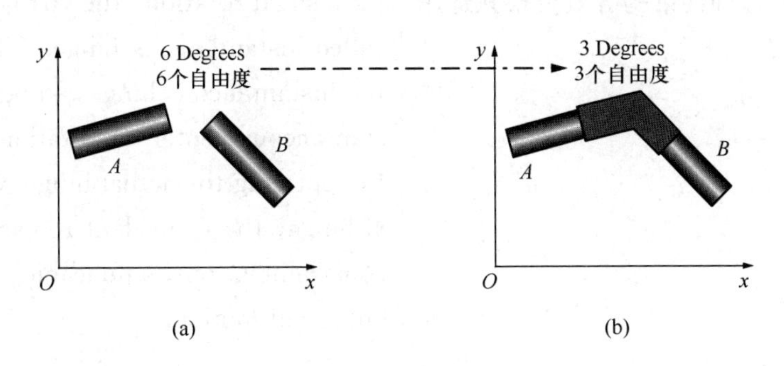

图2.6 刚节点

（a）2个刚片无连接；（b）1个刚节点连接的2个刚片

Figure 2.6 Single rigid joint

(a) 2 rigid bodies are not connected; (b) 2 rigid bodies are connected by 1 rigid joint

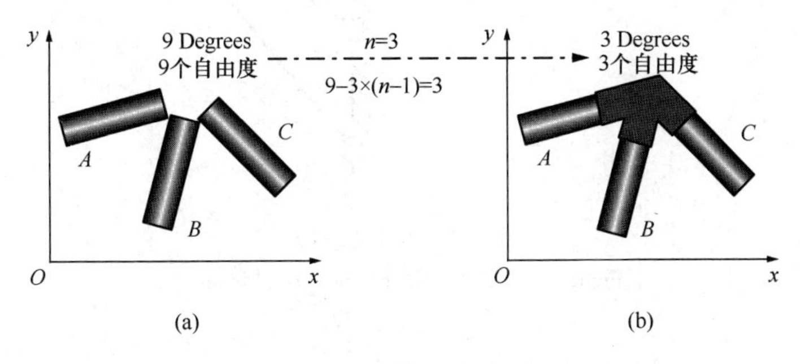

图2.7 复刚节点

（a）3个刚片无连接；（b）复刚节点连接的3个刚片

Figure 2.7 Multi-rigid joint

(a) 3 rigid bodies are not connected; (b) 3 rigid plates are connected by multi-rigid joint

2.1.3.4 实铰与虚铰

从提供的约束数目来看，连接2个刚片的1个单铰与2个不共线链杆的作用相当，所以可以进行等效代替。因此，将2个不共

2.1.3.4 Actual Hinge and Virtual Hinge

From the provided number of constraints, a single hinge connecting two rigid bodies has the same effect as two non-paral-

线链杆称为**虚铰**，如图 2.8 所示。

lel connection links, so they can be replaced equivalently. Therefore, the two non-parallel connection links are defined as **virtual hinge**, as shown in Figure 2.8.

从运动方式看，刚片只能绕实铰或虚铰进行转动。由于发生微小转动后，虚铰的位置是变化的，因此，虚铰也称之为瞬铰。瞬铰的位置可以称为瞬时转动中心。显然，用瞬铰替换实铰，这种约束的等效替换只适用于瞬时微小运动。

From the perspective of the motion mode, the rigid body can only rotate around the actual hinge or the virtual hinge. Since the position of the virtual hinge changes after a small rotation, the virtual hinge is also called instantaneous hinge. The location of the instantaneous hinge can be called the instantaneous center of rotation. Obviously, by replacing the actual hinge with the virtual hinge, the equivalent replacement of this constraint is only applicable to instantaneous small motion.

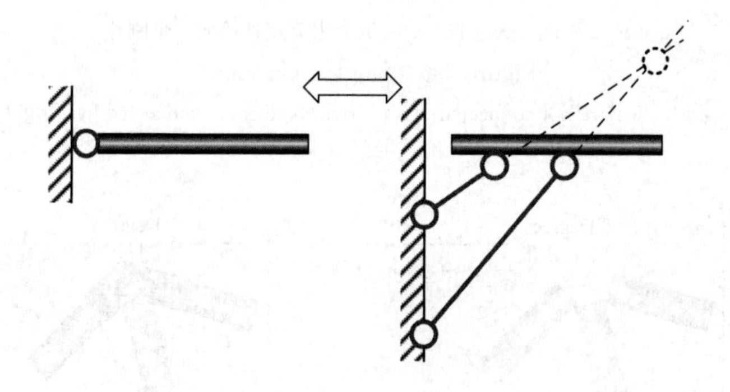

图 2.8 实铰与虚铰

Figure 2.8 Actual hinge and virtual hinge

2.1.3.5 无穷远铰

如果连接两个刚片的两根链杆平行，那么其约束作用相当于无穷远处的瞬铰。由于瞬铰在无穷远处，绕瞬铰的微小转动便转化为垂直于链杆方向的平动，如图 2.9 所示。

2.1.3.5 Infinity Hinge

If the two connection links are parallel, the constraint function is equal to the instantaneous hinge at infinity. Since the virtual hinge is at infinity, the rotation around the instantaneous hinge is transferred into the translation perpendicular to the connection link, as shown in Figure 2.9.

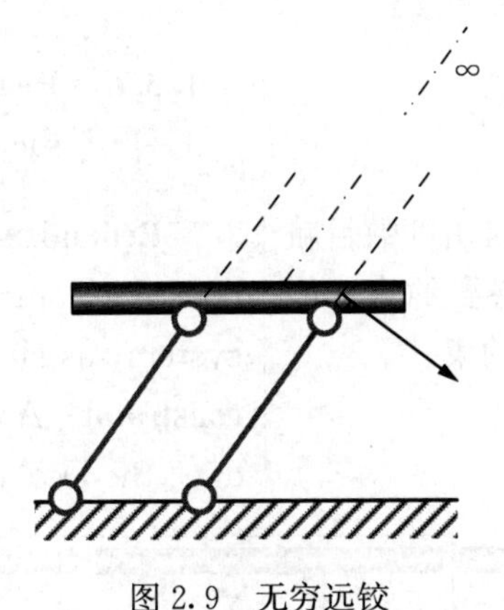

图 2.9 无穷远铰

Figure 2.9 Infinity hinge

在几何构造分析中应用无穷远处瞬铰的概念时，可采用射影几何中关于∞点和∞线的结论，如图 2.10 所示：

In geometrical formation analysis, when using the concept of instantaneous hinge at infinity, the conclusion of infinity point and infinity line in projection geometry can be used, as shown in Figure 2.10:

（1）每个方向有一个∞点，不同方向有不同的∞点；

(1) Each direction has an infinity point, different directions have different infinity points;

（2）各∞点都在同一条直线上，该直线称为∞线；

(2) Infinity points are in the same line, which is called infinity line;

（3）各有限点都不在∞线上。

(3) There are no finite point on infinity line.

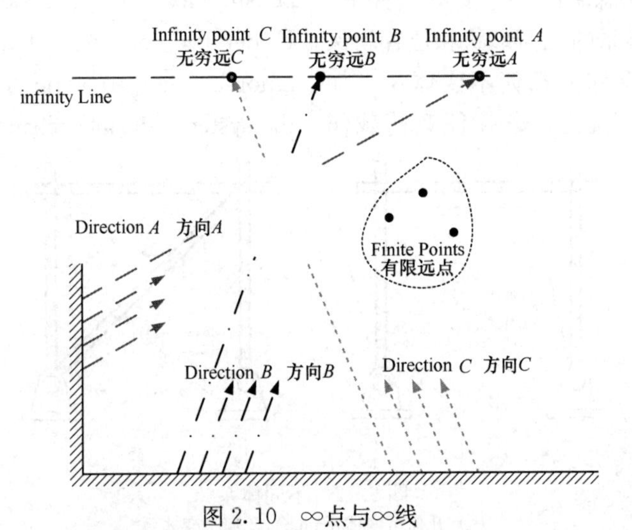

图 2.10 ∞点与∞线

Figure 2.10 Infinity point and infinity line

2.1.3.6 多余约束与必要约束

多余约束：不能使体系的自由度数目减少的约束，如图2.11所示；**必要约束**：能够使体系的自由度数目减少的约束。

2.1.3.6 Redundant Constraint and Necessary Constraint

Redundant constraint: A constraint does not contribute to reduce the DoF of a system, as shown in Figure 2.11; **necessary constraint**: A constraint contributes to reduce the DoF of a system.

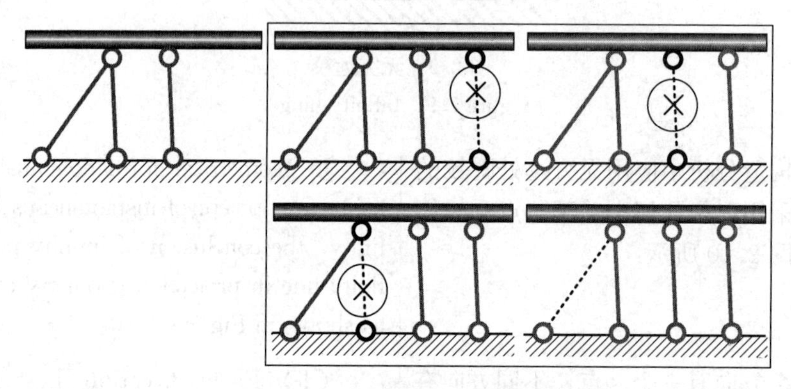

图2.11 多余约束与必要约束
Figure 2.11 Redundant constraint and necessary constraint

2.1.4 几何不变体系和几何可变体系

体系必须能够承受可能出现的各种荷载才能成为结构。在忽略材料应变的条件下，任意荷载作用下体系的几何形状和位置均保持不变，这种体系称为**几何不变体系**，如图2.12(b)所示。反之，若在任意荷载作

2.1.4 Geometric Stable System and Geometric Changeable System

A system has to be stable under any possible loads to be classified as a structure. Under the condition that material strain is ignored, the geometric shape and position of the system remain unchanged under any load,

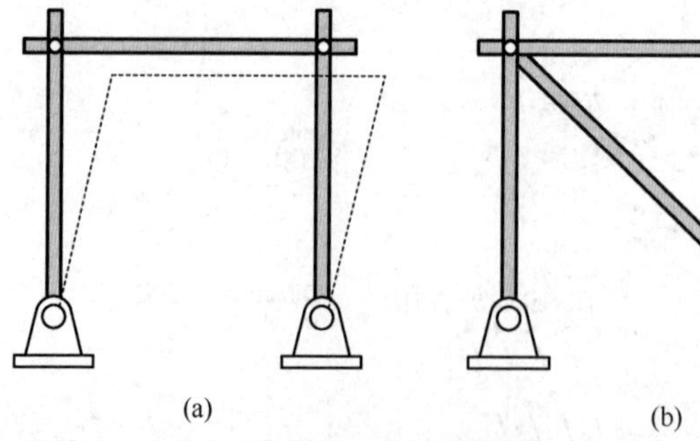

图2.12 不同体系
(a) 几何可变体系；(b) 几何不变体系
Figure 2.12 Different systems
(a) Geometric changeable system; (b) Geometric stable system

用下，体系的几何形状或位置可以改变，这种体系称为**几何可变体系**，如图 2.12(a) 所示。

which is called **geometric stable system**, as shown in Figure 2.12 (b). On the contrary, if the geometric shape or position of the system can be changed under any load, this system is called **geometric changeable system**, as shown in Figure 2.12(a).

特殊地，如图 2.13(a) 所示共线杆体系，初始状态下几何可变，但在 A 点发生微小竖向位移后，杆件 AB 和 AC 不再彼此共线，体系就成为几何不变的了。如果一个几何可变体系在发生微小位移后成为几何不变体系，这样的体系我们称为**瞬变体系**。

Specially, as the collinearity rod system shown in Figure 2.13(a), the original state is geometric changeable, but after the micro vertical displacement occurs at point A, the rods AB and AC are not parallel, and the system is turned into geometric stable system. Thus, we call this kind of system the **instantaneous unstable system.**

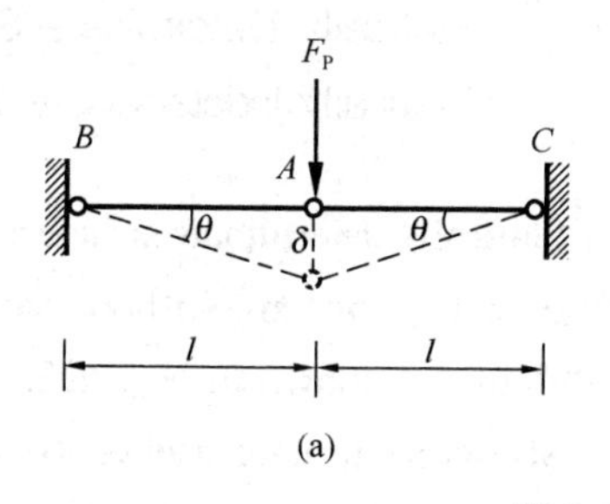

(a)

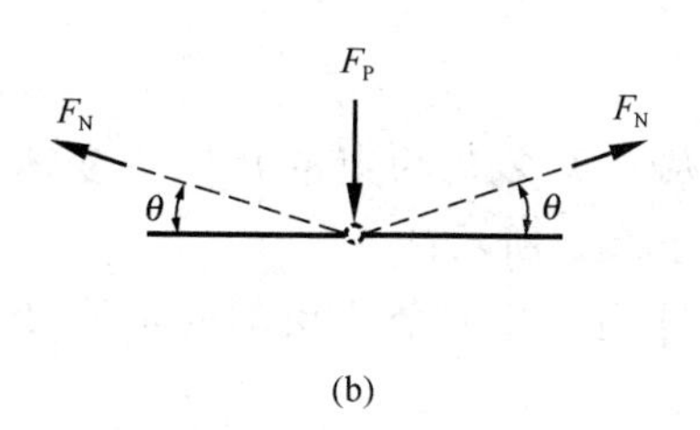

(b)

图 2.13 瞬变体系

Figure 2.13 Instantaneous unstable systems

瞬变体系经微小运动后即转化为几何不变体系，那么能否作为工程结构来使用呢？为此对体系施加集中力 F_P，如图 2.13(b) 所示。体系在初始位置是无法平衡的，当 A 点发生微小位移后，方可平衡。由平衡条件可知，此时，AB 和 AC 杆的轴力为：

The instantaneous unstable system is turned into geometric stable system after micro movement, so can it be used as an engineering structure? A concentrated force F_P is applied to the system, as shown in Figure 2.13 (b). The system cannot be equilibrated at the initial position, and can be equilibrated only after a micro displacement occurs at point A. According to the equilibrium condition, at this time, the axial forces of rod AB and rod AC are:

$$F_N = \frac{F_P}{2\sin\theta}$$

A 点发生微小位移，即 $\theta \to 0$，若 $F_P \neq 0$，则 $F_P \to \infty$。由此可见，瞬变体系在任意

Micro displacement occurs at point A, that is, $\theta \to 0$, if $F_P \neq 0$, then $F_P \to \infty$. It can

荷载作用下，通常都不能平衡，要待发生微小的移位后，才渐趋平衡。而瞬变体系在从不能平衡到平衡的过程中，会产生很大的内力，它远超过结构构件承载能力，使构件破坏。所以瞬变体系不能作为结构来使用，在结构设计中应避免采用瞬变体系或接近于瞬变的体系。

be seen that the instantaneous unstable system is usually not balanced under any load, and only becomes balanced after a micro displacement occurs. However, in the process of the instantaneous unstable system from unbalanced condition into balanced condition, it will produce a large internal force, which far exceeds the bearing capacity of the structural members and causes the members to fail. Therefore, instantaneous unstable system should not be used as a structure, and instantaneous unstable system or system close to it should be avoided in structural design.

2.1.5　静定结构与超静定结构

2.1.5　Statically Determinate Structure and Statically Indeterminate Structure

静定结构：无多余约束的几何不变体系，如图2.14(a)所示；

超静定结构：有多余约束的几何不变体，如图2.14(b)所示。

Statically determinate structure: It is a geometric stable system without any redundant constraint, as shown in Figure 2.14(a);

Statically indeterminate structure: It is a geometric stable system with at least one redundant constraint or more, as shown in Figure 2.14(b).

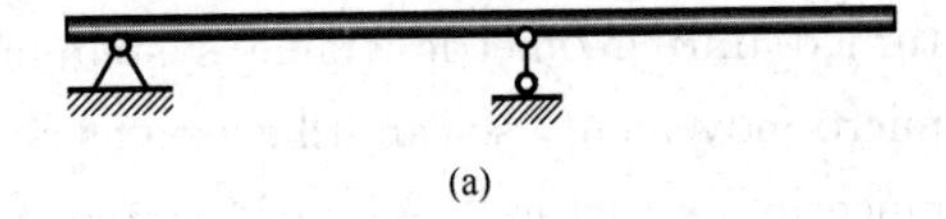

(a)　　(b)

图2.14　不同的结构

(a) 静定结构；(b) 超静定结构

Figure 2.14　Different structures

(a) Statically determinate structure; (b) Statically indeterminate structure

2.1.6　平面杆件体系的自由度

2.1.6　The Degrees of Freedom of Coplanar Structures

计算自由度W：组成体系的各构件自由度总和减去体系中全部约束的数目即为体系的计算自由度，即：

Computational degrees of freedom W: The number of DoF of all components of the system subtracts the number of all constraints in the system, i.e.:

W=体系的各构件自由度总和－体系中全部约束的数目

W = The number of DoF of all components－The number of all constraints

体系自由度 S：体系的各构件自由度总和减去体系中必要约束的数目，由于全部约束数与必要约束数的差值是多余约束数 n，因此，不难得出：

System's degrees of freedom S: The number of DoF of all components of the system subtracts the number of necessary constraints, since the difference between the number of all constraints and the number of necessary constraints equals the number of redundant constraints n, it is easy that:

S =体系的各构件自由度总和－体系中必要约束的数目
=体系的各构件自由度总和－(体系中全部约束的数目－体系中多余约束的数目)
=体系的各构件自由度总和－体系中全部约束的数目＋体系中多余约束的数目

$$S=W+n$$

S = The number of DoF of all components －The number of necessary constraints
= The number of DoF of all components －(The number of all constraints－The number of redundant constraints)
= The number of DoF of all components －The number of all constraints＋The number of redundant constraints

$$S = W + n$$

体系的自由度举例 表 2.1

The examples of DoF of systems Table 2.1

结构举例 Examples	计算自由度 W Computational degrees of freedom W	体系自由度 S System's degrees of freedom S	体系是否可变 Whether the system is changeable or not	多余约束 n Redundant constraints n
	1	1	可变 Geometric changeable system	0
	0	0	不变 Geometric stable system	0
	0	1	可变 Geometric changeable system	1
	−1	0	不变 Geometric stable system	1
	−1	1	可变 Geometric changeable system	2

表 2.1 的规律可进行如下总结：

The rules in Table 2.1 can be summarized as follows:

（1）若$W>0$，则$S>0$，体系约束的数目不够，体系是几何可变的；

（1）If $W>0$, then $S>0$, the constraints are scarce so the system is geometric changeable;

（2）若$W=0$，则$S=n$，如无多余约束，则为几何不变，如有多余约束，则为几何可变；

（2）If $W=0$, then $S=n$, if there is no redundant constraint, it is geometric stable; if there is redundant constraint, it is geometric changeable;

（3）若$W<0$，表明体系中必有多余约束，但体系是否几何不变还取决于约束的布置是否合理。

（3）If $W<0$, then it indicates that there must be redundant constraints in the system, but whether it is geometric stable or changeable depends on whether the arrangement of constraints is reasonable.

由此可见，几何不变体系的必要条件是：体系的计算自由度$W\leqslant 0$。若$W>0$，则体系一定几何可变。所以，计算自由度W就成为体系几何可变性判定的手段之一。

It can be seen that the necessary condition of a geometric stable system is: the computational freedom is $W\leqslant 0$ of the system. If $W>0$, then the system must be geometric changeable. Therefore, W becomes one of the means to determine the geometric variability of the system.

有时可不考虑体系与基础的连接，而只检查体系内部的几何不变性。这时，如果体系本身几何不变，则可以作为一个刚片，在平面内尚有3个自由度。因此，体系内部几何不变时，应有$W\leqslant 3$。

Sometimes the connection between the system and the foundation is not considered, and only the geometric invariance within the system is checked. At this time, if the geometry of the system itself is unchanged, it can be regarded as a rigid body, and there are still three degrees of freedom in the plane. Therefore, if the geometry of the redundant constraints system is stable, there should be $W\leqslant 3$.

计算自由度计算可按多种方法进行计算，对于某一体系：m为刚片数，g为单刚节点数，h为单铰节点数，a为地基约束数。计算自由度W可表示为：

The calculation of computational degrees of freedom can be carried out in various ways, for a certain system, m is the number of rigid bodies, g is the number of single rigid joint, h is the number of single hinge, and a is the number of foundation constraints. Computational degrees of freedom W can be calculated as:

$$W = 3m - (3g + 2h + a)$$

对于铰接体系：j 为节点数，b 为链杆数，a 为地基约束数。计算自由度 W 可表示为：

For hinge system, j is the number of joints, b is the number of connection links, and a is the number of foundation constraints. Computational degrees of freedom W can be calculated as:

$$W = 2j - (b + a)$$

计算举例如图 2.15 所示。

Calculation example is shown in Figure 2.15.

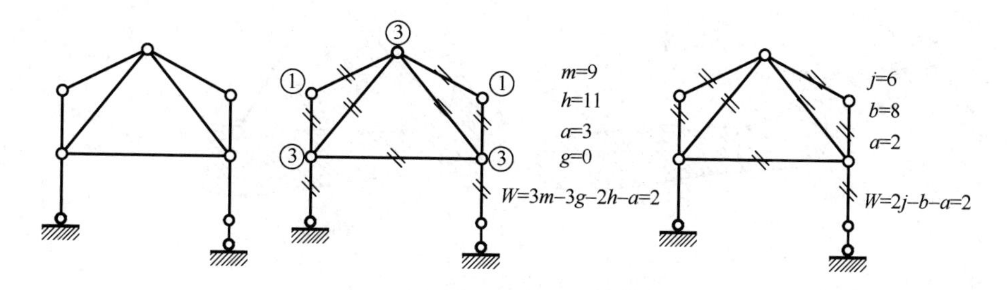

图 2.15 计算自由度的计算方法

Figure 2.15 Method for calculating degrees of freedom

特殊的，如果体系含有无铰封闭框（图 2.16），则需要减去每个无铰封闭框带来的 3 个约束：

In particular, if the system contains hingeless closed frames (Figure 2.16), the 3 constraints imposed by each hingeless closed frame need to be subtracted:

$$W = 3m - (3g + 2h + a + 3k)$$

式中，m 为刚片数；g 为单刚节点数；h 为单铰节点数；a 为地基约束数；k 为无铰封闭框数。

Where m is the number of rigid bodies, g is the number of single rigid joints, h is the number of single hinges, a is the number of foundation constraints, and k is the number of hingeless closed frames.

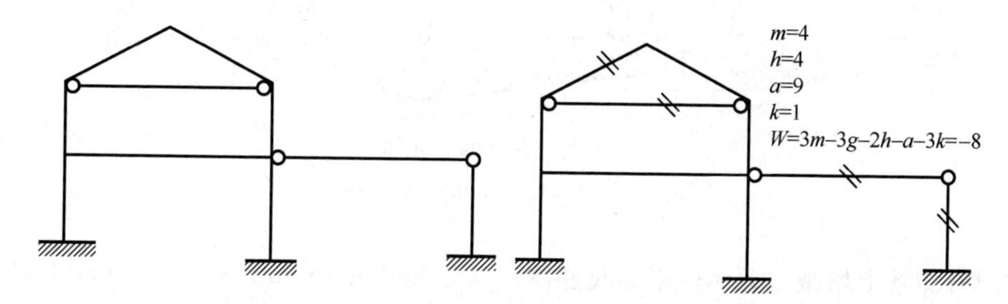

图 2.16 含无铰封闭框体系计算自由度的计算方法

Figure 2.16 Method of calculation degrees of freedom for hingeless closed frame system

2.2 平面几何不变体系的组成规则

2.2 Principles of Coplanar Geometric Stable System

2.2.1 固定一个点

2.2.1 Fix a Point

如图2.17所示，平面内如何固定一个点？

As shown in Figure 2.17, How do you fix a point in a plane?

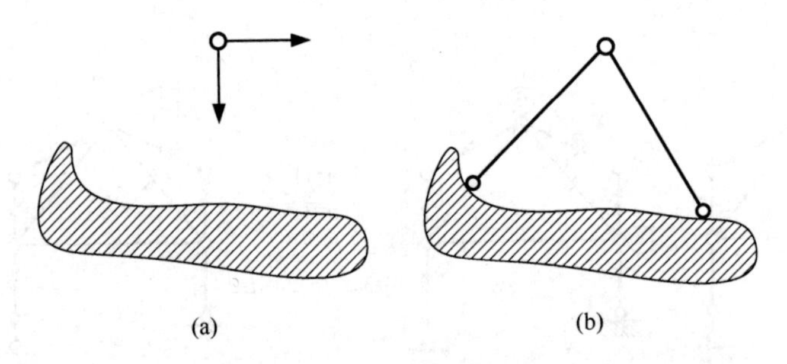

图2.17 一个点与刚片的连接

Figure 2.17 The connection of a point to a rigid body

一个刚片与一个点用不共线的两个链杆相连，体系几何不变，且没有多余约束。我们把这种由两个不共线链杆铰接于一点的构造称为**二元体**，如图2.18所示。

A rigid body is connected by two non-collinear connection links. This system is a geometric stable system without any redundant constraint. The constitution of two non-collinear links connecting to one hinge is called **dual body**, as shown in Figure 2.18.

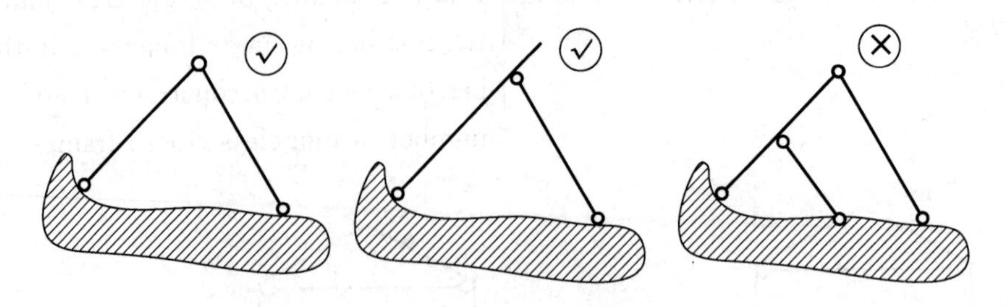

图2.18 常见的二元体

Figure 2.18 Common dual bodies

在原有体系上增减二元体，不会改变体系的几何组成特性。

Adding or subtracting a dual body from the original system does not change the geometric composition of the system.

2.2.2 两刚片规则

对上节得到的铰接三角形进行变换，可以得到两刚片规则，如图 2.19 所示。

2.2.2 Two-rigid-body Rule

Based on the hinge triangle obtained in the above section, the two-rigid-body rule can be obtained, as shown in Figure 2.19.

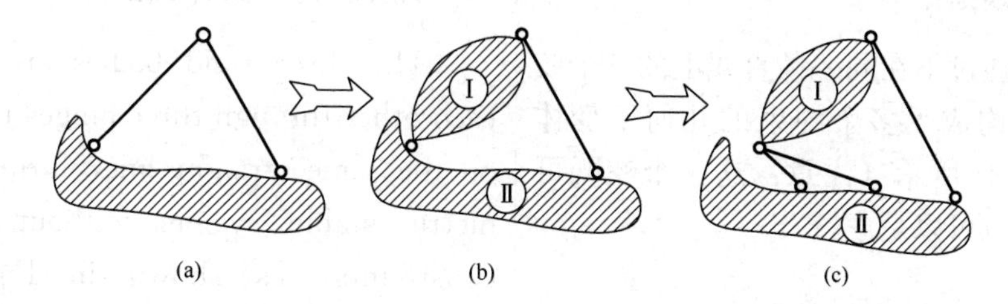

图 2.19 两刚片规则

Figure 2.19 Two-rigid-body rule

两刚片规则：

图 2.19(b)：两刚片通过一个铰和一个链杆连接，且链杆不通过铰，则体系几何不变，且没有多余约束。

Two-rigid-body rule:

Figure 2.19(b): Two rigid bodies are connected by a hinge and a connection link, and the connection link does not pass through the hinge, then the system is geometric stable and there are no redundant constraints.

图 2.19(c)：两刚片通过三个链杆连接，且三个链杆不交于同一点（注意：平行也视为交于同一点），则体系几何不变，且没有多余约束。

Figure 2.19(c): Two rigid bodies are connected by three connection links, and the three connection links do not intersect at the same point (Note: parallel condition is also considered as intersecting at the same point), then the system is geometric stable and there are no redundant constraints.

利用两刚片规则判断体系示例如图 2.20 所示。

Figure 2.20 shows the example of two-rigid-body rule.

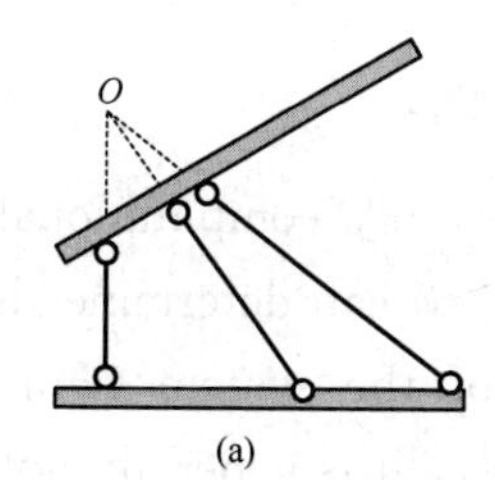

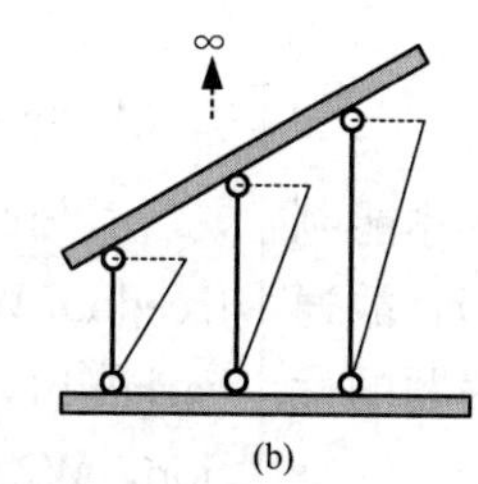

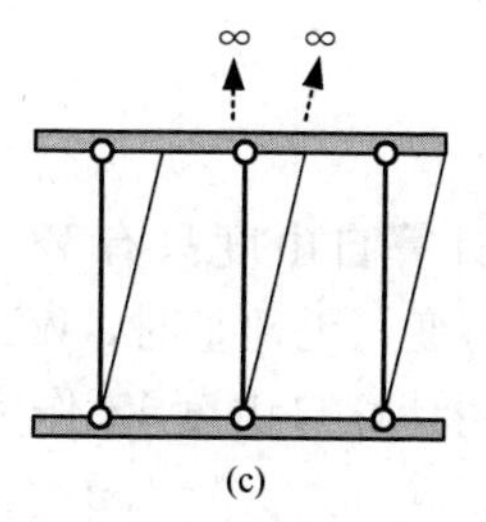

图 2.20 两刚片规则的应用

(a) 瞬变 1；(b) 瞬变 2；(c) 常变

Figure 2.20 Application of two-rigid-body rule

(a) Instantaneous unstable 1 ; (b) Instantaneous unstable 2; (c) Constantly changeable

2.2.3　三刚片规则

对铰接三角形变换，可以得到三刚片规则。

三刚片规则：

三刚片通过不在同一条直线上的三个铰两两相连，构成无多余约束的几何不变体系，如图2.21所示（注意：这三个铰既可以是实铰，也可以是虚铰）。

2.2.3　Three-rigid-body Rule

Based on hinged triangles, the three-rigid-body rule can be obtained.

Three-rigid-body rule:

The three rigid bodies are connected each other through three hinges that are not on the same straight line, forming a geometric stable system without redundant constraints, as shown in Figure 2.21 (Note: These three hinges can be either actual hinges or virtual hinges).

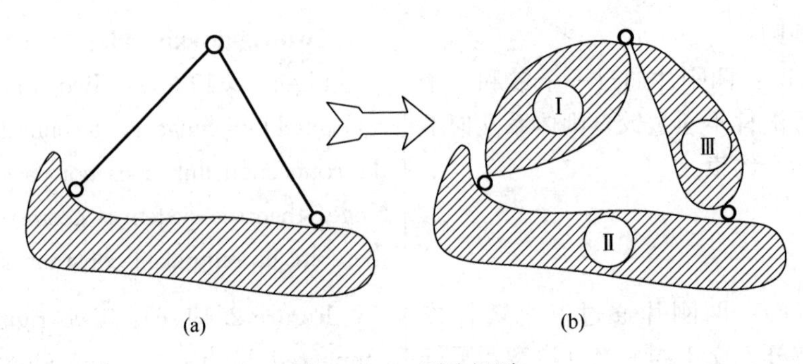

图2.21　三刚片规则

Figure 2.21　Three-rigid-body rule

2.3　几何构造分析方法举例

2.3　Examples of Geometric Formation Analysis Methods

2.3.1　概述

一般地：

1. 由于计算自由度只有 $W>0$ 才能判定体系几何可变（无地基时，$W>3$），故通常不建议采用计算自由度 W 作为解题的优先选择；

2.3.1　Overview

Generally:

1. Since only computational degrees of freedom $W>0$ can determine the geometric variability of the system (For no foundation, $W>3$), it is generally not recommended to use the computational degrees of freedom W as the preferred choice for solving problems;

2. 二元体规则操作简单、直观，建议解题时首先考虑；

2. The dual body rule is simple and intuitive to operate, and it is recommended to consider it first when solving problems;

3. 如体系与地基仅通过符合两刚片规则限制条件的三个约束连接，则可以先去掉地基；

3. If the system and the foundation are only connected by three constraints that meet the constraints of the two-rigid-body rule, the foundation can be removed first;

4. 体系中的折杆或曲杆可以在保持两头铰位置不变的前提下转换为直杆，如图 2.22 所示；

4. The folded rod or curved rod in the system can be converted into a straight rod under the premise of keeping the hinge positions of both ends unchanged, as shown in Figure 2.22;

5. 体系中若存在"直线二元体"（不再是二元体）构造，则可直接认定几何可变，如图 2.22 所示；

5. If there is a "straight line dual body" (no longer a dual body) constitution in the system, it can be directly determined as the geometric changeable, as shown in Figure 2.22;

6. 体系中杆件或约束不可遗漏，也不能重复使用。

6. The rods or constraints in the system cannot be omitted or reused.

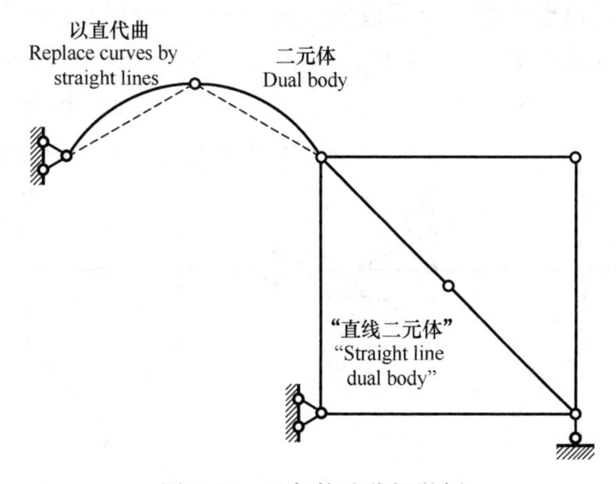

图 2.22 几何构造分析举例

Figure 2.22 Example of geometry analysis

2.3.2 先局部后整体的思路

2.3.2 The Idea of the Whole after the Part

这种思路常见于扩大地基法，如图 2.23 所示，逐渐扩大地基，从而简化体系。

This idea is commonly used in the expanded foundation method, as shown in Figure 2.23, the foundation is gradually enlarged, thereby simplifying the system.

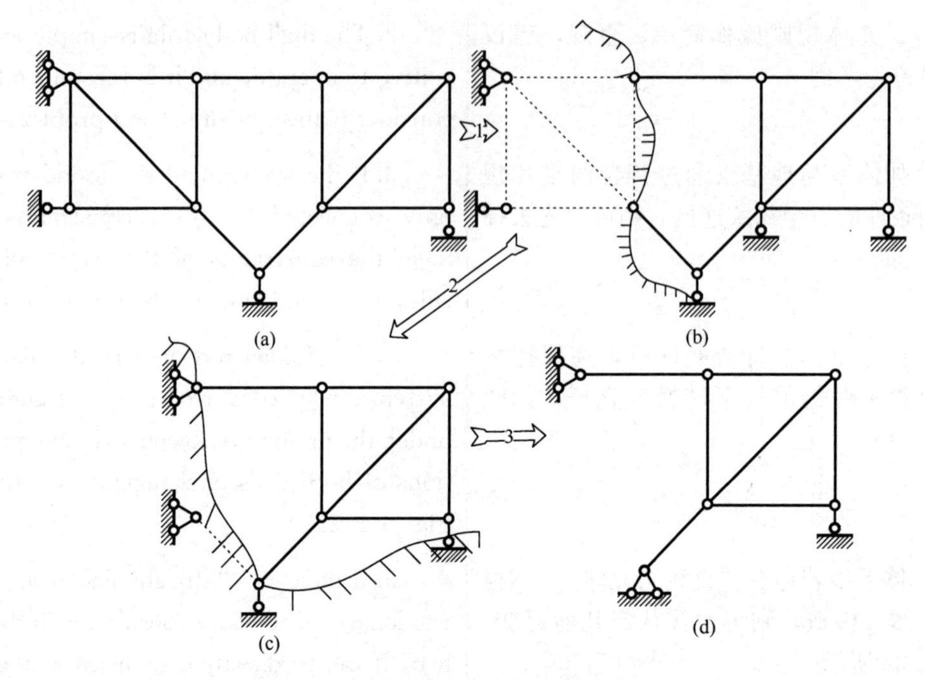

图 2.23　扩大地基法

Figure 2.23　Expanded foundation method

等效替换法：体系中任意一个几何不变部分，用若干个铰接三角形代替，且不改变其与其他部分的连接方式，如图 2.24 所示。

Equivalent substitution method: Any geometric stable part in the system is replaced by several hinged triangles, which does not change the connections with other parts, as shown in Figure 2.24.

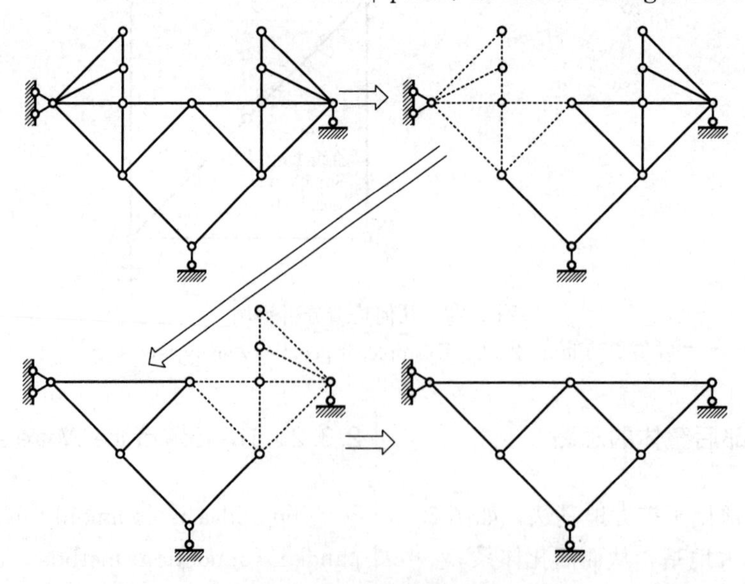

图 2.24　等效替换法

Figure 2.24　Equivalent substitution method

2.3.3 寻找三刚片规则中的刚片

可采用“顺藤摸瓜”的方法寻找刚片，如图 2.25 所示。

2.3.3 Find the Rigid Body in the Three-rigid-body Rule

The rigid body can be found by the method of “tracking down”, as shown in Figure 2.25.

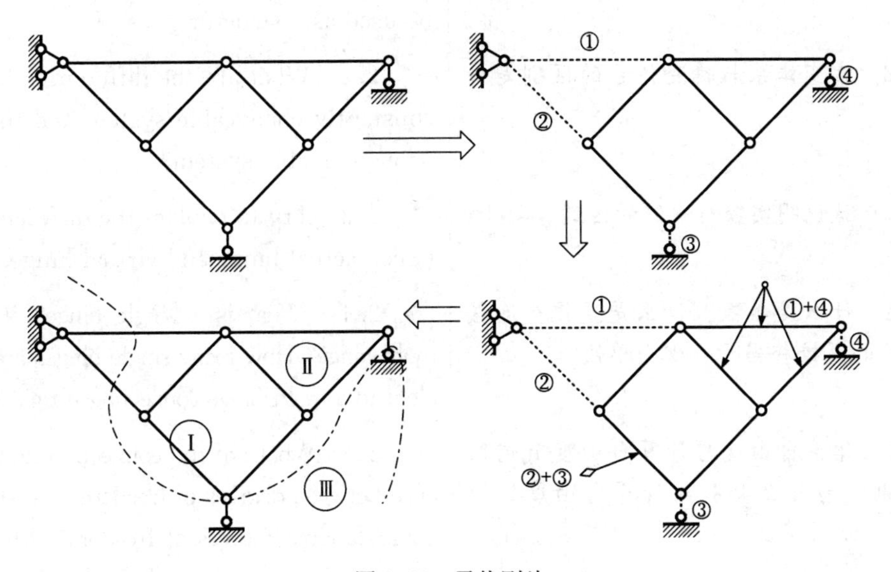

图 2.25 寻找刚片

Figure 2.25 Search for rigid body

2.3.4 化虚为实

某些复杂问题无法直接分析，可以将瞬时位置的虚铰转换为实铰进行分析，如图 2.26所示。

2.3.4 Change the Virtual Hinge into Actual Hinge

Some complex problems cannot be analyzed directly, but virtual hinges at instantaneous positions can be transformed into actual hinges for analysis, as shown in Figure 2.26.

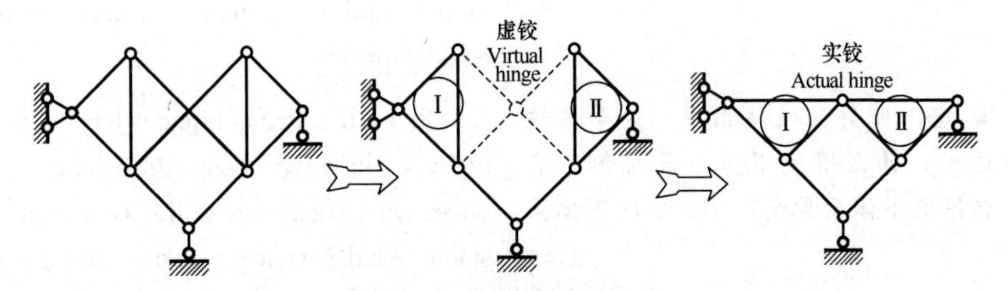

图 2.26 化虚为实示意图

Figure 2.26 Schematic diagram of changing virtual hinge into actual hinge

思 考 题

2.1 从不同角度归纳瞬变体系的特点，并举例说明瞬变体系不能作为结构使用的原因。

2.2 常变体系与瞬变体系的区别是什么？

2.3 试说明实铰与虚铰的区别。

2.4 什么是单铰？什么是复铰？连接 n 个刚片的复铰相当于多少个单铰？

2.5 体系自由度与计算自由度有何联系与区别？为什么要引入计算自由度的概念？

2.6 当体系计算自由度 $W \leqslant 0$ 时，体系一定几何不变吗？

2.7 什么是必要约束？什么是多余约束？瞬变体系一定存在多余约束吗？

2.8 无多余约束几何不变体系三个基本组成规则之间有何联系？

2.9 若三刚片三铰体系中，一个虚铰在无穷远处，什么情况下体系是几何不变的？什么情况下体系常变？什么情况下体系瞬变？

Questions

2.1 Summarize the characteristics of instantaneous unstable system from different aspects, and give examples to explain why instantaneous unstable system cannot be used as a structure.

2.2 What are the differences between constantly changeable system and instantaneous unstable system?

2.3 Try to explain the differences between actual hinge and virtual hinge.

2.4 What is a single hinge? What is multi-hinge? How many single hinges are equivalent to a multi-hinge connecting n rigid bodies?

2.5 What are the connection and difference between degrees of freedom of system and computational degrees of freedom? Why is the concept of computational degrees of freedom introduced?

2.6 When computational degrees of freedom $W \leqslant 0$, is the system geometric stable?

2.7 What are the necessary constraints? What are redundant constraints? Must there be redundant constraints in instantaneous unstable systems?

2.8 What are the connections among the three basic constituent rules of a geometric stable system without redundant constraints?

2.9 If a virtual hinge is located at infinity in a three-rigid-body three-hinge system, under what conditions is the system geometric stable? Under what conditions does the system constantly change? Under what conditions is the system instantaneous unstable?

2.10 若三刚片三铰体系中，两个虚铰在无穷远处，什么情况下体系是几何不变的？什么情况下体系常变？什么情况下体系瞬变？

2.10 If two virtual hinges are at infinity in a three-rigid-body three-hinge system, under what conditions is the system geometric stable? Under what conditions does the system constantly change? Under what conditions is the system instantaneous unstable?

2.11 若三刚片三铰体系中，三个虚铰在无穷远处，体系一定是几何可变吗？

2.11 If in a three-rigid-body three-hinge system with three virtual hinges at infinity, is the system geometric changeable?

2.12 在几何构造分析中可以进行哪些等效代换？如何保证变换的等效性？

2.12 What equivalent substitutions can be made in geometric formation analysis? How to ensure the equivalence of these substitution?

2.13 试总结平面体系几何构造分析的主要思路和需要注意的问题。

2.13 Try to summarize the main ideas and problems that are needed to pay attention to in geometric formation analysis of coplanar system.

习　题

Exercises

2.1 确定体系的计算自由度 W，见图 2.27。

2.1 Determine the computational degrees of freedom W of the systems in Figure 2.27.

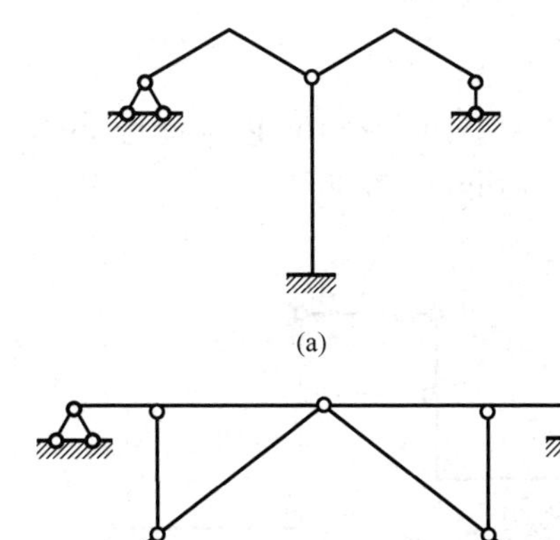

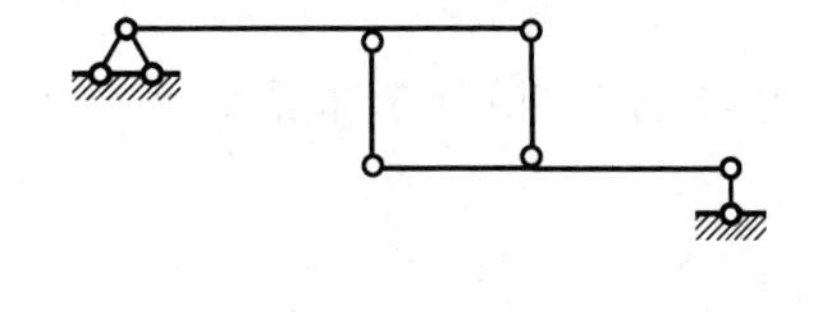

(a)　(b)

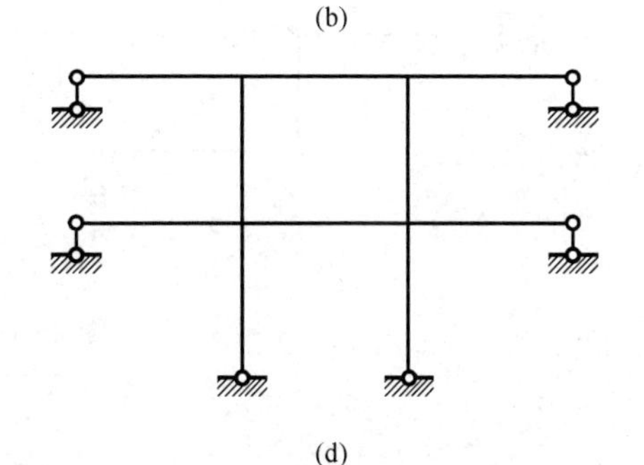

(c)　(d)

图 2.27 习题 2.1

Figure 2.27 Exercise 2.1

2.2　分析图2.28所示体系的几何构造。

2.2　Analyze the geometry of the systems in Figure 2.28.

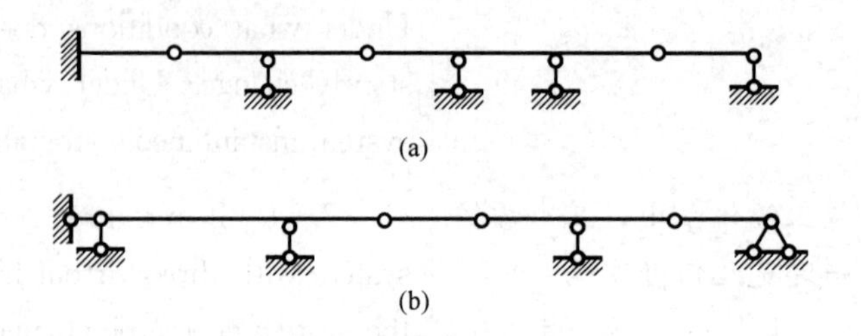

图2.28　习题2.2
Figure 2.28　Exercise 2.2

2.3　分析图2.29所示体系的几何构造。

2.3　Analyze the geometry of the systems in Figure 2.29.

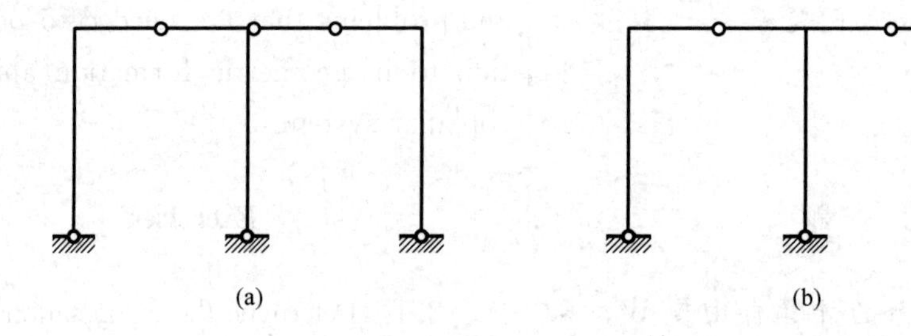

图2.29　习题2.3
Figure 2.29　Exercise 2.3

2.4　分析图2.30所示体系的几何构造。

2.4　Analyze the geometry of the systems in Figure 2.30.

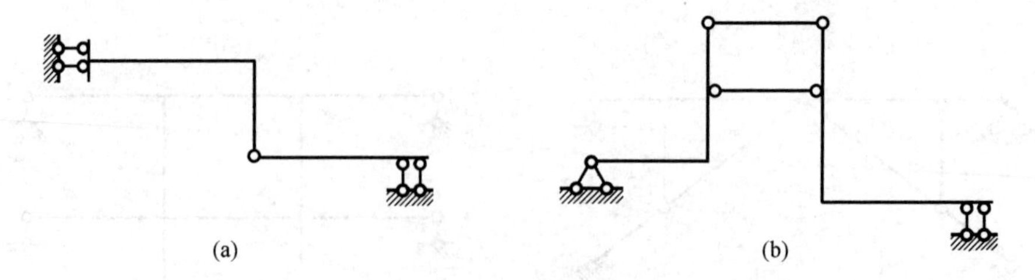

图2.30　习题2.4
Figure 2.30　Exercise 2.4

2.5　分析图2.31所示体系的几何构造。

2.5　Analyze the geometry of the systems in Figure 2.31.

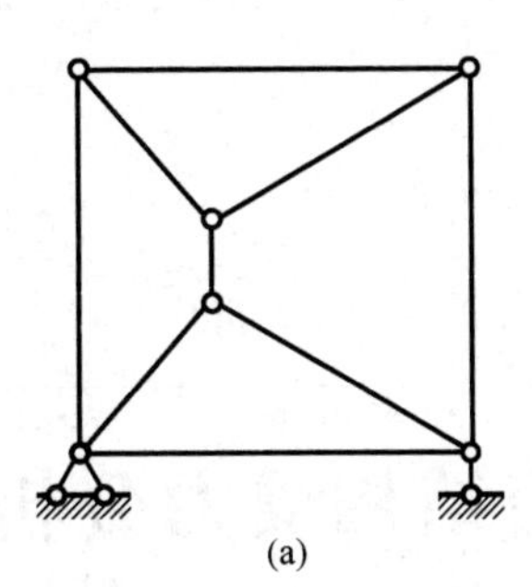
(a)

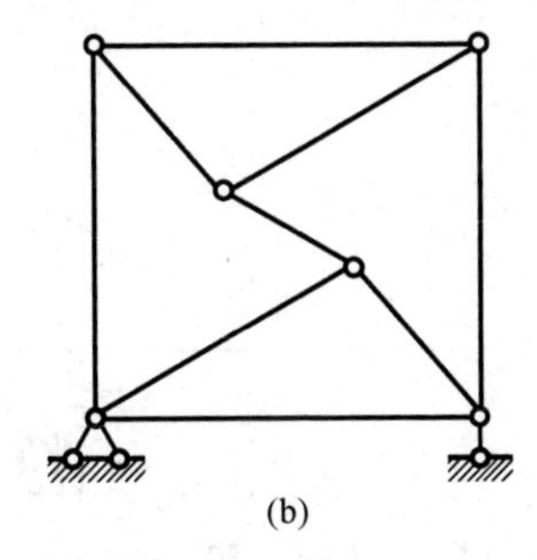
(b)

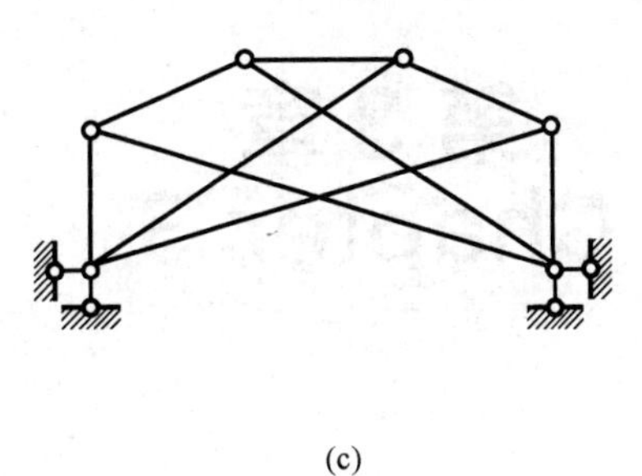
(c)

图 2.31 习题 2.5
Figure 2.31 Exercise 2.5

2.6 分析图 2.32 所示体系的几何构造。

2.6 Analyze the geometry of the systems in Figure 2.32.

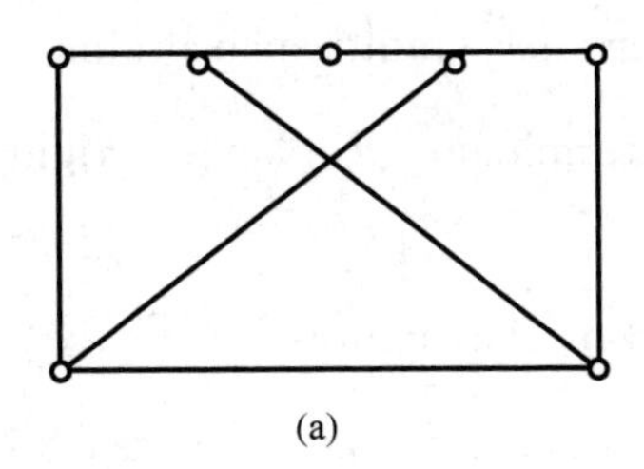
(a)

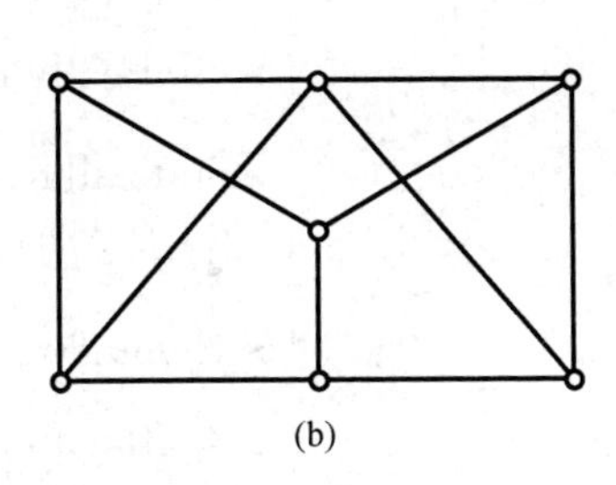
(b)

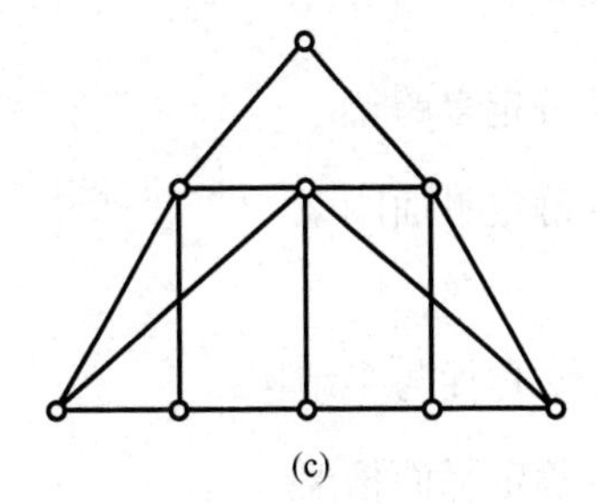
(c)

图 2.32 习题 2.6
Figure 2.32 Exercise 2.6

第3章 Chapter 3

静定结构受力分析 Analysis of Statically Determinate Structures

要点

➢截面法求解梁的内力

➢静定多跨梁

➢静定平面刚架

➢对称性与半结构

➢静定平面桁架

➢组合结构

➢三铰拱

Keys

➢ Solve internal forces by the method of sections

➢ Statically determinate multi-span beams

➢ Statically determinate coplanar rigid frames

➢ Symmetry and half structures

➢ Statically determinate coplanar trusses

➢ Composite structures

➢ Three-hinged arches

3.1 梁的内力计算及内力特征

3.1 Calculation and Characteristics of the Internal Forces for Beams

3.1.1 截面内力及其正负号规定

3.1.1 Internal Forces of Sections and Sign Convention

平面结构中的杆件，其横截面上一般有三个内力分量：轴力 F_N、剪力 F_Q和弯矩 M，其正负号规定如图 3.1 所示。

For members of coplanar structures, the internal forces acting at the section consist of an axial force F_N, a shear force F_Q, and a bending moment M. The positive direction is shown in Figure 3.1.

轴力：以拉力为正，压力为负；

Axial force: Positive axial force tends to elongate the segment.

剪力：以绕截面顺时针为正，逆时针为负；

Shear force: Positive shear force tends to rotate the segment clockwise.

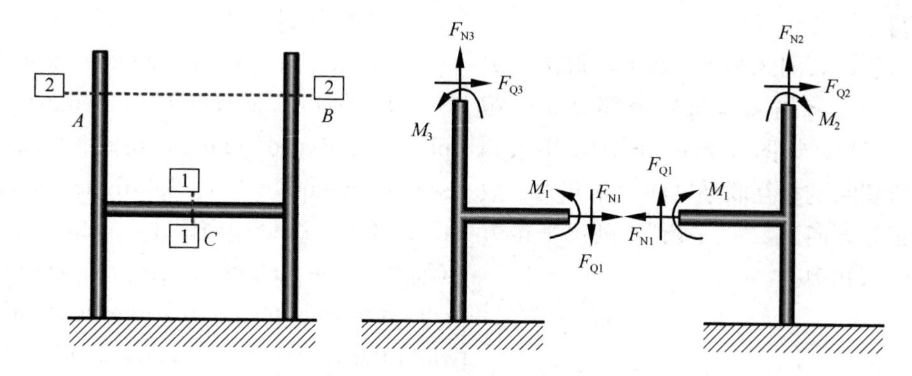

图 3.1 截面内力的正方向

Figure 3.1 Positive directions of sections

弯矩：对于水平杆件弯矩可假定杆件下方受拉为正，刚架中可假定内侧受拉为正。

Bending moment: Positive bending moment tends to stretch the bottom/interior of the beams/rigid frames.

弯矩图的弯矩竖标必须画在受拉侧，不标正负号，轴力图和剪力图必须标正负号。

It should be noted that the axial force diagrams and shear force diagrams should be marked with +/− signs, while the bending moment diagrams should be drawn at the stretch side of the member with no +/−signs.

3.1.2 截面法

3.1.2 Method of Sections

截面法是计算截面内力的基本方法，可分为三个步骤："一截二替三平衡"，下面进行举例说明：

The method of sections is fundamental to solve internal forces of members, by following steps of "cut, substitution, and equilibrium". The following shows the examples:

【例 3.1】求图 3.2(a) 所示结构支座反力。

［**Example 3.1**］Solve the reaction forces for structural supports in Figure 3.2(a).

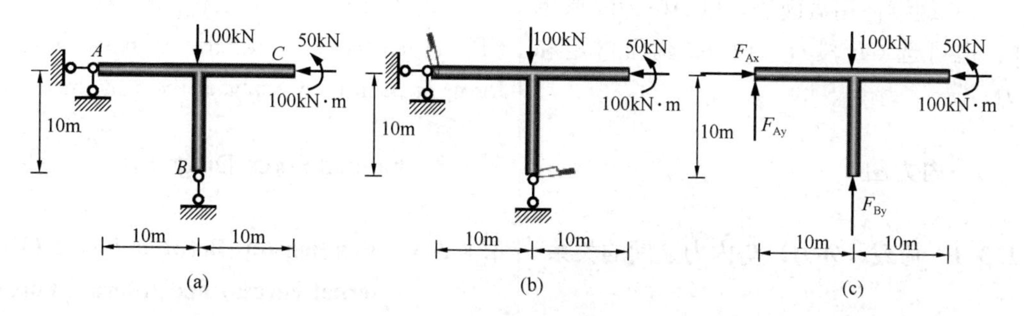

图 3.2 截面法求支座反力

Figure 3.2 Solve reaction forces by the method of sections

【解】

"一截"，如图3.2(b) 所示，取出隔离体（注：必须将隔离体与其余部分完全截断）；"二替"，如图3.2(c) 所示，用未知力代替所切断的相互联系（注：未知力一般假设为正号方向）；"三平衡"，列出平衡方程，求解未知力。

[Solution]

Cut first: Take the isolation body (Figure 3.2b) (Note: The isolation body must be completely truncated from the rest); Substitution second (Figure 3.2c): Substitute the connections from other parts by unknown forces (Note: The unknown force is generally assumed to be in the positive direction); Equilibrium third: Apply the equations of equilibrium to solve the unknown forces.

$$\left.\begin{array}{l}\sum F_x = 0\\ \sum F_y = 0\\ \sum M_A = 0\end{array}\right\} \rightarrow \left\{\begin{array}{c}F_{Ax} = 100\text{kN}\\ F_{Ay} + F_{By} = 100\text{kN}\\ 10F_{By} + 100 = 10 \times 100\text{kN} \cdot \text{m}\end{array}\right. \rightarrow \left\{\begin{array}{l}F_{Ax} = 100\text{kN}\\ F_{Ay} = 10\text{kN}\\ F_{By} = 90\text{kN}\end{array}\right.$$

【例3.2】求图3.3(a) 所示简支梁跨中 C 截面弯矩。

[Example 3.2] Solve the internal bending moment at point C, as shown in Figure 3.3(a).

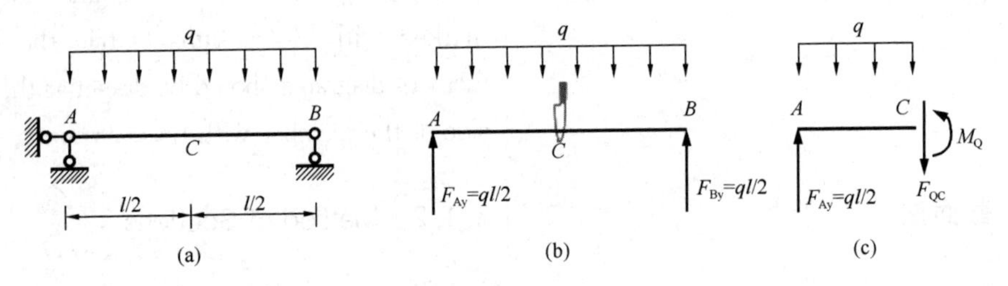

图3.3　截面法求截面内力

Figure 3.3　Solve internal forces by the method of sections

【解】

(1) 以 AB 为隔离体求出支座反力，如图3.3(b) 所示；

(2) 继续使用截面法，以 AC 为隔离体（注：也可选 CB 段），可得 C 截面弯矩和剪力。

[Solution]

(1) Isolate AB from supports as free body and solve the reactions, as shown in Figure 3.3(b);

(2) Cut at point C and then take AC (or CB) as free body. Solve the unknown internal forces at point C by applying the same method.

3.1.3　内力图

3.1.3　Internal Force Diagrams

3.1.3.1　荷载（外力）与内力之间的关系

3.1.3.1　Relationship Between Loads (External Forces) and Internal Forces

在荷载连续分布的直杆段内，取图3.4

As a distributed load acting on a straight

(a) 所示微段 dx 为隔离体。

member, we isolate a segment with a length of dx as a free body, as shown in Figure 3.4(a).

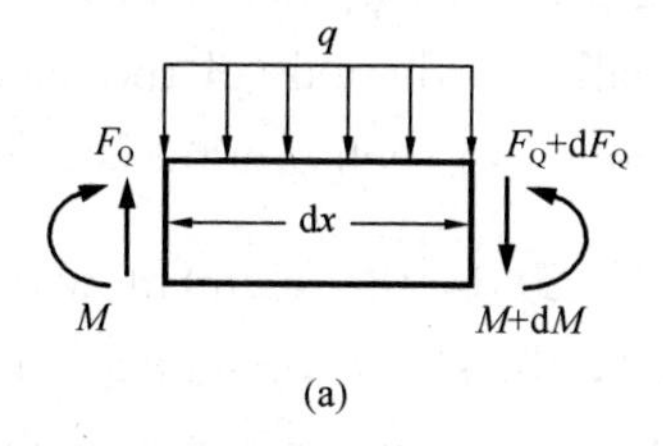

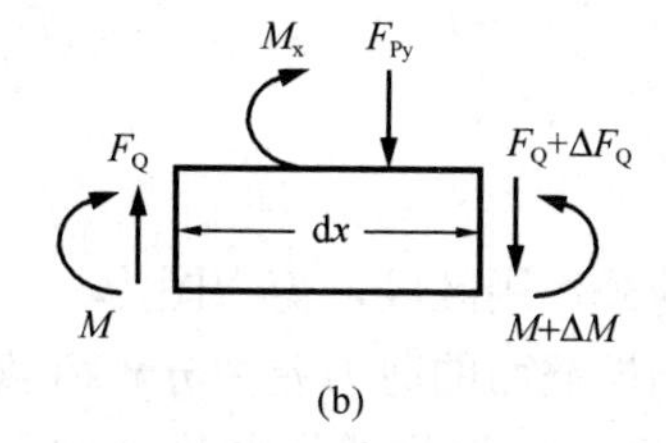

图 3.4 荷载与内力之间的关系

Figure 3.4 The relationship between loads and internal forces

其中分布荷载集度 q 向下取正，由平衡条件可导出微分关系如下：

Where the positive force of q is downward. Then by applying the equations of equilibrium, the differential relationship is as following:

$$\frac{dF_Q}{dx}=-q,\ \frac{dM}{dx}=F_Q,\ \frac{d^2M}{dx^2}=-q \tag{3.1}$$

在集中荷载作用处，取微段 dx 为隔离体，如图 3.4(b) 所示，其中竖向集中力 F_{Py} 向下取正，集中力偶 M 顺时针为正。由平衡条件可导出集中荷载与内力增量之间的关系如下：

At the point where the concentrated load locates, we also isolate a segment (Figure 3.4b) with a length of dx as a free body, where the positive concentrated load F_{Py} is downward and the positive bending moment M tends to rotate the segment clockwise. The relationship between the concentrated load and the increment of the internal force can be derived by using the equilibrium theory:

$$\Delta F_Q=-F_{Py},\ \Delta M=M_x \tag{3.2}$$

3.1.3.2 内力图的形状特征

3.1.3.2 The Shape Characteristics of the Internal Force Diagrams

由微分关系式 (3.1) 可以得到内力图在一段直杆上的形状特征（零、平、斜、抛）：

The shape characteristics for the internal force diagrams in straight rod can be drawn from the mathematic differential relationship of Equation (3.1) (zero, parallel, slope, parabola):

(1) 无分布荷载区段 ($q=0$)，剪力图平行轴线，弯矩图为一斜直线（即零、平、斜）。如果这一段内各截面的剪力均为零，

(1) In the segment with no loads ($q=0$), the shear force diagram is parallel to the axis of the member, the moment diagram is

弯矩图就平行轴线，即各截面弯矩为一常数。

a linear line (i. e. , zero, parallel, slope). In particular, if the shear force is zero in a segment, then the moment diagram yields parallel to the axis of the member, i. e. the moment of each section is a constant.

(2) 均布荷载作用区段，剪力图为一斜直线（斜直线两端截面的剪力差为分布荷载的合力），弯矩图为一抛物线（抛物线的凸向即荷载 q 的指向，即平、斜、抛）。剪力为零处，弯矩达到极值。请读者思考在线性分布荷载区段剪力图和弯矩图的形状。

(2) In the segment with a distributed load, the shear force diagram is a linear line (The total increment/decrement is the resultant value of the distributed load), then the moment diagram is a parabolic curve (The distributed load q goes towards the valley bottom of the parabolic curve, i. e. , parallel, slope, parabola). The moment reaches its maximum or minimum at the point where the shear force is zero。Please consider the shapes of shear force and bending moment diagrams in linearly distributed load sections.

由增量关系式（3.2）可以得到集中荷载作用处内力图的突变特征：

According to the incremental Equation (3.2), the abrupt characteristics of the internal force diagram at the place of concentrated load can be obtained:

(1) 横向集中力作用处，剪力图有突变，突变的值等于集中力的值。弯矩图连续，但发生拐折，形成尖点，尖角的指向与集中力的指向相同。

(1) At the point where the concentrated load acts, the shear force diagram will have a sudden increment/decrement which equals to the positive/negative value of the concentrated load. Meanwhile, the bending moment diagram will have a sharp corner, which points to the direction of the concentrated load.

(2) 集中力偶作用处，剪力图无变化。弯矩图有突变，突变的值为该集中力偶的值。因为集中力偶作用处两侧的剪力值相等，所以集中力偶作用处两侧弯矩图的切线应互相平行。

(2) At the point where the concentrated moment acts, it should be noted that the concentrated moment will not affect the shear force diagram (i. e. the shear force diagram will not change at that point), and the bending moment diagram will have a sudden increment/decrement which equals to the value of the concentrated moment.

Since the shear forces at both sides of the moment are equal, tangents of the bending moment diagrams at both sides of the point will be parallel.

上述荷载与内力图形状之间的对应关系可由图 3.5 直观地给出。

The above relationships can be seen in Figure 3.5.

由图 3.5 可以看到“弯矩图的弯折方向与外力的指向相同”，可参照“弓箭法则”（图 3.5b）进行快速判断：

From Figure 3.5,“The bending moment diagram bends in the same direction as the external force”, which can refer to “bow and arrow rule” for quick judgement as shown in Figure 3.5(b):

自由端、铰接的杆端、铰支杆端无集中力偶作用时，这些截面弯矩为零，有集中力偶作用时，这些截面弯矩就等于该集中力偶。

Free ends, hinge rod ends and hinge support rod ends commonly have no moments. Only if a concentrated moment acts, an internal moment with the same value as the external concentrated moment will be found in the moment diagram.

上述关于内力图的特征，对于绘制内力图和校核内力图都是十分有用的。

Above are the characteristics of the internal force diagrams which are very useful to draw and check the internal force diagrams.

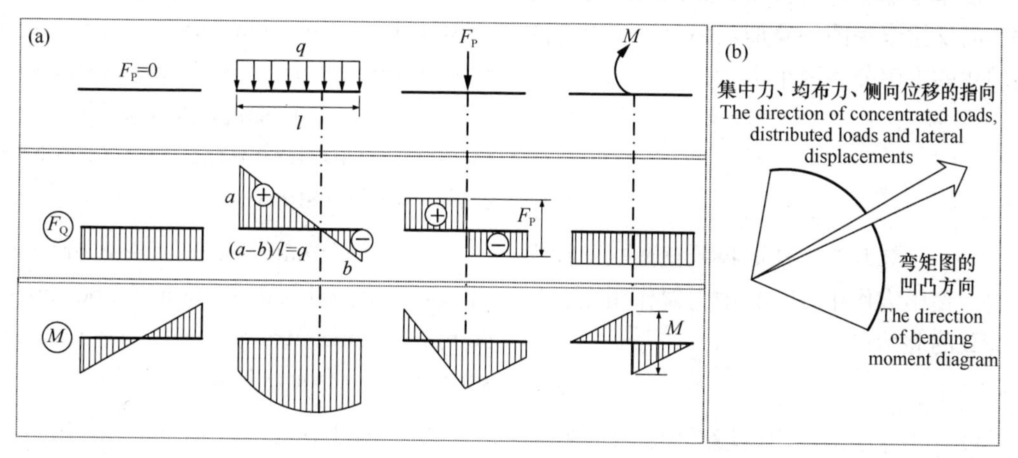

图 3.5 内力图

(a) 荷载与内力图之间的对应关系；(b) 弓箭法则

Figure 3.5 The internal force diagrams

(a) Relationship between loads and internal forces; (b) Bow and arrow rule

3.1.4 叠加法

3.1.4 The Principle of Superposition

内力图绘制中，叠加法是非常常用的方

The principle of superposition is very

法，其使用前提是：材料必须表现为线弹性方式，结构的几何形状在施加荷载时不能发生显著变化，其原理如图 3.6 所示。

useful for drawing the internal force diagrams. Two requirements must be imposed for the principle to apply: the material must behave in a linear-elastic manner and the geometry of the structure must not undergo significant changes when the loads are applied, as shown in Figure 3.6.

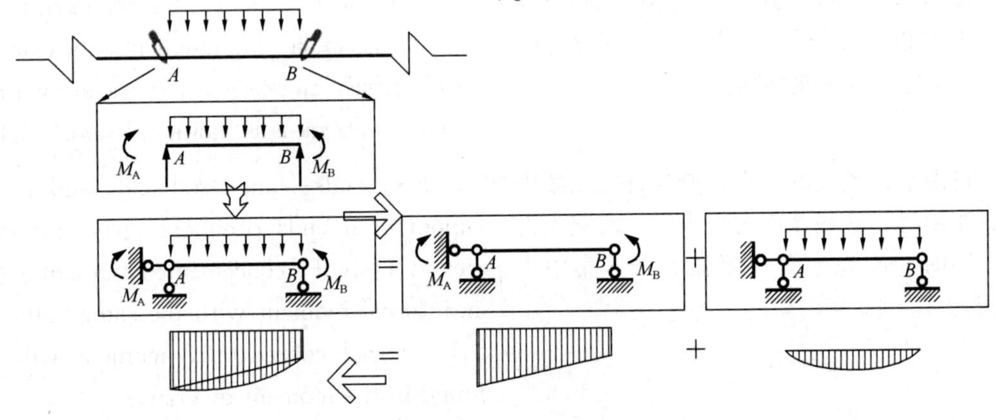

图 3.6 叠加法原理

Figure 3.6 The principle of superposition

使用叠加法可以将任意梁段的内力图分解为简支梁弯矩图的叠加，因此，可以把内力问题的求解分为两个步骤：

By applying the principle of superposition, a beam under arbitrary loads can be substituted as the sum of simple supported beams under some simple loads. Therefore, the solution to the internal forces can be divided into two steps:

（1）一算两头：选取控制截面（通常为节点、集中力作用点、集中力偶作用点、分布荷载的起始点）；

(1) Calculate both sides at first: Choose the critical section (The nodes, points of concentrated loads and moment, start points of distributed loads);

（2）二连中间：用区段对应简支梁在荷载下的弯矩图形状连接两头。

(2) Add them up to get the final results.

3.1.5 绘制弯矩图的注意点

3.1.5 Key Points for Drawing the Moment Diagrams

1. 弯矩图垂直于轴线绘制；

1. The moment diagrams are drawn perpendicular to the axes of the members.

2. 弯矩图画在杆件受拉侧，没有正负

2. Diagrams should be drawn at the

号；

side where the material is in tension.

3. 弯矩图的叠加是竖标的叠加，如图 3.7所示。

3. The superposition of the moment diagrams is the vertical values' adding, as shown in Figure 3.7.

4. 为了快速准确地绘制弯矩图，简支梁的基本弯矩图应牢记，如图 3.8 所示。

4. Some moment diagrams of simply supported beams under the certain loadings would be better to bear in mind, as shown in Figure 3.8.

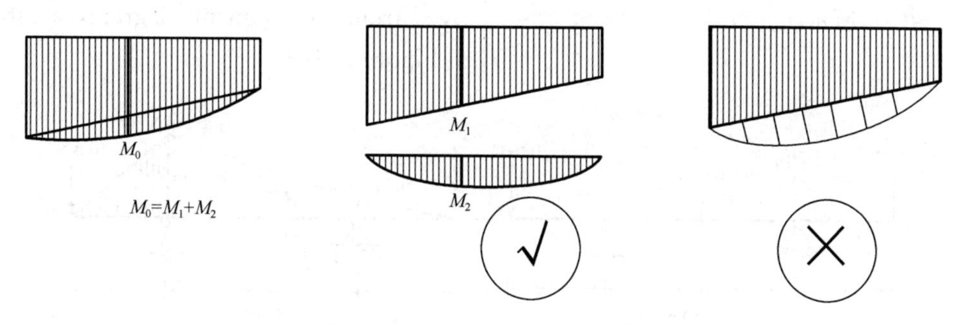

图 3.7 弯矩图叠加要注意的问题

Figure 3.7 Key points for applying the principle of superposition

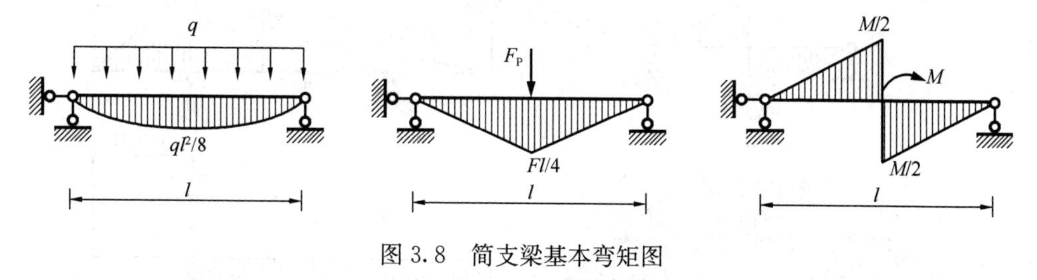

图 3.8 简支梁基本弯矩图

Figure 3.8 Some moment diagrams of simply supported beams under the certain loadings

3.1.6 静定梁内力求解举例

3.1.6 Examples of Solutions of Statically Determinate Beam Internal Forces

【例 3.3】 求图 3.9(a) 所示结构内力图。

[Example 3.3] Solve the internal force diagrams of the structure in Figure 3.9(a).

【解】

[Solution]

(1) 首先从两侧悬臂端开始，可求得两个支座处的弯矩如图 3.9(b) 所示（注：悬臂端开始求弯矩是非常推荐的快速计算技巧）。

(1) The process starts from the two free ends, the moment at the supports can be easily obtained, as shown in Figure 3.9 (b) (Note: Calculating the moment at the beginning of the cantilever end is a very rec-

ommended quick calculation technique).

（2）单独取出中间段，对应简支梁弯矩图如图3.9(c)所示。

(2) Take the middle part as a free body which is a simply supported beam, and then solve that case, as shown in Figure 3.9(c).

（3）采用叠加法，得到总体弯矩图如图3.9(d)所示。

(3) Apply the principle of superposition to get the whole moment diagram, as shown in Figure 3.9(d).

（4）剪力图直接由弯矩图得到，其方法如图3.9(e)所示。

(4) The shear force diagram can be derived from the moment diagram, as shown in Figure 3.9(e).

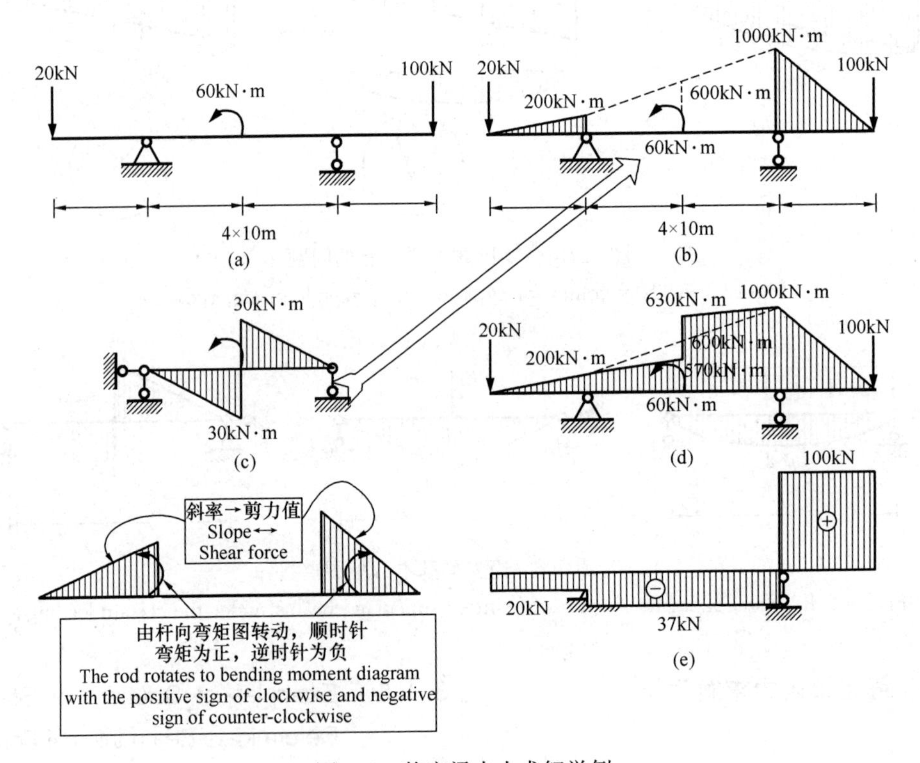

图3.9　静定梁内力求解举例

Figure 3.9　Examples of solving the internal force diagrams of a statically determinate beam

3.2　静定多跨梁

3.2　Statically Determinate Multi-span Beams

静定多跨梁是由若干根单跨静定梁（悬臂梁、简支梁、外伸梁）铰接而成的静定结

Statically determinate multi-span beams consist of several single-span statically de-

构，常见于桥梁结构（图 3.10a）等，其计算简图如图 3.10(b) 所示。

terminate beams which are connected by hinge joints (e. g. cantilever beams, simply supported beams, overhanging beams) . This type of beam is conventionally used in bridge structures (Figure 3.10a). Its calculation diagram is shown in Figure 3.10 (b).

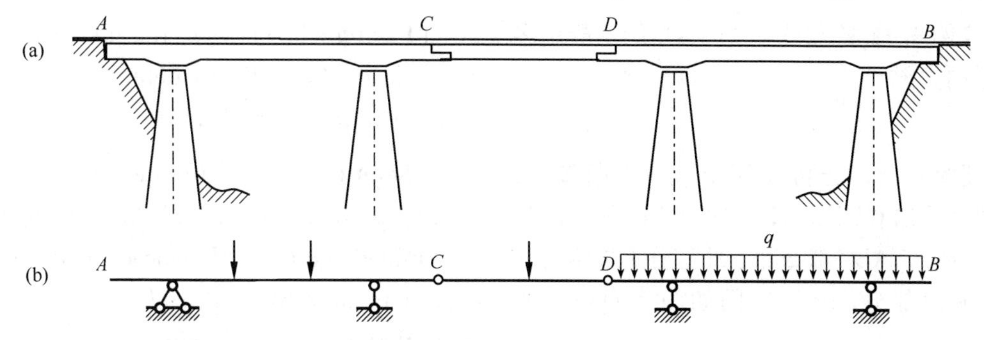

图 3.10 静定多跨梁

Figure 3.10 Statically determinate multi-span beam

静定多跨梁的几何组成特点是含有基本部分和附属部分，组成次序是先固定基本部分，后固定附属部分。

Statically determinate multi-span beams consist of main parts and secondary parts, the order of composition is to fix the main parts first, then the secondary parts.

静定多跨梁的受力特点是：当外力作用在基本部分上时，附属部分不受力，只有基本部分受力；当外力作用在附属部分上时，附属部分及其基本部分均受力。

The mechanical characteristics of statically determinate multi-span beams are as follows: Loads that act on the main parts will not induce any internal force in the secondary parts, while the secondary parts transfer loads to the main parts and cause internal forces there.

因此，计算静定多跨梁应遵循的原则是：先计算附属部分，将附属部分的支座反力反向加在基本部分上，再算基本部分。

Therefore, to solve the statically determinate multi-span beam, it is better to solve the secondary parts first and then apply the reactions of the secondary parts reversely to the main parts as the external loads.

必须指出，如果结构是由基本部分出发，逐次连接附属部分而组成的主从结构，应该反其组成次序进行这种结构的内力分析，即后装的先算，先装的后算。不仅对静定多跨梁，对以后要学到的其他主从结构也

If the structure is composed of the main parts and connecting the secondary parts successively, the internal force analysis of this structure should be carried out in reverse order of its composition, that is, the

是如此。

internal force of the first installation is calculated at last while the last installation is calculated first. Not only for statically determinate multi-span beams, this principle is also useful for other master-slave structures to be learned.

【例 3.4】 绘制图 3.11（a）所示静定多跨梁内力图。

[Example 3.4] Solve the internal force diagrams of the statically determinate multi-span beam in Figure 3.11(a).

【解】（1）分析几何组成。切断铰 C、F 即可看出 $CDEF$ 部分仍能维持平衡，是基本部分，ABC 部分和 FH 部分不能维持平衡是附属部分，其层次图如图 3.11(b) 所示。

[Solution] (1) Geometry analysis. Cutting at points C and F, then $CDEF$ is determined as a self-balanced main part, and parts of ABC and FH are secondary parts that cannot maintain self-balance, as shown in Figure 3.11(b).

（2）先算附属部分，求得支座反力及弯矩图，如图 3.11(c) 所示。

(2) Solve the secondary parts to get the reactions and moment diagrams, as shown in Figure 3.11(c).

（3）将附属部分的支座反力反向作用在基本部分上，求得基本部分的支座反力及弯矩图，如图 3.11(d) 所示。

(3) Apply the reactions as external forces on the main part reversely and then solve the main part, as shown in Figure 3.11(d).

（4）将各段梁的弯矩图合并，得到全梁的弯矩图，如图 3.11(e) 所示。注意，铰 C、F 处弯矩为零，BCD 段、EFG 段上无外荷载作用，弯矩图应为一斜直线。

(4) Apply the principle of superposition and add the internal forces up to get the final diagram, as shown in Figure 3.11(e). It is mentioned that the moments at points C and F are zero, while the moment diagrams tends to be straight slope lines in BCD and EFG with no external forces.

（5）多跨梁的剪力图，如图 3.11(f) 所示。先从 A 点开始向下突变 ql，形成平直线到 B 点；向上突变 $2ql$，形成平直线到 D 点；向下突变 $3ql/4$，DE 段剪力图为向右下斜直线，两端截面剪力差值为该段上分布荷载的合力，于是得到：$F_{QED}=F_{QDE}-2ql=ql/4-2ql=-7ql/4$；在 E 点向上突变 $9ql/4$，形成平直线到 G 点；向下突变 ql，

(5) The shear diagram is shown in Figure 3.11(f). It starts from point A and mutates ql downward to point B; mutates $2ql$ upward to reach the point D; $3ql/4$ is changed downwards, the shear force diagram of DE is slanted downward to the right, and the difference between the shear forces at both ends is the resultant force of

形成平直线到 H 点；向上突变 $ql/2$，正好回到基线，这表明所有外力满足竖向投影平衡。

the load distributed on this section. Thus, it can be obtained that $F_{QED}=F_{QDE}-2ql=ql/4-2ql=-7ql/4$; At point E, it mutates $9ql/4$ upward and flattens the line to point G; mutates ql downward and flattens line to point H; mutates $ql/2$ upward, just back to the baseline which indicates that all external forces satisfy vertical projection equilibrium.

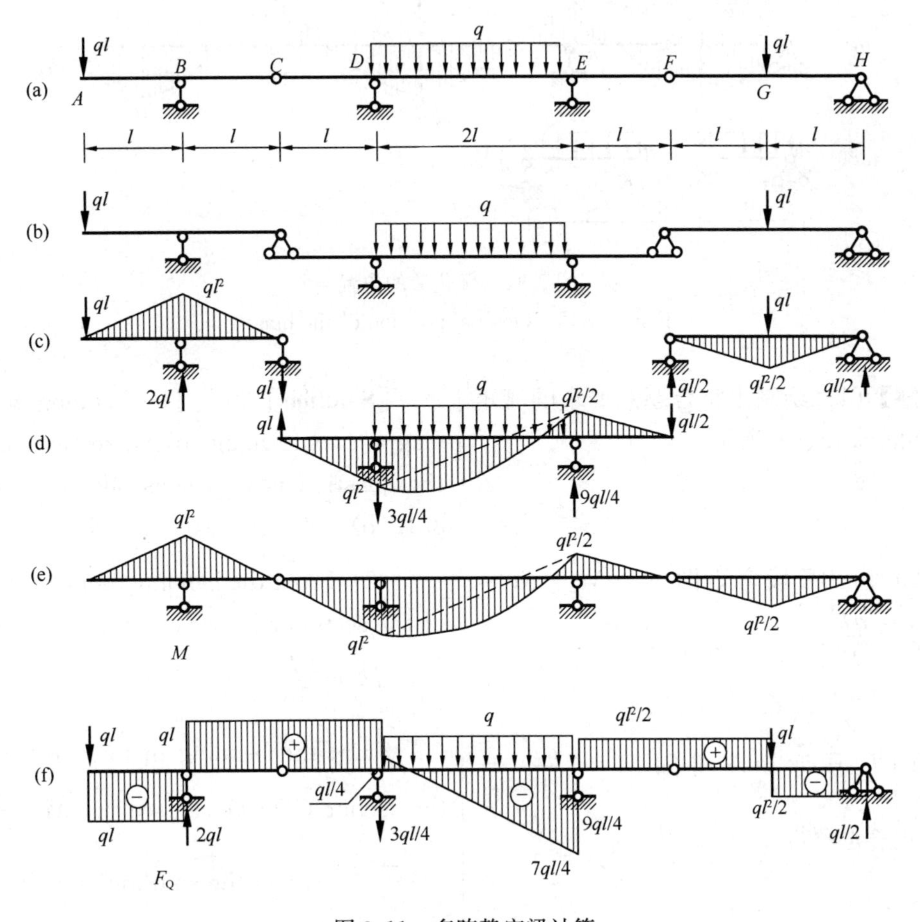

图 3.11　多跨静定梁计算

Figure 3.11　Calculation of the statically determinate multi-span beam

【例 3.5】三跨静定梁承受均布荷载 q，如图 3.12(a) 所示，铰 C 和铰 D 到支座的距离为 x。两边跨的跨度为 l_1，中跨的跨度为 l。试确定正负弯矩的峰值相等时 x，l_1，l 之间的关系。

[**Example 3.5**] The statically determinate three-span beam bears a distributed load q as shown in Figure 3.12(a). The distances from points C and D to supports are x. The lengths for side spans are l_1 and the length for the middle span is l. Try to calculate the relationships among x, l_1 and l when the peak values of positive and negative moments are equal.

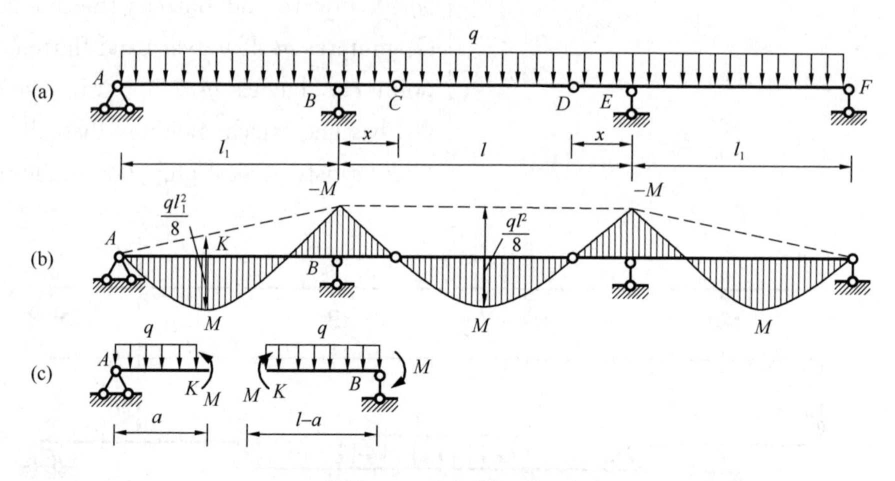

图 3.12　优化铰的位置

Figure 3.12　Optimal position of the hinge

【解】(1) 绘制出符合题意要求的弯矩图，如图 3.12(b) 所示。

[**Solution**] (1) The bending moment diagram conforming to the requirements of the question is drawn as shown in Figure 3.12(b).

(2) 由 BE 段弯矩图可知 $M+M=\frac{ql^2}{8}$, $M=\frac{ql^2}{16}$。

(2) From the bending moment diagram of the section BE, we can see that $M+M=\frac{ql^2}{8}$, $M=\frac{ql^2}{16}$.

由 CD 段弯矩图可知 $M=\frac{ql^2}{16}=\frac{q(l-2x)^2}{8}$，解得 $x=0.147l$。

From the bending moment diagram of the section CD, we can see that $M=\frac{ql^2}{16}=\frac{q(l-2x)^2}{8}$, and the solution is $x=0.147l$。

(3) 令 AB 跨的最大弯矩发生在 K 截面 ($AK=a$)，则 $M_K=M$，$F_{QK}=0$。如图 3.12(c) 所示，取隔离体 AK，由 $\Sigma M_A=0$，得到 $\frac{qa^2}{2}=M=\frac{ql^2}{16}$，求得：$a=\frac{l}{2\sqrt{2}}=0.354l$。

(3) Let the maximum bending moment of AB span occur at section K ($AK=a$), then $M_K=M$, $F_{QK}=0$。As shown in Figure 3.12(c), take the isolation body AK, by $\Sigma M_A=0$, get $\frac{qa^2}{2}=M=\frac{ql^2}{16}$, and the solution is $a=\frac{l}{2\sqrt{2}}=0.354l$。

取隔离体 KB，由 $\sum M_B = 0$，得到 $\frac{q(l_1-a)^2}{2} - M - M = 0$，求得：$l_1 - a = \frac{l}{2}$，$l_1 = 0.854l$。

Take the isolation body KB, by $\sum M_B = 0$, get $\frac{q(l_1-a)^2}{2} - M - M = 0$, and the solution is $l_1 - a = \frac{l}{2}$, $l_1 = 0.854l$.

当 $x = 0.147l$，$l_1 = 0.854l$ 时梁中的正负弯矩峰值均为 $ql^2/16$。如果改用三个简支梁，则最大弯矩为 $ql^2/8$，是本例最大弯矩的 2 倍。

When $x = 0.147l$ and $l_1 = 0.854l$, the peak values of positive and negative bending moments in the beam are both $ql^2/16$. If three simply supported beams are used, the maximum bending moment is $ql^2/8$, which is twice the maximum bending moment of this example.

由于静定多跨梁设置了带伸臂的基本梁，这不仅使中间支座处产生了负弯矩（它将降低跨中正弯矩），另外减少了附属梁的跨度。因此静定多跨梁比相应的多个简支梁弯矩分布均匀，节省材料，但其构造要复杂一些。

The statically determinate multi-span beam is equipped with a basic beam with cantilever beams, which not only generates a negative bending moment at the middle support (it will reduce the positive bending moment in the middle span), but also reduces the span of the secondary beam. Therefore, statically determinate multi-span beams have more uniform moment distribution than corresponding simple beams and save materials, but their structure is more complicated.

3.3 静定平面刚架

3.3 Statically Determinate Coplanar Rigid Frames

3.3.1 刚架概述

3.3.1 The Introduction for Rigid Frames

刚架：梁与柱主要采用刚节点连接的结构（注：连接可以含有铰节点，但不能全部是铰节点，全部铰接的是桁架），如图 3.13 所示，可以看出刚架内部围合空间更大，在房建中广泛使用。

Rigid frames are structures composed of beams and columns which are mainly connected by rigid joints (It should be noted that when all the joints are hinge joints, the structure can be classified as a truss). As shown in Figure 3.13, the large architecture available space is the main advantage of this type of structure and it is widely used in building structures.

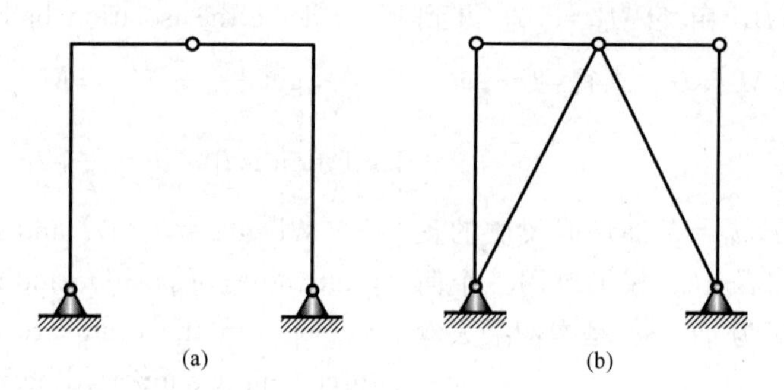

图 3.13 刚架和桁架
(a) 刚架；(b) 桁架
Figure 3.13 Rigid frames and trusses
(a) Rigid frames; (b) Trusses

从变形角度来看，刚节点连接处各杆间夹角保持不变；从受力角度来看，刚节点可以传递轴力、剪力和弯矩，刚架中的主要内力是弯矩。

From the perspective of deformation, the rigid joints will have no relative rotations between the members; while in the load-bearing perspective, such joints will transfer axial forces, shear forces, and bending moments. The main internal forces in rigid frames are moments.

基本静定刚架可分为：悬臂刚架、简支刚架、三铰刚架。基本形式的组合可以构成组合刚架，如图 3.14 所示。

Three main statically determinate rigid frames are cantilever rigid frames, simply supported rigid frames, and three-hinged rigid frames. Composite rigid frames are the combinations of the three above, as shown in Figure 3.14.

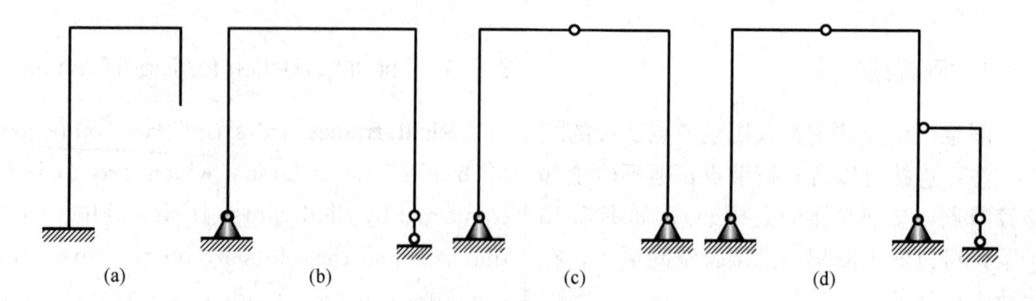

图 3.14 刚架的类型
(a) 悬臂刚架；(b) 简支刚架；(c) 三铰刚架；(d) 组合刚架
Figure 3.14 The types of the rigid frames
(a) Cantilever rigid frames; (b) Simply supported rigid frames; (c) Three-hinged rigid frames;
(d) Composite rigid frames

3.3.2 静定平面刚架的内力计算

3.3.2 Internal Forces Calculation of the Statically Determinate Coplanar Rigid Frames

3.3.2.1 方法概述

3.3.2.1 General Method

1. 对于复杂的组合刚架，可参照多跨静定梁的解题思路，拆分为基本结构和附属结构，按照“先附属再基本”的顺序求解。

1. To solve the complex composite rigid frames, the same idea of solving the statically determinate multi-span beams can be applied. The main parts and secondary parts should be determined at first, and then the order of “secondary parts first and then main parts” should be followed for calculation.

2. 对悬臂刚架，可不求解支座反力，直接由悬臂端开始计算。

2. For cantilever rigid frames, we can start from the free ends without solving the reactions.

3. 对简支刚架，可通过整体平衡条件直接计算支座反力，然后取控制截面依次计算。

3. For simply supported rigid frames, reactions should be solved at first by equilibrium condition, and then calculations can be taken for specific sections.

4. 对三铰刚架，由于支座反力超过三个，因此支座反力的计算必须拆铰联立方程组求解。

4. For three-hinged rigid frames, there are more than three unknowns for reactions which can be solved by simultaneous equations of structural parts after removing the hinge joints.

3.3.2.2 静定刚架计算举例

3.3.2.2 Statically Determinate Rigid Frame Case Study

【例 3.6】 定性画出图 3.15(a) 所示结构弯矩图。

[**Example 3.6**] Solve the bending moment diagram shown in Figure 3.15 (a) qualitatively.

【解】 求解步骤如图 3.15(b)～(d) 所示。

[**Solution**] The solving process is shown in Figures 3.15(b)-(d).

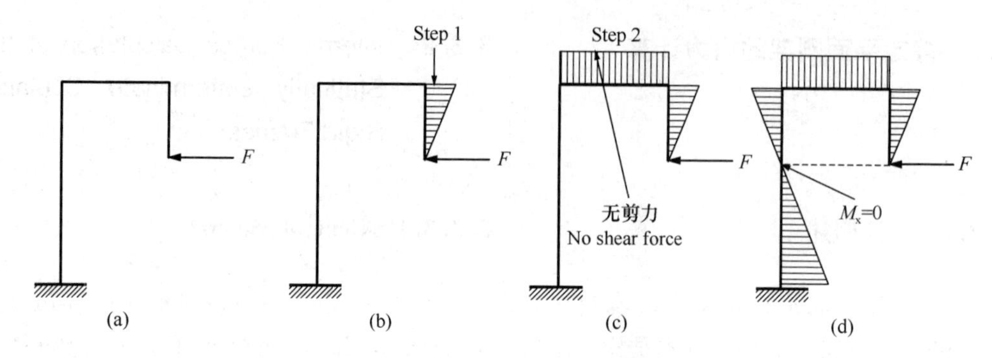

图 3.15 悬臂刚架弯矩图求解过程

Figure 3.15 Solving process for the cantilever rigid frame

【例 3.7】计算图 3.16(a) 所示简支刚架的内力。

[Example 3.7] Solve the internal forces for the simply supported rigid frame shown in Figure 3.16 (a).

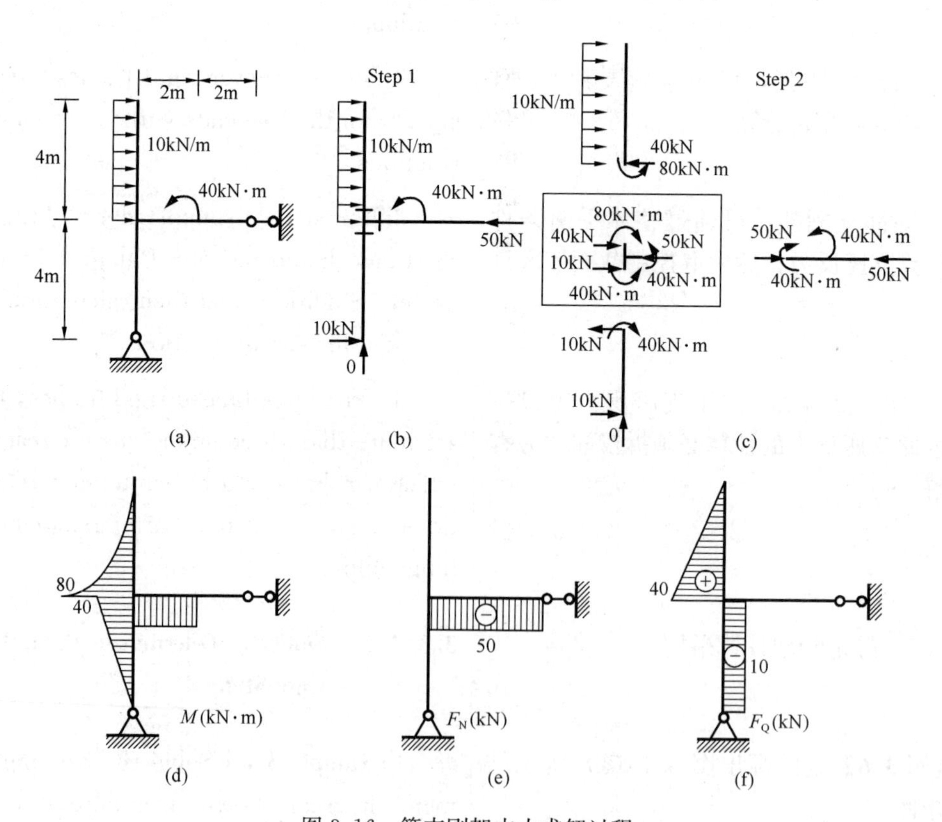

图 3.16 简支刚架内力求解过程

Figure 3.16 Solving process for the simply supported rigid frame

【解】解题过程如图 3.16 所示。需要指出，这一过程为常规解法，即求解支座反力，分解结构，取各隔离体为研究对象，求

[Solution] The problem solving process is shown in Figure 3.16. It should be noted that the process above is a conventional pro-

解各截面内力，绘制内力图。

cedure, i.e., solve reactions, decompose structure, take free bodies, solve internal forces, and draw diagrams.

【例 3.8】求图 3.17 所示三铰刚架支座反力。

［**Example 3.8**］As shown in Figure 3.17, solve the reactions for the three-hinged rigid frame.

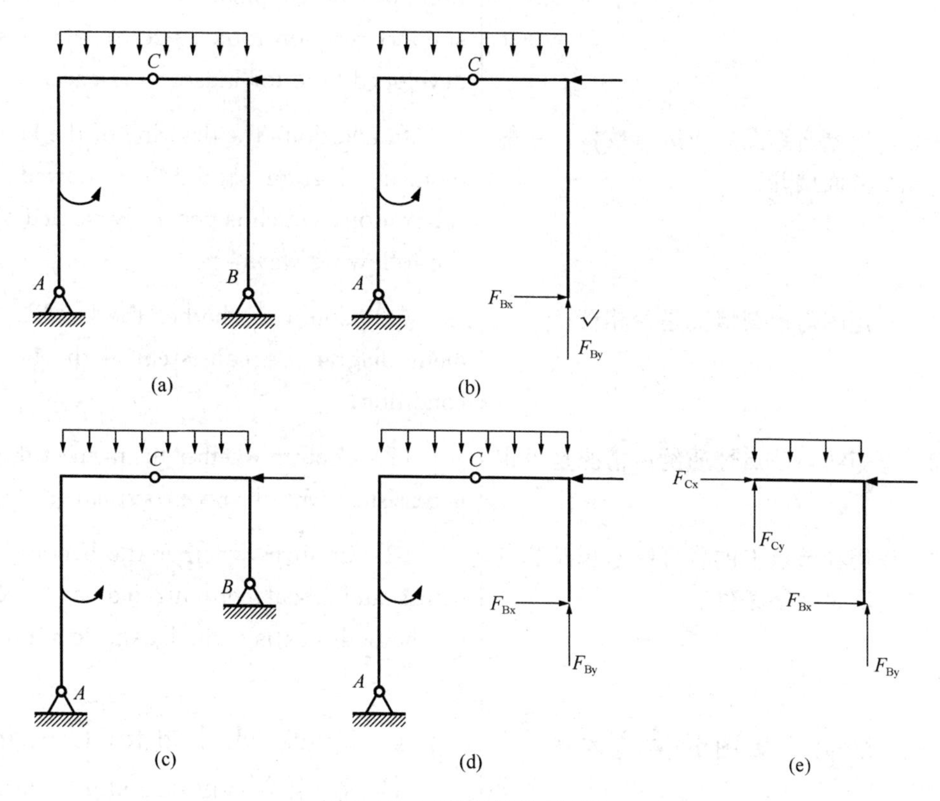

图 3.17 三铰刚架的支座反力

Figure 3.17 The support reaction forces of three-hinged rigid frame

【解题思路】

［**Ideas of solution**］

目标：*B* 支座反力；

Objective: The support reaction force of *B*;

方法：对另外两个铰 *A*、*C* 分别取力矩平衡方程，联立求解（图 3.17d、e 所示）。

Method: Take the moment balance equations for the other two hinges *A* and *C* respectively and solve them jointly (Figures 3.17d、e).

图 3.17(a) 为特殊情况，*A*、*B* 两铰平齐，可直接对 *A* 取力矩平衡方程，计算 F_{By}。

Figure 3.17(a) shows a special case where the two hinges *A* and *B* are flush and the moment balance equation can be taken directly for A to calculate F_{By}.

注意，复杂情况下，另外两铰可以是有限远的虚铰；若三铰刚架中的三个铰存在无穷远铰，则可补充 x 或 y 方向合力为零求解支座反力，限于篇幅，此处不再展开。

It should be mentioned that in complex cases, the other two hinges can be finitely distant virtual hinges; if they are infinitely distant hinges in the three-hinged rigid frames, the combined forces in x or y directions can be supplemented to solve for the support reaction force as zero, which is not developed here for reasons of space.

此外，绘制弯矩图后应进行校核，一般从以下几个方面展开：

In addition, the drawing of the bending moment diagram should be followed by a calibration, which is generally carried out in the following ways:

（1）弯矩图与荷载情况是否相符；

(1) Confirm whether the bending moment diagram is consistent with the load condition;

（2）弯矩图与节点性质约束情况是否相符；

(2) Confirm whether the moment diagram is consistent with the node constraints;

（3）作用在节点上的各杆端弯矩及节点集中力是否满足平衡条件。

(3) Confirm whether the bending moments and nodal concentrated loads acting on the nodes satisfy the balance conditions.

3.4　绘制弯矩图的快速方法

3.4 Quick Method for Drawing Bending Moment Diagrams

1. 对无剪力杆进行快速判定，如图3.18所示。

1. As in Figure 3.18, quick determination of shearless bars is taken.

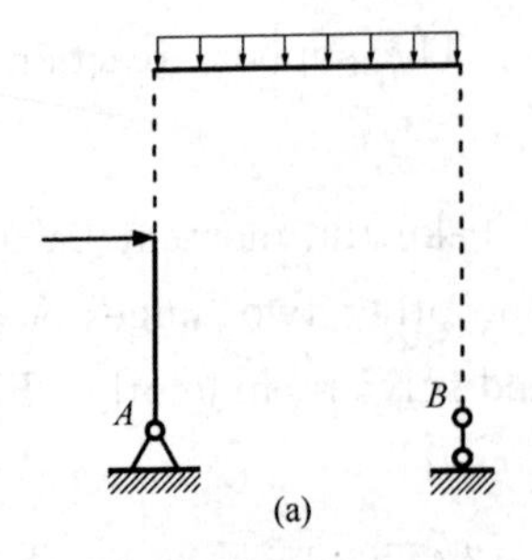

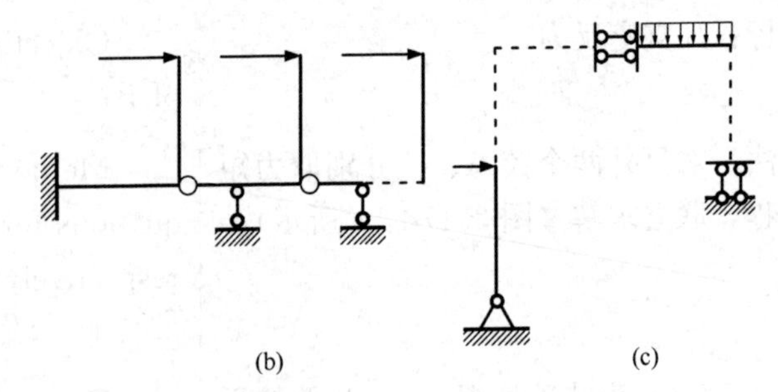

图3.18　无剪力杆

Figure 3.18　Shearless bars

2. 寻找"局部突破口"，如图 3.19 所示。

2. As in Figure 3.19, find "local breakthroughs".

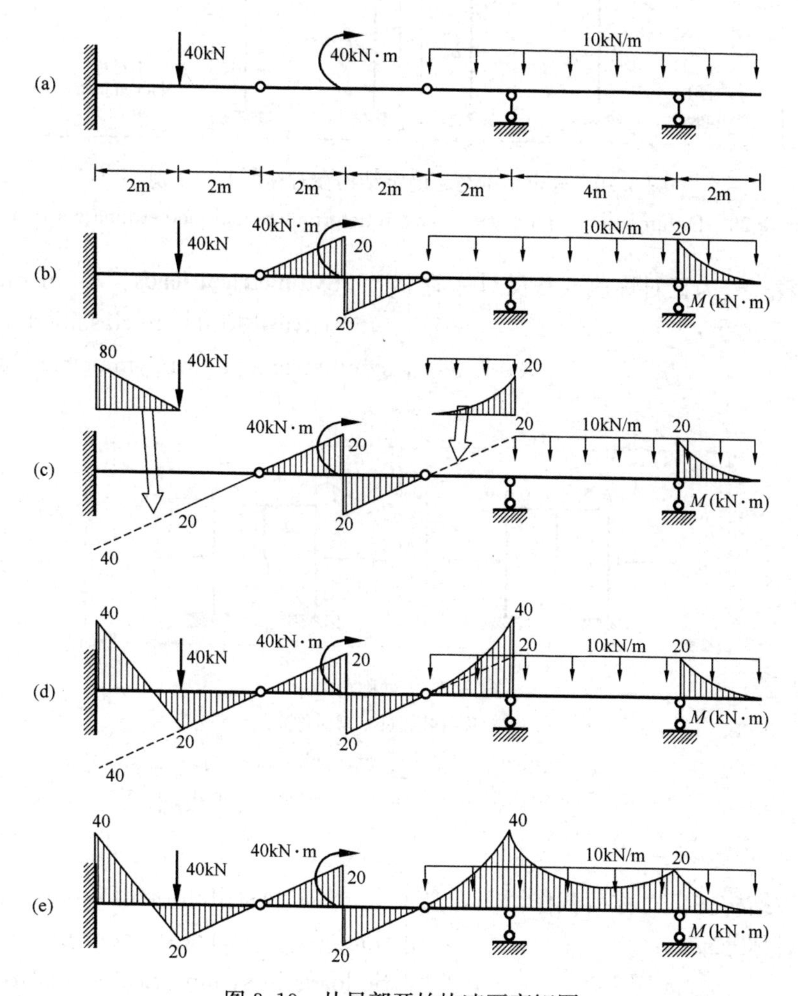

图 3.19 从局部开始快速画弯矩图

Figure 3.19 Quick drawing of bending moment diagrams starting from the local part

3.5 对称性与半结构

3.5 Symmetry and Semi-structure

对称结构：几何形状、支承情况、刚度分布关于某轴对称的结构（由于静定结构的内力与刚度无关，故只需要几何形状与支承情况对称即可，如图 3.20 所示）。

Symmetrical structure: Structures in which the geometry, supports and stiffness distribution are symmetrical of an axis. (As in Figure 3.20, internal forces are independent of stiffness, the geometry only needs to be symmetrical to the support situation).

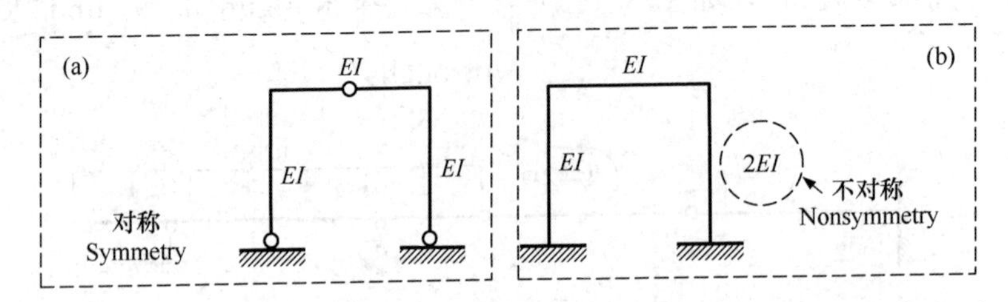

图 3.20 静定结构与超静定结构在对称性上的区别

Figure 3.20 Differences in symmetry between determinate and indeterminate structures

对称荷载：分为正对称荷载和反对称荷载，如图 3.21 所示：

Symmetrical loads: As in Figure 3.21, symmetrical loads are classified into positive symmetric and anti-symmetric loads:

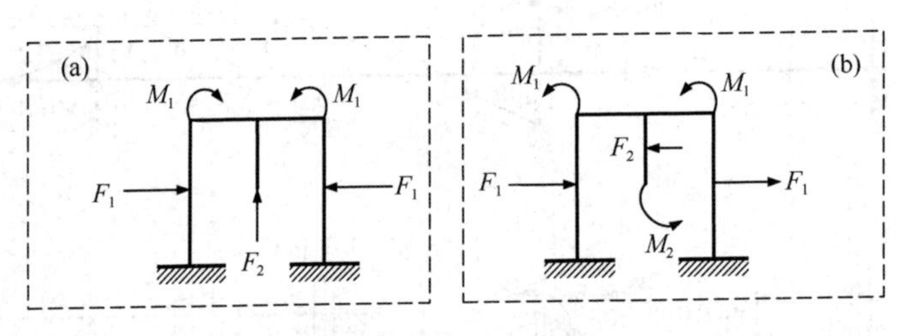

图 3.21 对称荷载

(a) 正对称荷载；(b) 反对称荷载

Figure 3.21 Symmetrical loads

(a) Asymmetric loads; (b) Anti-symmetric loads

(1) 对称结构在正对称荷载作用下，受力、变形呈对称分布；

(1) Distribution of forces and deformations is symmetrical under positive symmetric loads in symmetrical structure;

(2) 对称结构在反对称荷载作用下，受力、变形呈反对称分布；

(2) Distribution of forces and deformations is anti-symmetrical under anti-symmetric loads in symmetrical structure;

(3) 正对称荷载作用下对称轴经过截面只有对称内力，即轴力、弯矩；

(3) There are only symmetrical internal forces (axial forces, bending moments) in the section that axis of symmetry passes through under positive symmetric loads;

(4) 反对称荷载作用下对称轴经过截面只有反对称内力，即剪力。

(4) There is only anti-symmetrical internal forces (shear forces) in the section that axis of symmetry passes through under anti-symmetric loads.

根据对称轴经过截面内力特点及对称性（图 3.22），可得常见的半结构如图 3.23 所示。

According to the internal force characteristics and symmetry in the cross section that the axis of symmetry passes through (Figure 3.22), the common semi-structures can be obtained, as shown in Figure 3.23.

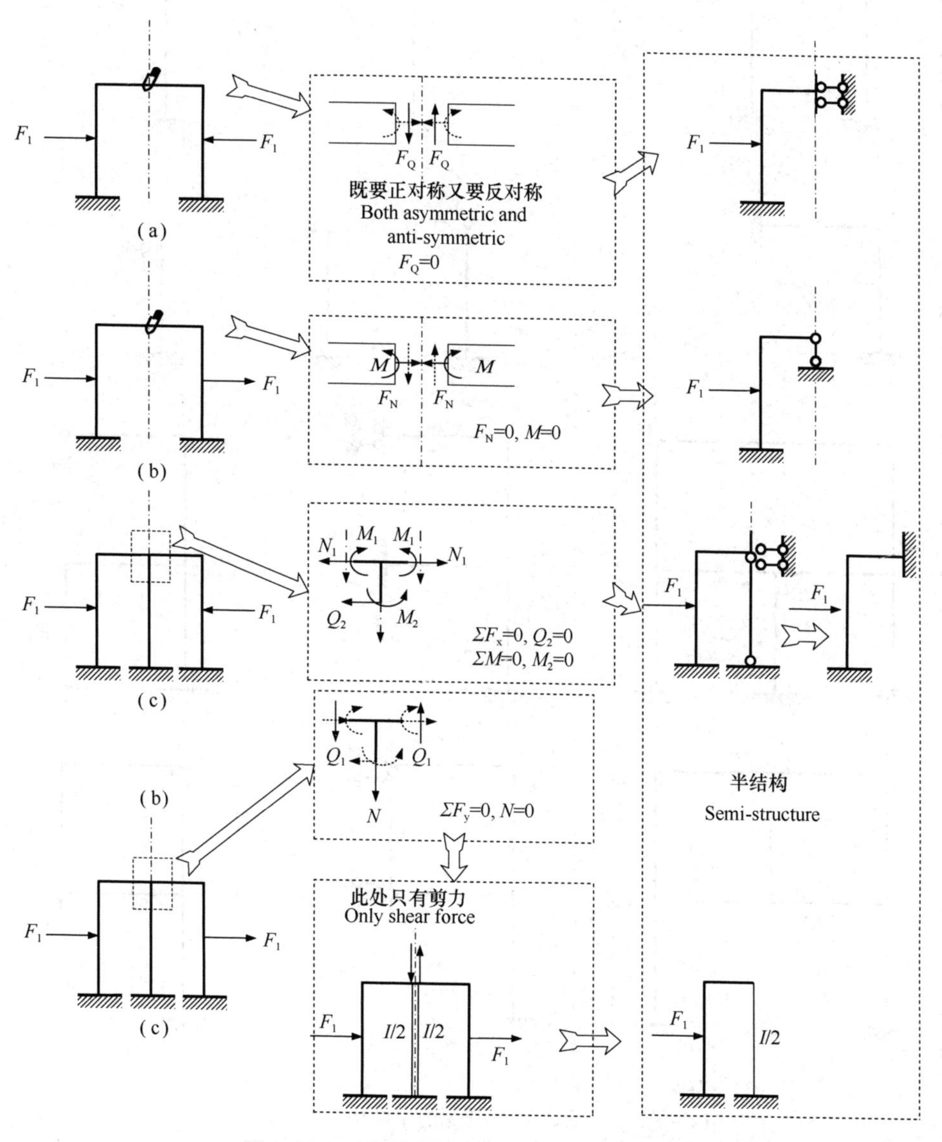

图 3.22　对称轴经过截面内力特点及对称性

Figure 3.22　The internal force characteristics of the section that the axis of symmetric passes through and symmetry

如果作用于对称结构的荷载不具有对称性，可以将荷载进行分组，如图 3.24 所示。

If the loads acting on a symmetrical structure are not symmetrical, the loads can be divided, as shown in Figure 3.24.

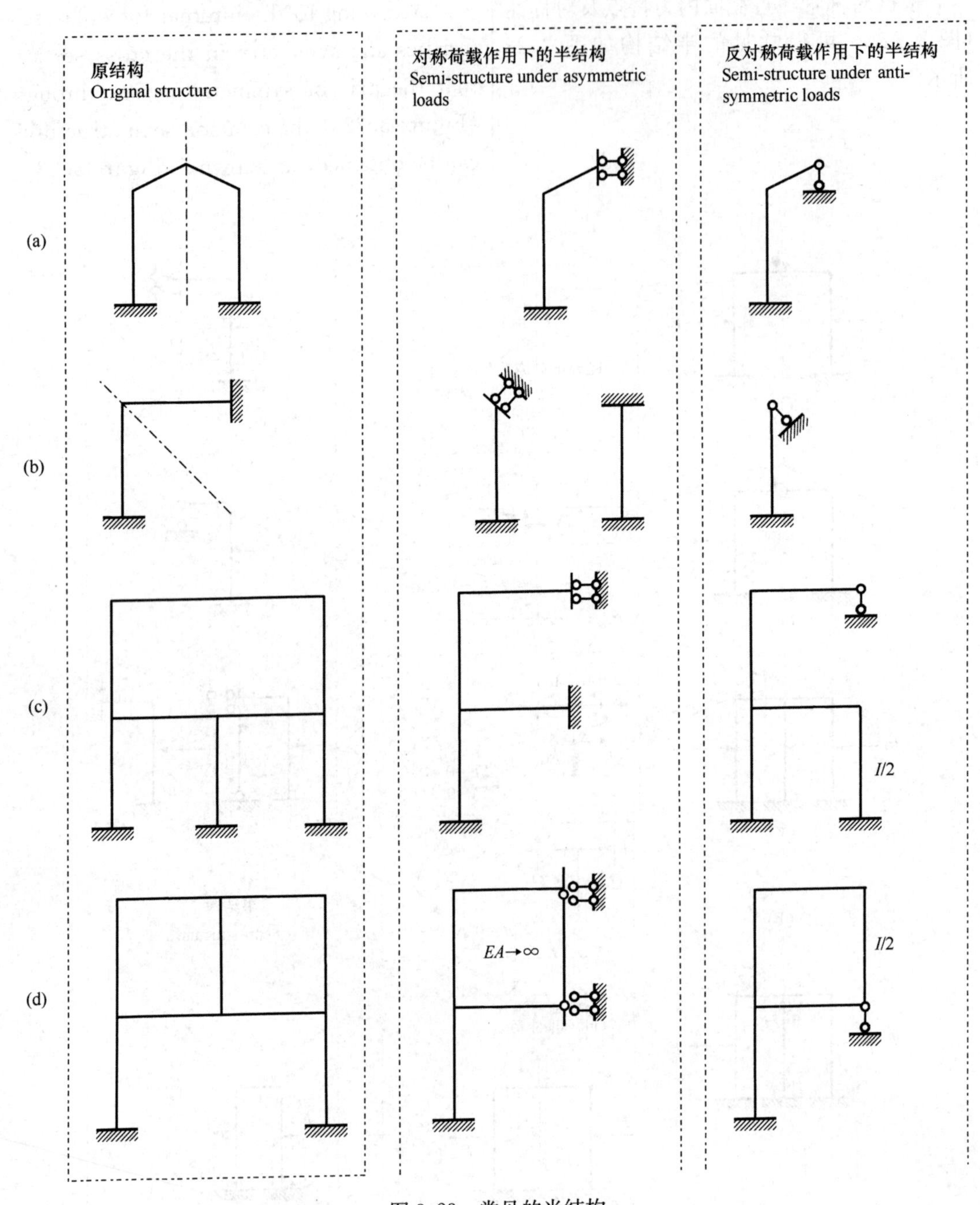

图 3.23　常见的半结构

Figure 3.23　Common semi-structures

对称性应用举例如图 3.25 所示（注：第5步中，正对称荷载无弯矩，所以舍弃）。

The example of symmetry application can be shown in Figure 3.25 (Note: In step 5, the positive symmetric load has no bending moment, so it is discarded).

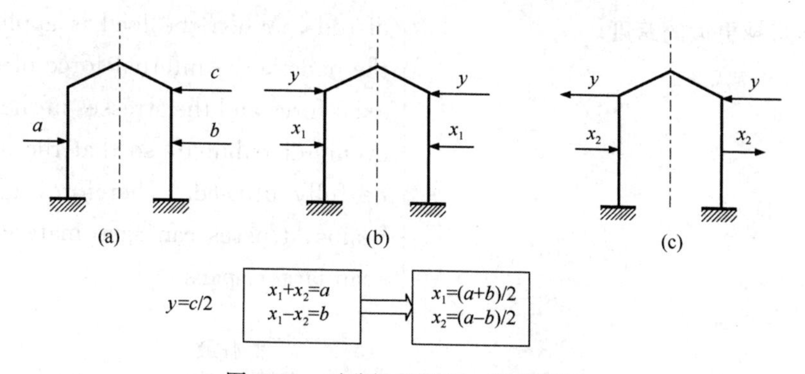

图 3.24　对称性的利用：荷载分组
Figure 3.24　Use of symmetry: load division

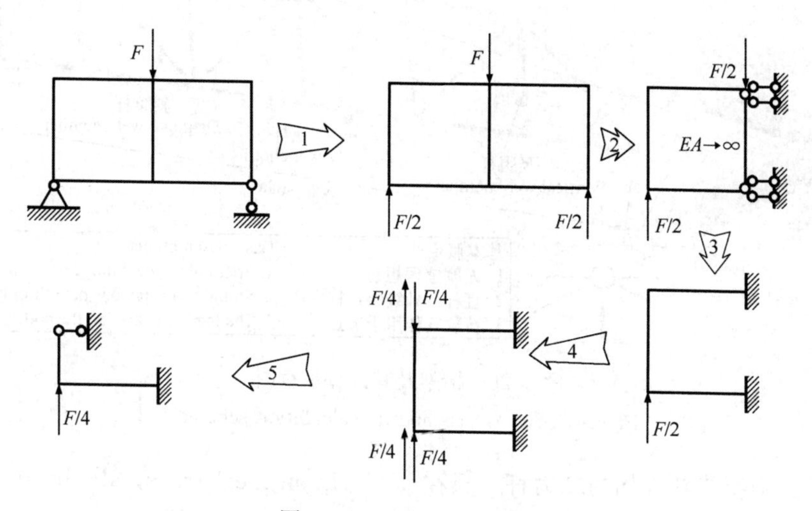

图 3.25　对称性应用举例
Figure 3.25　The example of symmetry application

3.6 静定平面桁架

3.6 Statically Determinate Coplanar Truss

3.6.1 桁架的组成及特点

3.6.1 The Composition of Truss and Characteristics

梁和刚架承受荷载后，主要产生弯曲内力，截面上应力分布是不均匀的，其边缘处应力最大，而中部的材料并未充分利用。桁架（图 3.26）是由杆件组成的格构体系，当荷载只作用在节点上时，各杆内力只有轴力，截面上的应力均匀分布，可以充分发挥材料的作用。因此，与梁相比，桁架的用料

When beams and rigid frames are loaded, they mainly produce bending internal forces and the stress distribution in the cross section is uneven, with the highest stresses at their edges and the material in the middle not fully utilised. The truss (Figure 3.26) is a lattice system composed

较省，并能跨越更大的跨度。

of rods. When the load is applied only on the nodes, the internal force of each rod is axial force and the stresses in the section are evenly distributed, so that the material can be fully utilised. Therefore, compared to beams, trusses can save materials and can span larger spans.

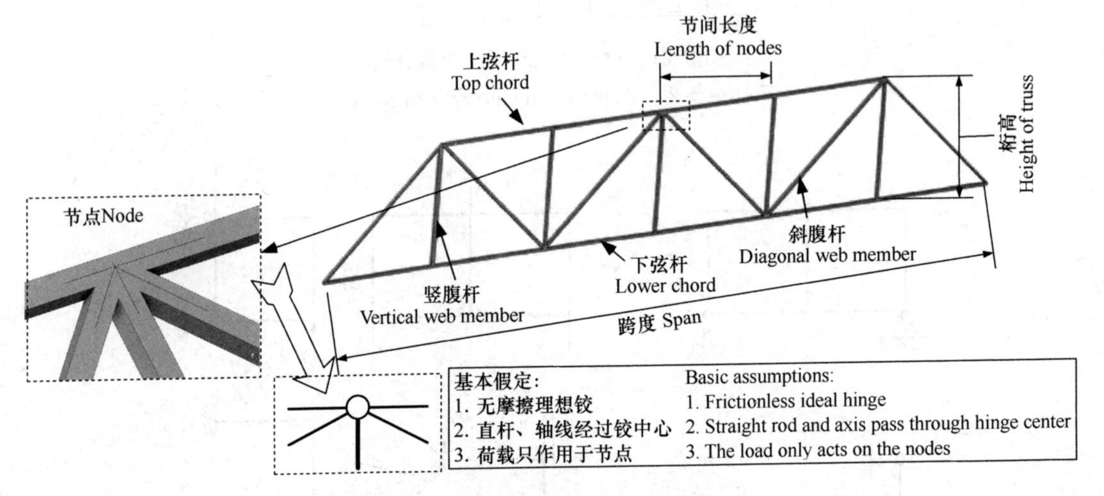

图 3.26 桁架及其计算示意图

Figure 3.26 Truss and its calculation scheme

理想桁架中所有杆件均为**二力杆**（只在两端受力，此二力必然大小相等，方向相反，并沿杆轴线作用，故杆件只产生轴力），然而，实际工程的桁架与上述假定并不完全吻合。

In an ideal truss, all the rods are **two-force rods** (The forces are applied only at the ends, which are necessarily equal in magnitude and opposite in direction and act along the rod axis, so the rods only produce axial force). However, the actual engineering trusses do not exactly match the above assumptions.

(1) 除了木桁架的榫接节点比较接近于铰节点外，钢桁架和钢筋混凝土桁架的节点都有很大的刚性。有些杆件在节点处是连续不断的。

(1) Steel and reinforced concrete trusses have very rigid knots, while wood trusses where the tongue and groove knots are more like the hinge knots. Some rods are continuous at the knots.

(2) 各杆的轴线不可能绝对平直，在节点处也不可能完全交于一点。

(2) The axes of the rods cannot be abosolutely straight, nor can they intersect precisely at a point at the junction.

(3) 桁架不可能只受节点荷载作用（如风荷载、杆件自重）等等。

(3) The truss cannot be only subjected to nodal loads (e.g., wind loads, self-weight of rods), etc.

以上这些情况都可能使杆件在产生轴力的同时还产生其他附加内力，如弯矩。

All of the above situations may cause the rod to generate axial forces as well as other internal forces such as bending moments.

通常将按上述假定计算得到的桁架内力称为主内力。由于实际情况与上述假定不同而产生的附加内力称为次内力。理论分析和实验结果表明，一般情况下，次内力的影响是不大的，可忽略不计。本节只研究主内力的计算。

The internal forces in the truss calculated according to the above assumptions are usually referred to as primary internal forces. The additional internal forces resulting from the situations different from the above assumptions are called secondary internal forces. Theoretical analysis and experimental results show that, in general, the effect of secondary internal forces is insignificant and can be ignored. In this section, only the calculation of the primary internal forces is studied.

根据几何构造的特点，静定平面桁架可分为三类：

According to the geometrical configuration, stafically determinate coplanar trusses can be divided into three categories:

(1) 简单桁架。由基础或一个基本铰接三角形开始，依次增加二元体而组成的桁架，如图 3.27(a)、(b) 所示。

(1) Simple truss. Simple truss is consisting of a base or a basic hinged triangle, with successively adding dual bodies, as in Figures 3.27 (a)、(b).

(2) 联合桁架。由几个简单桁架按几何不变体系的基本组成规则而联合组成的桁架，如图 3.27(c) 所示。

(2) Combined truss. Combined truss is composed of several simple trusses by the basic rules of composition of geometrically invariant systems, as in Figure 3.27 (c).

(3) 复杂桁架。不属于前两类的其他静定桁架，如图 3.27(d) 所示。

(3) Complex trusses. Any other static trusses other than the above two categories, as in Figure 3.27 (d).

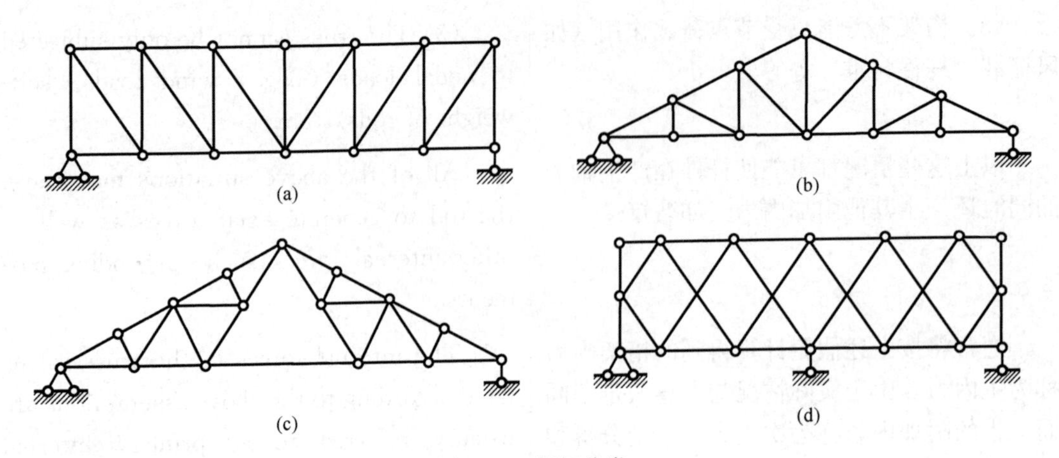

图 3.27 桁架分类

Figure 3.27 Truss classification

3.6.2 桁架内力计算方法

3.6.2.1 节点法

桁架内力 F_N 以拉力为正，对于桁架中的斜杆，通常采用力三角形与长度三角形的相似关系进行对比（如图 3.28 所示），免去正交坐标系中三角函数的换算，以简化计算。

3.6.2 Method of Calculating Internal Forces of Trusses

3.6.2.1 Nodal Method

The truss internal force F_N is positive in tension. For diagonal rods in trusses, the similarity rules between force triangles and length triangles is usually used for comparison (as shown in Figure 3.28), dispensing with the conversion of trigonometric functions in the orthogonal coordinate system to simplify calculations.

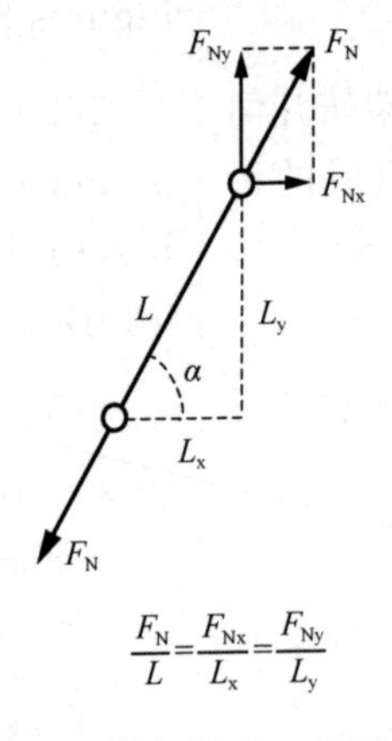

$$\frac{F_N}{L}=\frac{F_{Nx}}{L_x}=\frac{F_{Ny}}{L_y}$$

图 3.28 力三角形与长度三角形的相似关系

Figure 3.28 Similarity rules of force triangles to length triangles

【例 3.9】利用节点法计算图 3.29(a)中 AC、AD 杆轴力。

[Example 3.9] Use the nodal method to calculate the axial forces of rod AC and rod AD in Figure 3.29 (a).

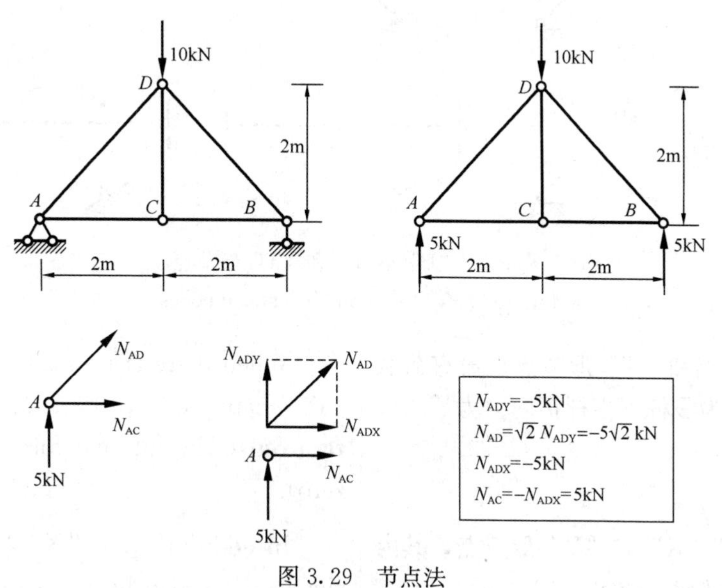

图 3.29 节点法

Figure 3.29 Nodal method

【解题要点】节点法的解题关键在于合理选择求解节点的顺序，如果选择节点所连未知轴力杆件数量过多则无法求解，所以，通常选择连接 2 根杆件的节点求解。

[Key points of the solution] The key point of the nodal method is to choose the order of the nodes. If the number of unknown axial force rods connected to the nodes is too big, it is impossible to solve the problem, so the nodes connected to 2 rods are usually chosen to solve the problem.

3.6.2.2 零杆

3.6.2.2 Zero Rods

单杆：如果在同一节点的所有 n 个内力未知杆件中，有 $n-1$ 个杆件共线，那么不共线那 1 根杆件称为此节点的单杆。单杆常出现于“L”形和“T”形节点中，如图 3.30中虚线所示。

Single rod: If $n-1$ rods are collinear in n rods of the same node with unknown internal forces, then the one that is not collinear with the other rods is called the single rod of this node. Single rod is often found in “L” and “T” shape nodes, as the dotted lines shown in Figure 3.30.

“L”形节点和“T”形节点上有荷载时，单杆是可以优先计算的杆件；

When there are loads on the “L” and “T”shape nodes, the single rod is in priority for calculation.

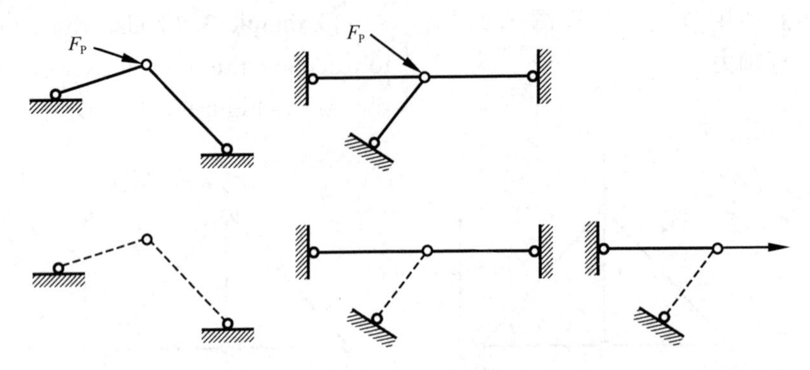

图 3.30 “L”形节点和“T”形节点

Figure 3.30 “L” and “T” shape nodes

“L”形节点和“T”形节点上没有荷载时，单杆内力为**零杆**（零杆指内力为零的杆件）。

When there is no load on the “L” and “T” shape nodes, the single rod is called **zero rod** (The internal force of zero rod is zero).

此外，“X”“K”和“Y”形节点，其内力也有一定的规律，如图 3.31 所示。

In addition, the “X” “K” and “Y” shape nodes also have a specific pattern of internal forces, as shown in Figure 3.31.

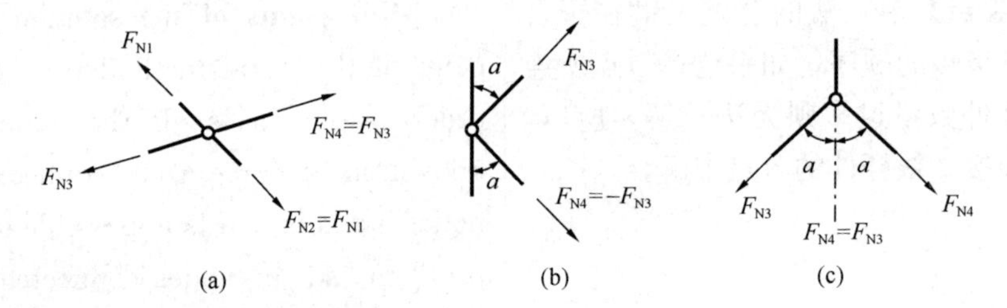

图 3.31 其他类型节点

(a)“X”形节点；(b)“K”形节点；(c)“Y”形节点

Figure 3.31 Other shapes of nodes

(a) “X” shape nodes; (b) “K” shape nodes; (c) “Y” shape nodes

结合对称性，有如下规律，如图 3.32 所示：

Combined with symmetry, the rules are the following, as shown in Figure 3.32:

(1) 对称桁架在正对称荷载作用下，对称轴上的“K”形节点的两斜杆是零杆。

(1) For symmetrical truss under positive symmetric loadings, the two diagonal rods at the “K” shape node on the axis of symmetry are zero rods.

(2) 在反对称荷载作用下，与对称轴垂直贯穿的杆是零杆；与对称轴重合的杆也是零杆。

(2) Under anti-symmetric loadings, a rod that runs perpendicular to the axis of symmetry is a zero rod and a rod that coin-

cides with the axis of symmetry is also a zero rod.

（3）在反对称荷载作用下，对称轴上的“Y”形节点的两斜杆是零杆。

(3) Under anti-symmetric loadings, the two diagonal rods at the “Y” shape node on the axis of symmetry are zero rods.

（4）内接三角形，如果节点上无荷载，则都是零杆，如图 3.33 所示。

(4) In internally connected triangles, if there is no load on the node, the rods connected to the node are all zero rods, as shown in Figure 3.33.

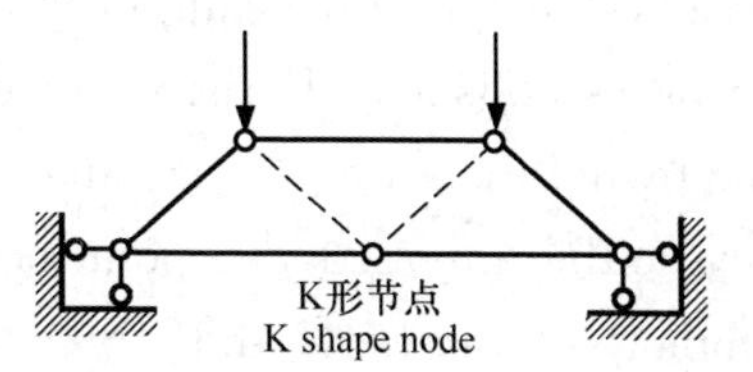

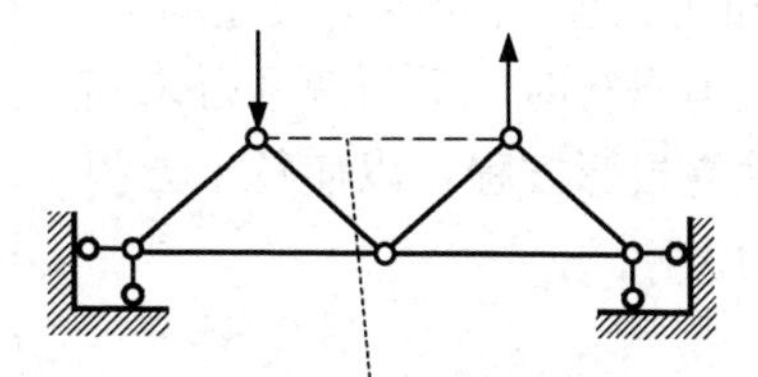

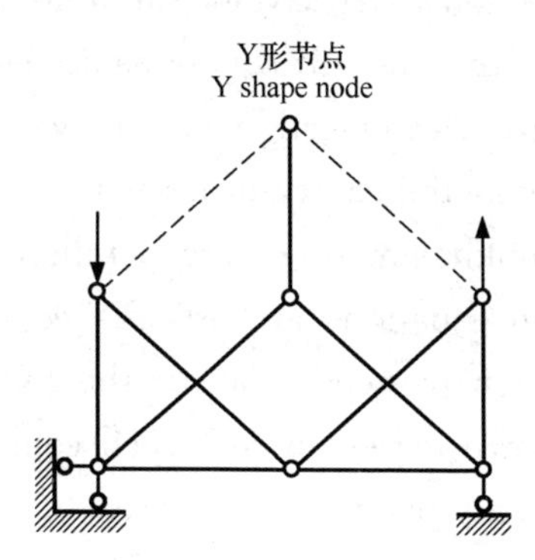

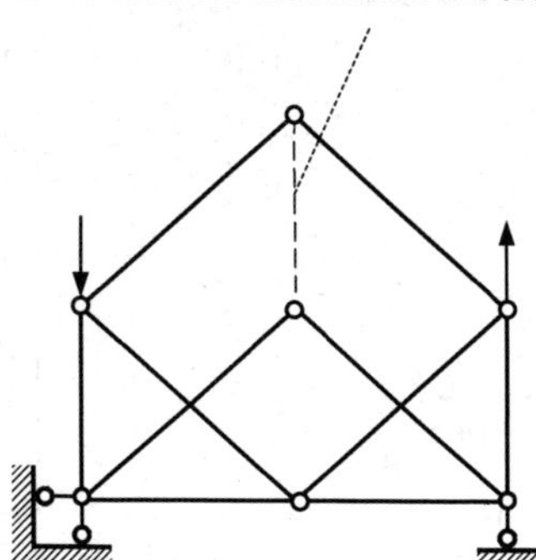

图 3.32 利用对称性找零杆

Figure 3.32 Using symmetry to find zero rods

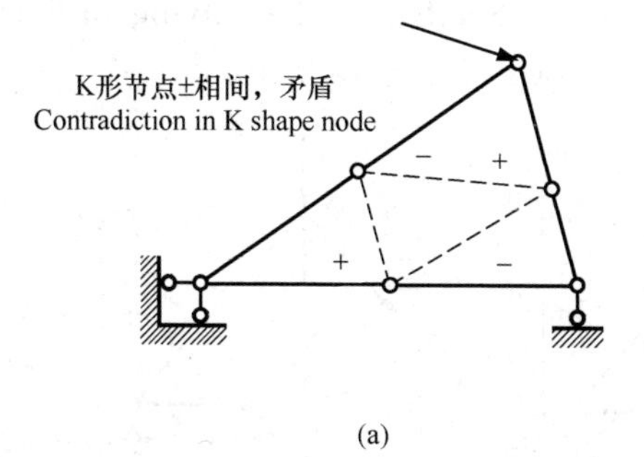

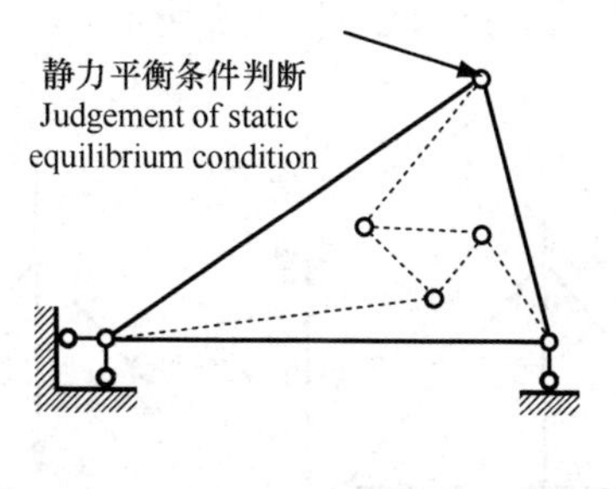

图 3.33 铰接三角形中的零杆

Figure 3.33 Zero rods in a hinged triangle

3.6.3 截面法

截面法是用截面切断拟求内力的杆件，从桁架中取出一部分为隔离体（至少包括两个节点），根据平衡方程来计算所截杆件的内力。通常作用在隔离体上的各力属于平面一般力系，故可建立三个平衡方程。如果隔离体上的未知力只有三个，且它们既不相交于一点，也不完全平行，则用截面法就可以直接求出这三个未知力。为了避免联立求解，以两个未知力的交点为矩心建立矩方程。当两个未知力平行时，沿与两平行未知力垂直的方向建立投影方程，以使每个方程只含一个未知力。

3.6.3 Method of Sections

Method of sections is to cut off the rod for the proposed internal forces by the section, then a part of the truss is removed as an isolated body (including at least two nodes) and calculate the internal forces of the cut rod according to the equilibrium equations. Usually, the forces acting on the isolated body belong to the plane general force system, so three equilibrium equations can be established. If there are only three unknown forces on the isolated body and they neither intersect at a point nor are they completely parallel, the three unknown forces can be found directly by the method of sections. To avoid simultaneous solutions, the moment equations are established using the intersection of the two unknown forces as the center of moments. When the two unknown forces are parallel, the projection equations are established along the direction perpendicular to the two parallel unknown forces so that each equation contains only one unknown force.

【例 3.10】求解图 3.34 中 *DA*、*DC*、*BC* 杆轴力。

[Example 3.10] Solve the axial forces of rods *DA*, *DC*, and *BC* in Figure 3.34.

【解】求解过程如图 3.34 所示。

[Solution] The solving process can be shown in Figure 3.34.

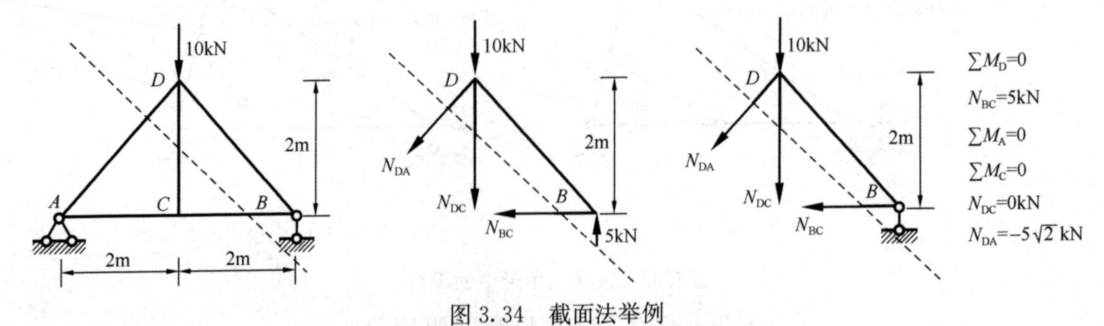

图 3.34 截面法举例

Figure 3.34 Example of the method of sections

【例 3.11】 求解图 3.35 中杆件轴力时，应如何选择合适的截面？

[Example 3.11] How to choose a suitable section to solve for the axial force of the rods in Figure 3.35?

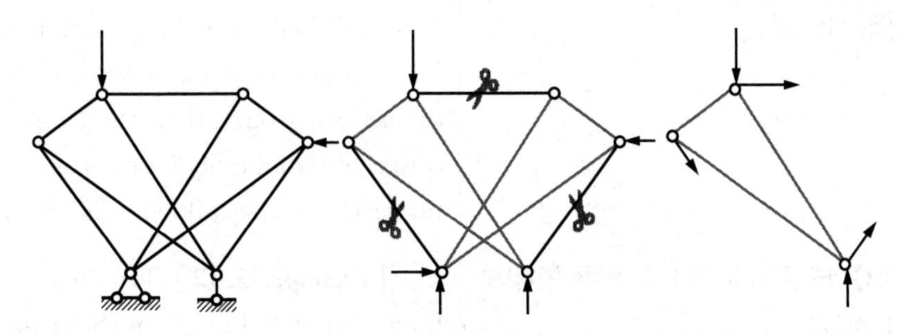

图 3.35 基于“两刚片法则”，选取合适的截面

Figure 3.35 Based on the “two-rigid-bodyrule” to select the appropriate section

【解题思路】 截面法并非只能“一刀切”，而是根据几何组成规律，截取其中一部分进行巧妙解题，图 3.35(b) 中的内部三角形，也是基于这一思路选取出来的。

[The ideas of solution] The method of sections is not just a “one-size-fits-all” approach but a clever solution based on the laws of geometric composition, intercepting a part of it. The inner triangle in Figure 3.35(b) is chosen based on this idea.

此外，有时所作截面可能切断三根以上的杆件，但如果被切断各杆中，除一杆外，其余均交于一点或均平行，则该杆内力仍可由力矩方程或投影方程求出，如图 3.36 所示。

In addition, sometimes the section may cut more than three rods, but if all but one of the rods cut intersect at a point or are parallel, the forces within that rod can still be found by the moment equation or the projection equation, as shown in Figure 3.36.

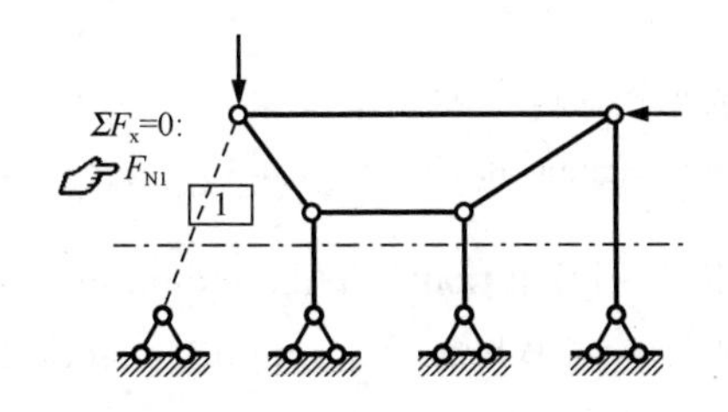

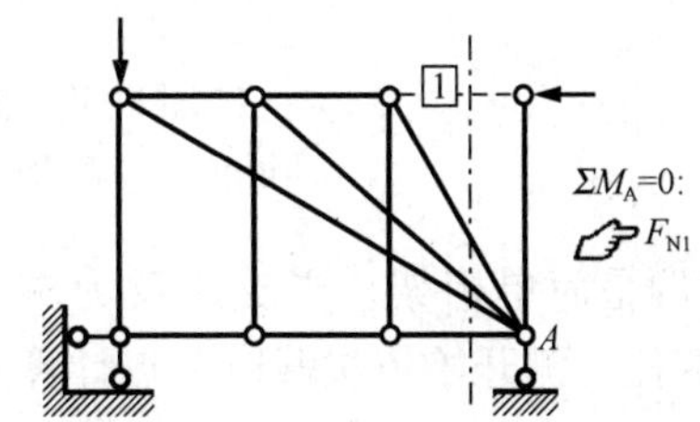

图 3.36 寻找截面法中的单杆

Example 3.36 Find the single rod by the method of sections

3.6.4 节点法和截面法的联合应用

3.6.4 Combined Application of the Nodal Method and the Method of Sections

节点法和截面法是静定平面桁架内力计算的两种基本方法。在某些桁架计算中，若

The nodal method and the method of sections are two basic methods for calculat-

只需求解几根指定杆件的内力，而单独应用节点法或截面法又不能一次求出结果时，则联合应用节点法与截面法，常可获得较好的效果。下面举例说明。

ing the internal forces of static plane trusses. In some calculations, if the internal forces of only a few specified rods need to be solved and the results cannot be obtained by applying the nodal method or the method of sections alone, then the combined application of two methods can often achieve better results. The following is an example.

【例 3.12】 试求图 3.37(a) 所示桁架中 1 杆和 2 杆的轴力。

[Example 3.12] Try to find the axial forces for rods 1 and 2 in the truss shown in Figure 3.37(a).

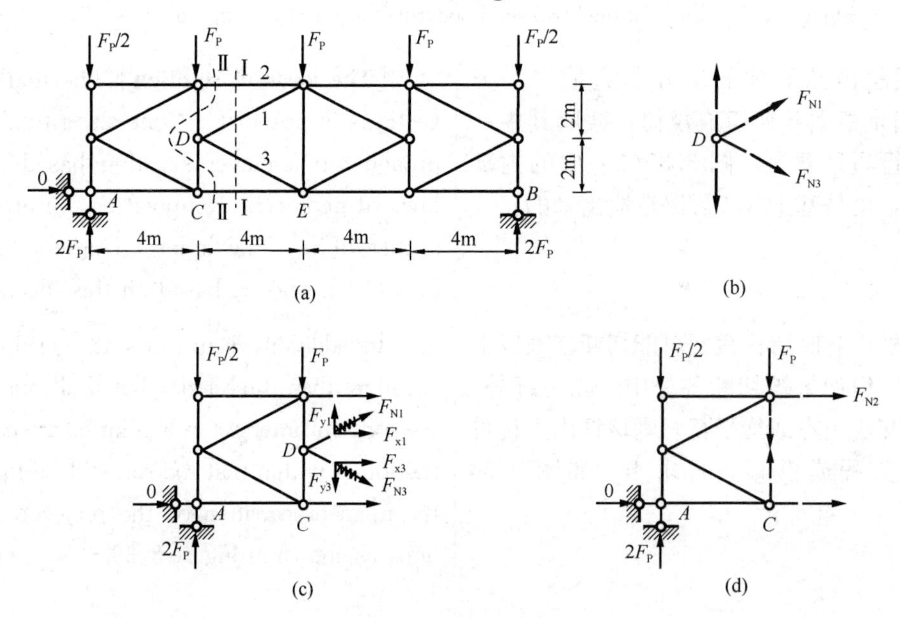

图 3.37　例 3.12 求解

Figure 3.37　Solve Example 3.12

【解】 本例是简单桁架。当支座反力求得后，从两侧任一侧开始依次截取节点计算即可，但要多次截取节点。若仅用截面法截取任一截面，则超出所要求的未知量数，即要解联立方程。为了减少计算步骤，采取节点法和截面法联合应用。

[Solution] This example is a simple truss. When the support reactions are obtained, the calculation can be done by intercepting the nodes in sequence starting from either side, but it requires many times of interceptions of the nodes. If only the method of sections is used to intercept any section, the number of unknowns is exceeded, i. e. , the combined equation is needed. In order to reduce the number of calculation steps, the nodal method and the method of sections are combined.

（1）求支座反力。

(1) Calculate support reaction force.

$$F_{yA}=F_{yB}=2F_P$$

（2）求1杆轴力，取节点 D 为隔离体，如图3.37(b) 所示。由水平投影平衡方程可知：

(2) Calculate the axial force of rod 1, taking the node D as the isolator, as shown in Figure 3.37(b). From the horizontal projection equilibrium equation:

$$F_{x1}=-F_{x3}$$
$$F_{y1}=-F_{y3},\ F_{N1}=-F_{N3}$$

作截面Ⅰ-Ⅰ，取桁架的左半部分为隔离体，如图3.37(c) 所示。

Take section Ⅰ-Ⅰ and the left half of the truss as the isolated body, as shown in Figure 3.37(c).

Based on (基于) $\Sigma F_y=0$, $F_{y1}-F_{y3}+2F_P-\frac{F_P}{2}-F_P=0$, $2F_{y1}+\frac{F_P}{2}=0$,

$$F_{y1}=-\frac{F_P}{4},\ \frac{F_{N1}}{\sqrt{5}}=\frac{F_{y1}}{1},\ F_{N1}=\frac{\sqrt{5}}{1}\times\left(-\frac{F_P}{4}\right)=-\frac{\sqrt{5}}{4}F_P \text{ (Pressure，受压)}$$

（3）求2杆轴力，作截面Ⅱ-Ⅱ，取桁架的左半部分为隔离体，如图3.37(d) 所示。由于除了2杆外，其余三杆都通过 C 点，故可用力矩平衡方程求得2杆内力。

(3) To get the axial force of rod 2, make section Ⅱ-Ⅱ, take the left half of the truss as the isolated body, as shown in Figure 3.37(d). Since all the three rods except the rod 2 pass through point C, the moment balance equation can be used to get the internal force of the rod 2.

$$\Sigma M_C=0\ ,F_{N2}\times 4+2F_P\times 4-\frac{F_P}{2}\times 4=0,\ F_{N2}=-\frac{3}{2}F_P \text{ (Pressure，受压)}$$

3.7 组合结构

3.7 Composite Structure

组合结构是指由链杆（两端为铰接的直杆且杆上无荷载作用）和梁式杆组成的结构，其中链杆只受轴力作用，梁式杆除受轴力作用外，还受弯矩和剪力作用。图3.38为桥梁中的组合结构。

Composite structure refers to the structure composed of chain rods (both ends are hinged straight rod and no load is acted on the rod) and beam rod, in which the chain rod only bears the axial force, beam rod bears not only the axial force, but also bending moment and shear force. Figure 3.38 shows the composite structure in bridges.

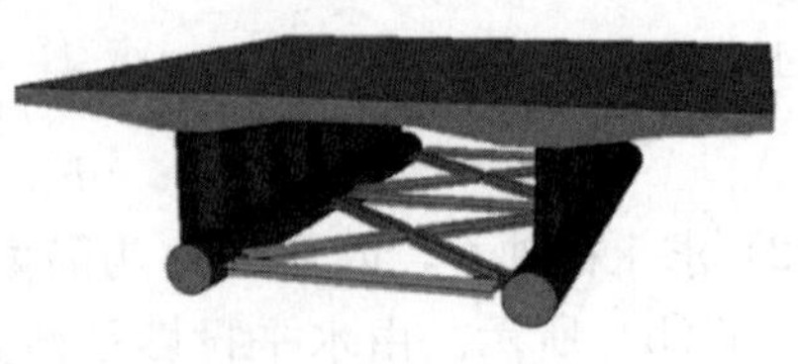

图 3.38 桥梁中的组合结构

Figure 3.38 Composite structure in bridges

下撑式五角形屋架（图 3.39）是较为常见的组合结构。

The pentagonal roof truss with lower support (Figure 3.39) is a typical composite structure.

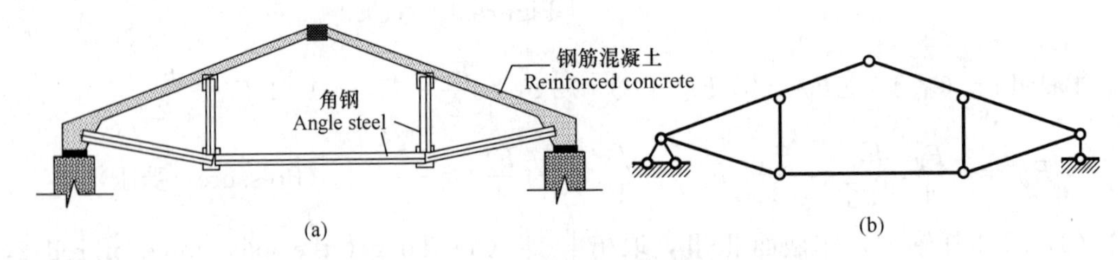

图 3.39 常见组合结构之一

(a) 下撑式五角形屋架；(b) 计算简图

Figure 3.39 One of the common composite structures

(a) The pentagonal roof truss with lower support; (b) Computing model

由于梁式杆的截面有三个内力，为了使隔离体上的未知力不致过多，应尽量避免截断梁式杆。因此，计算组合结构时，一般是先求出各链杆的轴力，然后再计算梁式杆的内力并作其 M、F_Q、F_N图。

Due to the beam rod section with three internal forces, in order to make the number of unknown forces on the isolation body not too much, cutting off the beam rod should be avoided. Therefore, when calculating the composite structure, the axial force of each chain rod is usually calculated first, then the internal forces of the beam rod is calculated, and its M, F_Q, F_N diagrams are made.

在计算时，必须特别注意区分链杆和梁式杆。截断链杆，截面上只有轴力；截断梁式杆，截面上一般作用有三个内力，即弯矩、剪力和轴力。如图 3.40(a) 中，截取 F 节点为隔离体时（图 3.40b），由于杆 FA 和 FC 是梁式杆，两端各存在三个内力。当

During the calculation, special attention must be paid to distinguish between chain rods and beam rods. When cutting a chain rod, there is only axial force in the cross section while cutting a beam rod, there are generally three internal forces act-

取截面Ⅰ-Ⅰ左部分为隔离体时（图3.40c），链杆 DE 中只有轴力，而梁式杆 FC 在铰旁边截断，除了有杆端轴力外，不要忘记还有杆端剪力。

ing in the cross section, namely bending moment, shear force, and axial force. As in Figure 3.40(a), when taking out the node F as an isolated body (Figure 3.40b), the rods FA and FC are beam rods, and there are three internal forces at each end. When taking the left part of section Ⅰ-Ⅰ as the isolated body (Figure 3.40c), there is only axial force in the chain rod DE, while the beam rod FC is cut next to the hinge, the beam rod FC not only bears the axial force at the rod end, but also the shear force.

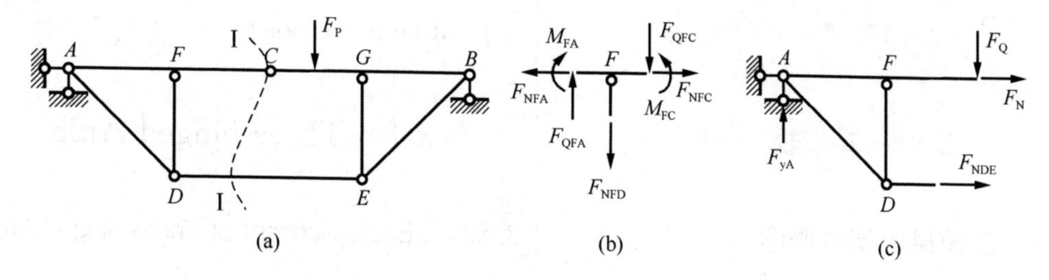

图3.40 截面法取隔离体的关键位置示意图

Figure 3.40 Schematic diagram of the key position of the isolated body taken by the method of sections

【例3.13】求解图3.41所示组合结构时，如何选取截面？

[Example 3.13] How to choose the section when solving the composite structure shown in Figure 3.41?

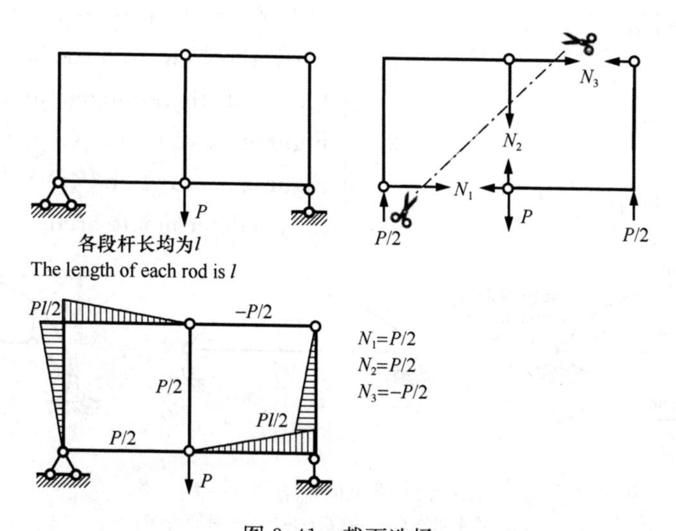

图3.41 截面选择

Figure 3.41 Cross section selection

【思路总结】求解组合结构内力时，应先求解桁架杆，方法上优先考虑截面法，所以合理地选择截面是解题关键。

[**Summary of ideas**] When solving the internal forces of the composite structure, the truss rod should be solved first, and the method of sections should give priority, so a reasonable choice of the section is the key to solving the problem.

求解完成后，梁式杆绘制弯矩图（画在受拉侧），桁架杆无需绘制轴力图，把所求轴力值标于桁架杆旁即可（符号以拉力为正）。

After the solution is completed, the bending moment diagram is drawn for the beam rod (drawn on the tension side), while the axial force diagram is not required for the truss rod, only the axial force value is marked next to the truss rod (the symbol is positive in tension).

3.8 三 铰 拱

3.8 Three-hinged Arch

3.8.1 三铰拱的基本概念

3.8.1 Basic Concept of Three-hinged Arch

三铰拱是一种静定的拱式结构，在桥梁和屋盖中都得到应用。拱结构的特点是：在竖向荷载作用下，支座将产生水平反力（或称推力）。水平推力的存在与否是区别拱和梁的重要标志。三铰拱常见的形式如图3.42(a)、(b)、(c)所示，图3.42(e)、(f)则属于超静定拱。

The three-hinged arch is a static arch structure that is used in both bridges and roofs. The arch structure is characterized by the horizontal reaction force (or thrust) that will be generated by the support under the vertical load. The presence or absence of horizontal thrust is a vital sign to distinguish the arch from the beam. The common forms of three-hinged arch are shown in Figures 3.42 (a), (b) and (c), while Figures 3.42(e) and (f) belong to the statically indeterminate arch.

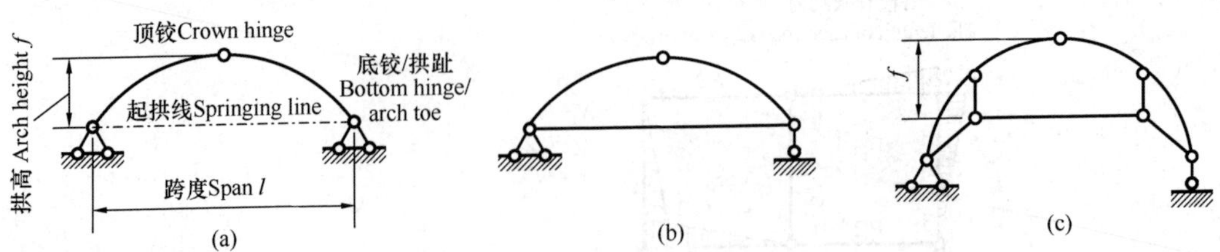

图3.42 常见的拱结构（一）

(a) 三铰平拱；(b) 带拉杆拱；(c) 带链杆拱；

Figure 3.42 Common arch structures (One)

(a) Three-hinged flat arch; (b) Arch with ties; (c) Arch with chain rods;

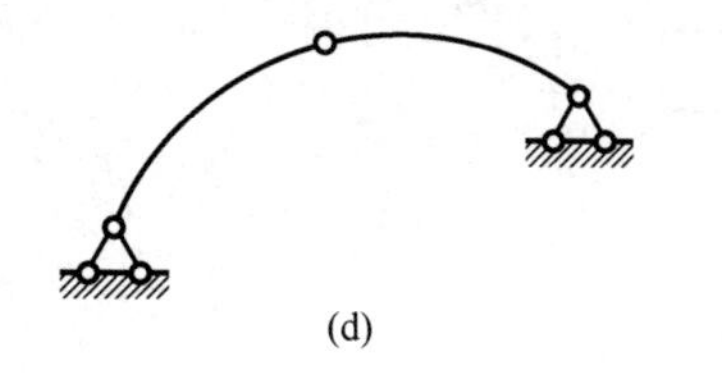
(d)

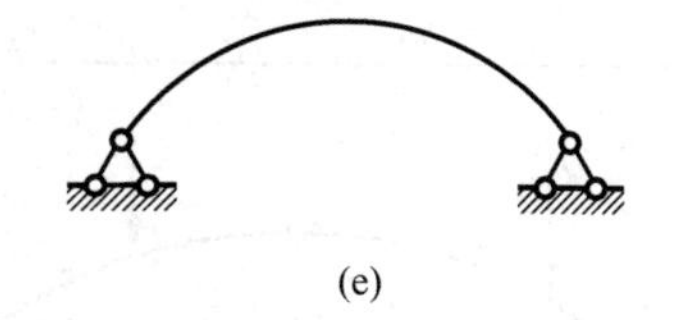
(e)

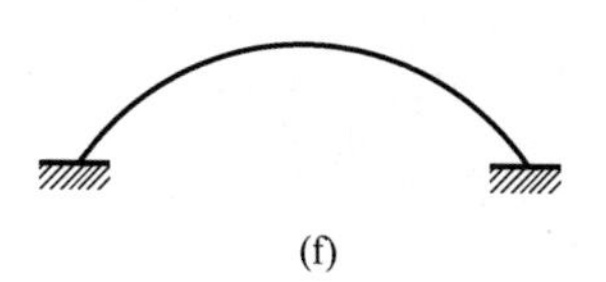
(f)

图 3.42 常见的拱结构（二）
（d）三铰斜拱；（e）两铰拱；（f）无铰拱
Figure 3.42 Common arch structures (Two)
(d) Three-hinged inclined arch; (e) Two-hinged arch; (f) Unhinged arch

3.8.2 三铰拱的计算

3.8.2 Calculation of Three-hinged Arch

3.8.2.1 支座反力的计算

3.8.2.1 Calculation of Support Reaction Forces

图 3.43(a) 所示三铰拱，支座反力共有四个。求反力时，除了取整体为隔离体可建立三个平衡方程外，还需取左（或右）半拱为隔离体，以中间铰 C 为矩心，根据平衡条件 $\sum M_C = 0$ 建立一个方程，从而求出所有的反力。

Figure 3.43(a) shows the three-hinged arch, there are four support reaction forces. To find the reaction forces, in addition to taking the whole body as an isolator to establish three equilibrium equations, we also need to take a left (or right) half arch as an isolator, with the middle hinge C as the moment center, to establish an equation according to the equilibrium condition $\sum M_C = 0$, to obtain all reaction forces.

为了便于比较，在图 3.43(b) 中画出了与该三铰拱具有相同跨度、相同荷载的简支梁。

For comparison, a simply supported beam with the same span and the same load with this three-hinged arch is drawn in Figure 3.43 (b).

首先，考虑整体平衡。由 $\sum M_B = 0$ 和 $\sum M_A = 0$ 可求出两支座的竖向反力为：

First, consider the overall equilibrium. By $\sum M_B = 0$ and $\sum M_A = 0$, we can find the vertical reaction forces of two bearings:

$$F_{VA} = \frac{1}{l}(F_{P1} b_1 + F_{P2} b_2) \tag{3.3}$$

$$F_{VB} = \frac{1}{l}(F_{P1} a_1 + F_{P2} a_2) \tag{3.4}$$

$$\because \sum F_x = 0 \quad \therefore F_{HA} = F_{HB} = F_H$$

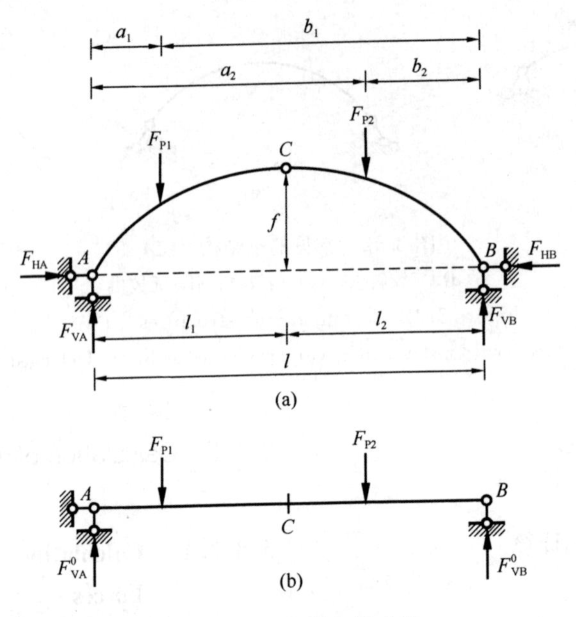

图 3.43 三铰拱和简支梁

Figure 3.43 Three-hinged arch and simply supported beam

式中，F_H表示三铰拱在竖向荷载作用下的水平推力。

In the equation, F_H refers to the horizontal thrust of three-hinged arch under vertical loads.

然后，取左半拱为隔离体：

Then, the left half arch is taken as the isolated body:

$$\because \sum M_C = 0 \quad \therefore F_{VA}\,l_1 - F_{P1}(l_1 - a_1) - F_H f = 0, \text{Then (于是):}$$

$$F_H = \frac{F_{VA}\,l_1 - F_{P1}(l_1 - a_1)}{f} \tag{3.5}$$

观察式（3.3）和式（3.4）的右边项，可知其恰好等于相应简支梁的竖向支座反力F_{VA}^0和F_{VB}^0。注意到式（3.5）的分子恰等于简支梁相应截面C的弯矩M_C^0，则可将以上各式写为：

Observe the right terms of Equations (3.3) and (3.4), they are equal to the vertical support reactions F_{VA}^0 and F_{VB}^0 of the corresponding simply-supported beam. It is noticed that the numerator of Equation (3.5) is equal to the bending moment M_C^0 of the corresponding section C in the simply-supported beam. The all of above can be written as:

$$\left.\begin{aligned} F_{VA} &= F_{VA}^0 \\ F_{VB} &= F_{VB}^0 \\ F_H &= \frac{M_C^0}{f} \end{aligned}\right\} \tag{3.6}$$

由式（3.6）可知，三铰拱的竖向反力与相应简支梁相同，水平推力 F_H 等于相应简支梁截面 C 的弯矩 M_C^0 除以拱高 f。三铰拱的反力只与荷载及三个铰的位置有关，而与各铰间的拱轴线形状无关。当荷载及跨度 l 不变时，水平推力 F_H 与拱高 f 呈反比，拱越低推力越大。若 $f=0$，则 $F_H=\infty$，此时三铰在一条直线上，属于瞬变体系。

From Equation (3.6), the vertical reaction force of three-hinged arch is the same as the corresponding simply supported beam, and the horizontal thrust F_H is equal to the bending moment M_C^0 of corresponding simply supported beam's section C divided by the arch height f. The reaction forces of three-hinged arch are only related to the loads and the position of three hinges, and are irrelevant with the shape of arch axis between each hinge. When the load and span l are constant, the horizontal thrust F_H is inversely proportional to the arch height f, and the lower the arch, the greater the thrust. When $f=0$, $F_H=\infty$, the three hinges are in a straight line, which is the instantaneously unstable system.

3.8.2.2 内力的计算

3.8.2.2 Calculation of Internal Forces

支座反力求出后，用截面法即可求得拱上任一横截面的内力。注意到拱轴为曲线，任一截面 K 的位置取决于其形心坐标 x、y，以及该处拱轴切线的倾角 φ（图 3.44）。

After the support reactions are obtained, the internal force of any cross section on the arch can be found by the method of sections. It should be noted that the arch axis is a curve, and the position of any section K depends on the arch's shape center coordinates x, y, and the tangent angle φ of the arch axis at the section (Figure 3.44).

在拱中，通常规定弯矩以使拱内侧受拉为正，反之为负。取 AK 段为隔离体，如图 3.44(b) 所示。

In the arch, bending moment is usually positive that strains the inside of the arch tension, and vice versa. Take section AK as the isolator, as shown in Figure 3.44(b).

$$\because \Sigma M_K = 0 \qquad \therefore M = [F_{VA}x - F_{P1}(x-a_1)] - F_H y$$

由于 $F_{VA}=F_{VA}^0$，且相应简支梁(图 3.44c) K 截面处的弯矩为：

Due to $F_{VA} = F_{VA}^0$, and the bending moment at section K of the corresponding simply supported beam (Figure 3.44c) is:

$M^0 = F_{VA}^0 x - F_{P1}(x-a_1)$，故上式可写为：

$M^0 = F_{VA}^0 x - F_{P1}(x-a_1)$, so the expression can be written as:

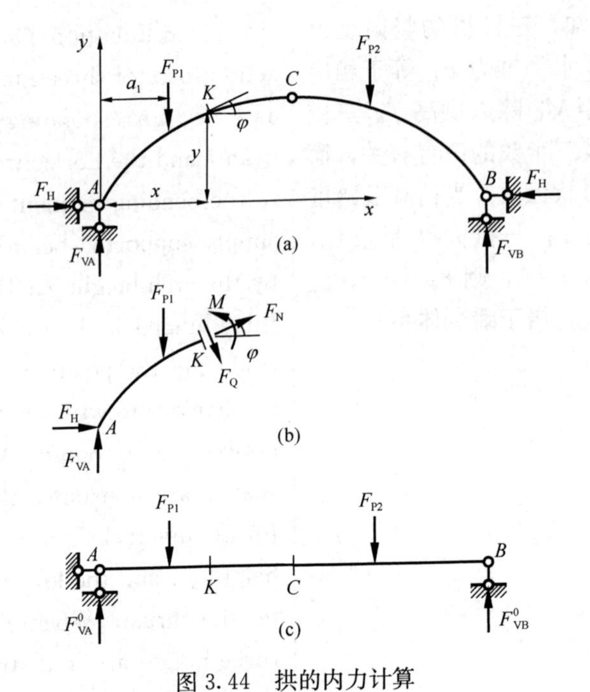

图 3.44 拱的内力计算

Figure 3.44 The internal forces of the arch

$$M = M^0 - F_{H} y$$

即拱内任一截面的弯矩 M 等于相应简支梁对应截面的弯矩 M^0 减去水平推力所引起的弯矩 $F_H y$。可见，由于水平推力的存在，拱的弯矩比相应简支梁的弯矩要小。

That is, the bending moment M of any section in the arch is equal to the bending moment M^0 of the corresponding section of the corresponding simply supported beam minus the bending moment $F_H y$, which is caused by the horizontal thrust. It can be seen that due to the horizontal thrust, the bending moment of the arch is smaller than the bending moment of the corresponding simply supported beam.

剪力是以绕隔离体顺时针转动为正，反之为负。取 AK 段为隔离体，如图 3.44(b) 所示，将所有力沿 F_Q方向投影，由平衡条件知：

Shear force is positive that rotates clockwise around the isolator, and vice versa. Take section AK as the isolator, as shown in Figure 3.44(b), project all forces along the F_Q' s direction, it can be known by the equilibrium condition:

$$\begin{aligned} F_Q &= F_{VA}\cos\varphi - F_{P1}\cos\varphi - F_H\sin\varphi \\ &= (F_{VA} - F_{P1})\cos\varphi - F_H\sin\varphi \\ &= F_Q^0\cos\varphi - F_H\sin\varphi \end{aligned}$$

式中，F_Q^0 为相应简支梁 K 截面处的剪力，$F_Q^0 = F_{VA} - F_{P1}$。在图 3.44(a) 坐标系中，左半拱 φ 取正，右半拱 φ 取负。

Where F_Q^0 is the shear force at the section K of the corresponding simply supported beam, $F_Q^0 = F_{VA} - F_{P1}$. In the coordinate system of Figure 3.44(a), φ is positive in the left half arch, φ is negative in the right half arch.

轴力以拉力为正，反之为负。取 AK 段为隔离体，如图 3.44(b) 所示，将所有力沿 F_N 方向投影，由平衡条件知：

Axial force in tension is positive, and vice versa. Take section AK as the isolator, as shown in Figure 3.44 (b), project all forces along the F_N' s direction, it can be known by the equilibrium condition:

$$
\begin{aligned}
F_N &= -(F_{VA} - F_{P1})\sin\varphi - F_H\cos\varphi \\
&= -F_Q^0\sin\varphi - F_H\cos\varphi
\end{aligned}
$$

综上所述，三铰平拱在竖向荷载作用下的内力计算公式为：

To sum up, calculation for the internal forces of three-hinged flat arch under vertical loads is:

$$
\left.
\begin{aligned}
M &= M^0 - F_H y \\
F_Q &= F_Q^0\cos\varphi - F_H\sin\varphi \\
F_N &= -F_Q^0\sin\varphi - F_H\cos\varphi
\end{aligned}
\right\} \tag{3.7}
$$

由上式可知，三铰拱的内力不但与荷载及三铰的位置有关，而且与各铰间拱轴线的形状有关。

From the above equation, it can be seen that the three-hinged arch' s internal forces are not only related to the loads and the positions of the three hinges, but also the shape of the arch axis between each hinge.

【例 3.14】 试作图 3.45(a) 所示三铰拱的内力图，拱轴线为抛物线：$y = \frac{4f}{l^2}x(l-x)$。

[Example 3.14] Try to draw the internal force diagram of the three-hinged arch shown in Figure 3.45 (a), the shape of the arch axis is a parabola: $y = \frac{4f}{l^2}x(l-x)$.

【解】(1) 求支座反力。由式 (3.6) 可得：

[Solution] (1) Calculate the support reaction force. From Equation (3.6):

$$F_{VA} = F_{VA}^0 = \frac{4\times4+1\times8\times12}{16} = 7\text{kN}$$

$$F_{VB} = F_{VB}^0 = \frac{1\times8\times4+4\times12}{16} = 5\text{kN}$$

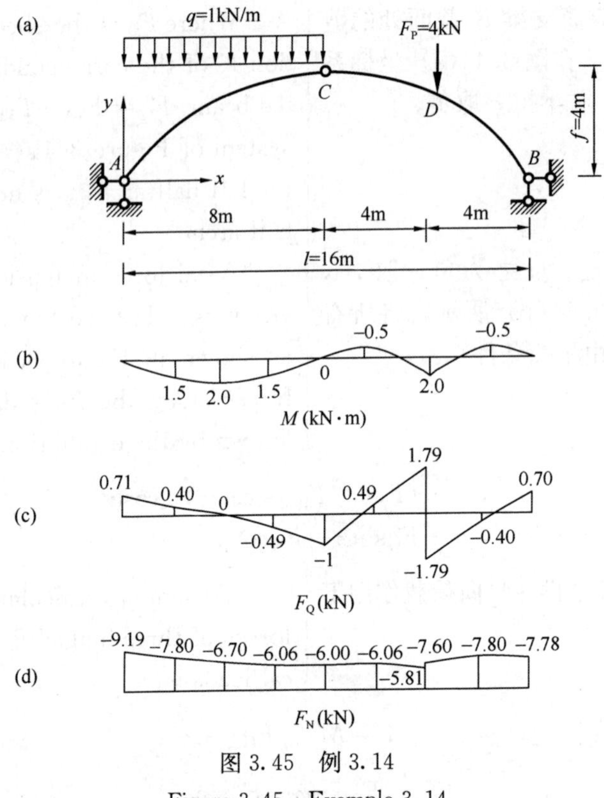

图 3.45 例 3.14

Figure 3.45 Example 3.14

$$F_H = \frac{M_C^0}{f} = \frac{5\times8-4\times4}{4} = 6\text{kN}$$

(2) 指定截面的内力计算

(2) Calculate the internal forces of the specific section

反力求出后，即可按式（3.7）计算各截面的内力。为了绘制内力图，可将拱轴沿水平方向分为八等份，计算每个截面的弯矩、剪力和轴力的数值。现以 $x=12\text{m}$ 的截面 D 为例说明计算步骤。

After the support reaction forces are obtained, the internal forces of each section can be got by Equation (3.7). In order to draw the internal force diagram, the arch axis along the horizontal direction can be divided into eight equal parts, to calculate the bending moment, shear force and axial force of each section. The calculation steps are now illustrated with section D at $x=12\text{m}$ as an example.

① 求截面 D 的几何参数

① Find out the geometric parameters of section D

将 $x=12\text{m}$，$l=16\text{m}$ 及 $f=4\text{m}$ 代入拱轴方程，得：

Take $x=12\text{m}$, $l=16\text{m}$, and $f=4\text{m}$ into the arch axis equation:

$$y=\frac{4f}{l^2}x(l-x)=\frac{4\times4}{16^2}\times12\times(16-12)=3\text{m}$$

$$\tan\varphi=\frac{\mathrm{d}y}{\mathrm{d}x}=\frac{4f}{l^2}(l-2x)=\frac{4\times4}{16^2}\times(16-2\times12)=-0.5$$

$$\therefore\varphi=-26°34',\ \sin\varphi=-0.447,\ \cos\varphi=0.894$$

三铰拱内力计算 **表 3.1**

Internal forces calculation of the three-hinged arch **Table 3.1**

截面几何参数（x，y 单位：m） Section geometry parameters （x，y in "m"）					F_Q^0 (kN)	弯矩计算（kN·m） Bending moment calculation (kN·m)			剪力计算（kN） Shear force calculation (kN)			轴力计算（kN） Axial force calculation (kN)		
x	y	$\tan\varphi$	$\sin\varphi$	$\cos\varphi$		M^0	$-F_Hy$	M	$F_Q^0\cos\varphi$	$-F_H\sin\varphi$	F_Q	$-F_Q^0\sin\varphi$	$-F_H\cos\varphi$	F_N
0	0	1	0.707	0.707	7	0	0	0	4.95	−4.24	0.71	−4.95	−4.24	−9.19
2	1.75	0.75	0.600	0.800	5	12	−10.5	1.5	4.00	−3.60	0.40	−3.00	−4.80	−7.80
4	3.00	0.50	0.447	0.894	3	20	−18	2	2.68	−2.68	0	−1.34	−5.36	−6.70
6	3.75	0.25	0.243	0.970	1	24	−22.5	1.5	0.97	−1.46	−0.49	−0.24	−5.82	−6.06
8	4.00	0	0	1	−1	24	−24.0	0	−1.00	0	−1.00	0	−6.00	−6.00
10	3.75	−0.25	−0.243	0.970	−1	22	−22.5	−0.5	−0.97	1.46	0.49	−0.24	−5.82	−6.06
12	3.00	−0.50	−0.447	0.894	−1 −5	20	−18	2	−0.89 −4.47	2.68	1.79 −1.79	−0.45 −2.24	−5.36	−5.81 −7.60
14	1.75	−0.75	−0.600	0.800	−5	10	−10.5	−0.5	−4.00	3.60	−0.40	−3.00	−4.80	−7.80
16	0	−1	−0.707	0.707	−5	0	0	0	−3.54	4.24	0.70	−3.54	−4.24	−7.78

② 求截面 D 的内力

由式（3.7）可知：

② Find the internal force of section D

By Equation (3.7):

$$M=M^0-F_Hy=5\times4-6\times3=2\text{kN}\cdot\text{m}$$

在集中荷载作用处，F_Q^0 发生突变，因此 F_Q 和 F_N 都要发生突变，需算出左、右两边的剪力 F_{QL}、F_{QR} 和轴力 F_{NL}、F_{NR}。

At the point of the concentrated load, F_Q^0 mutates, so F_Q and F_N mutate as well, we need to calculate shear forces F_{QL}, F_{QR} and axial forces F_{NL}, F_{NR}, respectively.

$$\begin{cases} F_{QL} = F_{QL}^0\cos\varphi - F_H\sin\varphi = -1\times0.894 - 6\times(-0.447) = 1.79\text{kN} \\ F_{NL} = -F_{QL}^0\sin\varphi - F_H\cos\varphi = -(-1)\times(-0.447) - 6\times0.894 = -5.81\text{kN} \end{cases}$$

$$\begin{cases} F_{QR} = F_{QR}^0\cos\varphi - F_H\sin\varphi = -5\times0.894 - 6\times(-0.447) = -1.79\text{kN} \\ F_{NR} = -F_{QR}^0\sin\varphi - F_H\cos\varphi = -(-5)\times(-0.447) - 6\times0.894 = -7.6\text{kN} \end{cases}$$

其他各截面的内力计算与上相同，可列表进行，详见表3.1。根据表中算得的结果绘出M、F_Q、F_N图，如图3.45(b)～(d)所示。

The other sections of the internal force calculation are the same with the above procedure, which can be seen in Table 3.1. According to the results in Table 3.1, M, F_Q, and F_N can be plotted, as shown in Figures 3.45(b)-(d).

3.8.3 三铰拱的合理拱轴线

3.8.3 Reasonable Arch Axis of the Three-hinged Arch

由前已知，当荷载及三个铰的位置给定时，三铰拱的反力就可确定，而与各铰间拱轴线形状无关；三铰拱的内力则与拱轴线形状有关。当拱上所有截面的弯矩都等于零（剪力也为零）而只有轴力时，截面上的正应力是均匀分布的，材料的使用最经济。在固定荷载作用下使拱处于无弯矩状态的轴线称为合理拱轴线。

As can be seen, when the loads and the positions of three hinges are given, the reaction forces of three-hinged arch can be determined, and it is not related to the shape of the arch axis between each hinge; but the internal forces of the three-hinged arch are related to the shape of the arch axis. When the bending moments of all sections on the arch are equal to zero (shear forces are also zero) and only the axial forces exist, the positive stress on the section is uniformly distributed, and the use of materials is the most economical. Under the action of fixed loads, the axis that makes no bending moment in each section is called reasonable arch axis.

合理拱轴线可以由弯矩为零的条件来确定。在竖向荷载作用下，三铰拱任一截面的弯矩为$M = M^0 - F_H y$。当为合理拱轴时，则$M = 0$，故有：

Reasonable arch axis can be determined by the condition that the bending moment is zero. Under vertical loads, the bending moment for any section of the three-hinged arch is $M = M^0 - F_H y$. When at a reasonable arch axis, $M=0$, therefore:

$$y(x) = \frac{M^0(x)}{F_H} \tag{3.8}$$

其中，$y(x)$和$M^0(x)$是x的函数，F_H

Where $y(x)$ and $M^0(x)$ are functions

是常数。上式表明，在竖向荷载作用下，三铰拱合理拱轴线的纵坐标 y 与相应简支梁弯矩图的竖标呈正比。

of x, F_H is a constant. The above equation indicates that the vertical coordinate y of the reasonable arch axis of the three-hinged arch is in direct proportion to the vertical coordinate of the bending moment diagram of the corresponding simply supported beam under vertical loads.

【例 3.15】 试作图 3.46(a) 所示三铰拱在图示满跨竖向均布荷载作用下的合理拱轴线。

[Example 3.15] Try to make the reasonable arch axis for the three-hinged arch shown in Figure 3.46(a) under the full-span vertical uniform loads.

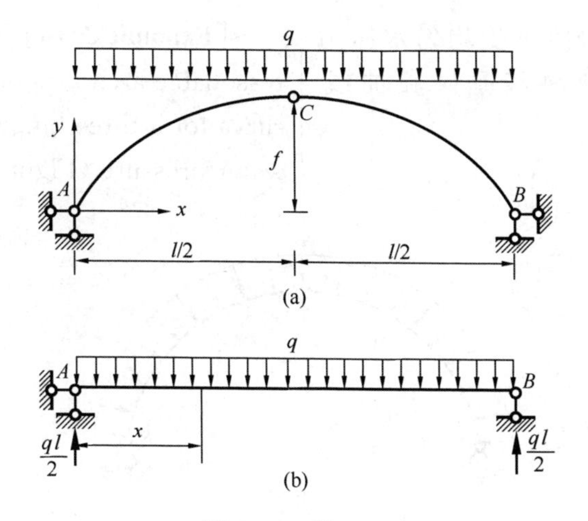

图 3.46 例 3.15

Figure 3.46 Example 3.15

【解】 由式（3.8）知：

[Solution] From the Equation (3.8):

$$y = \frac{M^0}{F_H}$$

图 3.46(b) 所示简支梁的弯矩方程为：

The bending moment equation of the simply supported beam in Figure 3.46(b) is:

$$M^0 = \frac{ql}{2}x - \frac{qx^2}{2} = \frac{1}{2}qx(l-x)$$

由式（3.6）求得推力为：

The thrust force can be obtained from Equation (3.6):

$$F_H = \frac{M_C^0}{f} = \frac{ql^2}{8f}$$

所以：

Hence:

$$y=\frac{4f}{l^2}x(l-x)$$

可见，在满跨竖向均布荷载作用下，三铰拱的合理拱轴线是抛物线。由于在合理拱轴线方程中，拱高 f 没有确定，因此具有不同高跨比的一组抛物线都是合理轴线。

It can be seen that the reasonable arch axis of the three-hinged arch is parabolic under the full-span vertical uniform loads. Since the arch height f is not determined in the equation of the reasonable arch axis, a group of parabolic lines with different height-to-span ratios are all reasonable arch axes.

【例 3.16】设三铰拱承受均匀水压力作用，试证明其合理轴线是圆弧曲线（图 3.47a）。

［**Example 3.16**］Try to prove that the reasonable arch axis of the arch is a circular curve for a three-hinged arch under uniform water pressure (Figure 3.47a).

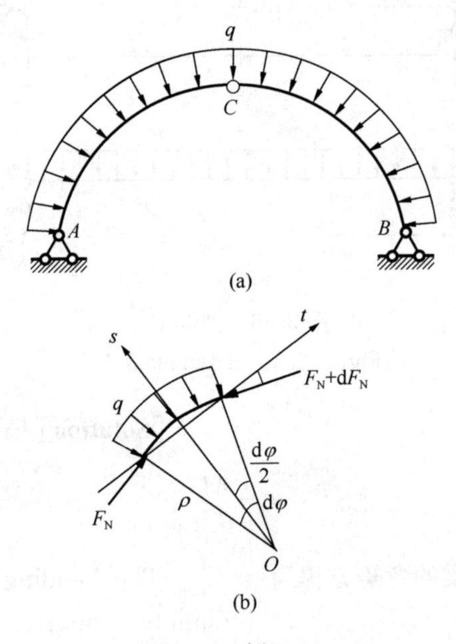

图 3.47　例 3.16

Figure 3.47　Example 3.16

【证】可以先假定拱处于无弯矩状态，然后根据平衡条件推求合理拱轴线方程。从拱中截取一微段为隔离体，如图 3.47(b) 所示。设微段两端横截面上弯矩、剪力均为零，而只有轴力 F_N 和 F_N+dF_N。

［**Proof**］It can be assumed that that arch is in a state without bending moment, and then the equation of the reasonable arch axis can be deduced according to the equilibrium conditions. A micro-segment is cut from the arch as the isolator, as shown in

Figure 3.47(b). It is assumed that the bending moment and shear force in the cross sections of both ends of the micro-segment are zero, with only the axial forces F_N and F_N+dF_N.

由 $\Sigma M_O = 0$ 得：

By $\Sigma M_O = 0$:

$$F_N\rho - (F_N + dF_N)\rho = 0$$

式中 ρ 为微段的曲率半径。由上式可知：

Where ρ is the radius of curvature of the micro-segment. From the above equation:

$$dF_N = 0$$

则 F_N =常数。

Then, F_N =constant.

再沿 s 轴列投影方程得：

And then project along the s axis and list the equation:

$$F_N\sin\frac{d\varphi}{2} + (F_N + dF_N)\sin\frac{d\varphi}{2} - q\rho d\varphi = 0$$

由于 $d\varphi$ 角极小，故可取 $\sin\frac{d\varphi}{2} = \frac{d\varphi}{2}$，并略去高阶微量得：

Since the angle of $d\varphi$ is extremely small, $\sin\frac{d\varphi}{2} = \frac{d\varphi}{2}$ can be taken and the higher differentiation can be omitted:

$$F_N - q\rho = 0$$

其中 F_N 为常数，q 亦为常数，故 $\rho = \frac{F_N}{q}$ =常数。

Where F_N is a constant and q is also a constant, $\rho = \frac{F_N}{q}$ = constant.

这就证明了合理轴线是圆弧曲线。

This proves that the reasonable arch axis of the arch is a circular curve.

【例 3.17】 试求图 3.48 所示在填土重量下三铰拱的合理拱轴线。拱上荷载集度按 $q = q_0 + \gamma y$ 变化，其中 q_0 为拱顶处的荷载集度，γ 为填土重度。

[**Example 3.17**] Try to find the reasonable arch axis of the three-hinged arch under the fill weight as shown in Figure 3.48. The load set on the arch varies according to $q = q_0 + \gamma y$, where q_0 is the load set at the arch vault and γ is the fill unit weight.

【解】 由式 (3.8) 知：

[**Solution**] From Equation (3.8):

$$y(x) = \frac{M^0(x)}{F_H}$$

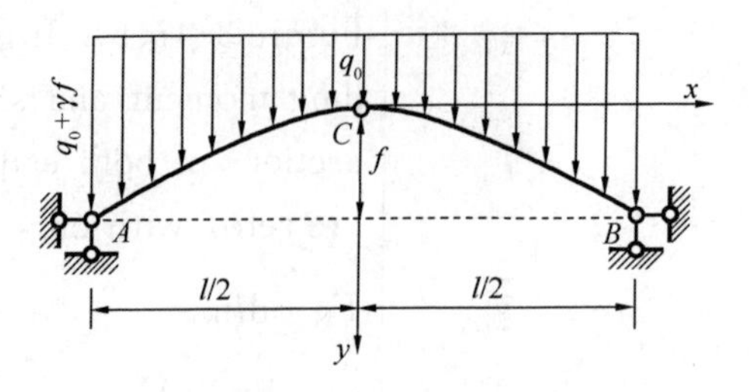

图 3.48　例 3.17

Figure 3.48　Example 3.17

上式对 x 微分两次，得：

Differentiate the above equation of x twice:

$$\frac{\mathrm{d}^2 y}{\mathrm{d}x^2}=\frac{1}{F_{\mathrm{H}}}\frac{\mathrm{d}^2 M^0}{\mathrm{d}x^2}$$

由于 $\frac{\mathrm{d}^2 M^0}{\mathrm{d}x^2}=-q(x)$，其中 $q(x)$ 表示沿水平线单位长度的荷载值，故：

Due to $\frac{\mathrm{d}^2 M^0}{\mathrm{d}x^2}=-q(x)$, where $q(x)$ is the load of unit length along the horizontal line, therefore:

$$\frac{\mathrm{d}^2 y}{\mathrm{d}x^2}=-\frac{q(x)}{F_{\mathrm{H}}} \tag{3.9}$$

这就是竖向荷载作用下合理拱轴线的微分方程，式中规定 y 向上为正。但在图 3.48 中，y 轴是向下的，因此式（3.9）右边应改为正号，即：

This is the differential equation of the reasonable arch axis under vertical loads, where y is positive when upward. However in Figure 3.48, the y axis is downward, so the right side of Equation (3.9) should be changed to be positive, i. e. :

$$\frac{\mathrm{d}^2 y}{\mathrm{d}x^2}=\frac{q(x)}{F_{\mathrm{H}}} \tag{3.10}$$

在本题中，由于荷载集度 q 随拱轴线纵坐标 y 而变，故相应简支梁的弯矩方程 M^0 无法事先求得，因而求合理轴线时，不用式（3.8）而用式（3.10）。

In this example, because q varies with the vertical coordinate y of the arch, so the bending moment equation M^0 of corresponding simply supported beam cannot be obtained in advance, therefore the solution for reasonable arch axis should not use Equation (3.8) but Equation (3.10).

将 $q=q_0+\gamma y$ 代入式（3.10），可得：

Substitute $q=q_0+\gamma y$ into Equation (3.10), we can get:

$$\frac{\mathrm{d}^2 y}{\mathrm{d}x^2}-\frac{\gamma}{F_{\mathrm{H}}}y=\frac{q_0}{F_{\mathrm{H}}}$$

这是二阶常系数线性非齐次微分方程，它的一般解可用双曲线函数表示：

This is a second-order linear non-homogeneous differential equation with constant coefficients, the general solution of which can be expressed by the hyperbolic function:

$$y = A \cdot \cosh\sqrt{\frac{\gamma}{F_{\mathrm{H}}}}x + B \cdot \sinh\sqrt{\frac{\gamma}{F_{\mathrm{H}}}}x - \frac{q_0}{\gamma}$$

常数 A、B 可由边界条件确定：

The constants A, B may be determined by boundary conditions:

当 $x = 0$ 时，$y = 0$，得：

When $x = 0$, $y = 0$:

$$A = \frac{q_0}{\gamma}$$

当 $x = 0$ 时，$y' = 0$，得：

When $x = 0$, $y' = 0$:

$$B = 0$$

于是，合理拱轴线方程为：

Thus, the reasonable arch axis equation is:

$$y = \frac{q_0}{\gamma}\left(\cosh\sqrt{\frac{\gamma}{F_H}}x - 1\right)$$

上式表明，在填土重量作用下，三铰拱的合理拱轴线是悬链线。

The above equation shows that the reasonable arch axis of the three-hinged arch is the catenary under the fill weight.

3.9 静定结构的特性综述

3.9 Overview of Properties of Statically Determinate Structures

静定结构在静力学方面有以下几个特性，掌握这些特性，对了解静定结构的性能和内力计算都是有益的。

Statically determinate structures have the following properties in terms of statics, which help understand the performance and internal force calculations of statically determinate structure.

（1）静力解答的唯一性。

(1) The uniqueness of the static force solution.

在给定的荷载作用下，静定结构的全部反力和内力都可以由静力平衡条件求出，而且得到的解答是唯一的有限值。这就是静定结构静力解答的唯一性。根据这一特性，在静定结构中，凡是能够满足平衡条件的内力

Under a given load, all the reaction and internal forces of a statically determinate structure can be found by the static equilibrium condition, and the solution obtained is a unique finite value. This is the uniqueness

解答就是唯一真正的解答，并可确信除此之外没有其他任何解答存在。

of the static solution of the statically determinate structure. According to this property, in a statically determinate structure, any internal force solution that can satisfy the equilibrium condition is the only true solution, and it can be assured that no other solution exists.

静力解答的唯一性是静定结构的基本静力特性。下面提到的一些特性，都是在此基础上派生出来的。

The uniqueness of the static force solution is the basic static property of the statically determinate structure. Some of the properties mentioned below are derived from this.

（2）在静定结构中，除荷载外，其他任何原因如温度改变、材料胀缩、支座移动、制造误差等均不引起内力。

（2）In statically determinate structure, any causes other than load, such as temperature change, material expansion and contraction, support movement, manufacturing error, etc., do not cause internal force.

如图 3.49（a）所示悬臂梁，当上侧温度升高 t_1，下侧温度升高 t_2时（设 $t_1>t_2$），梁将会发生自由的伸长和弯曲，不会产生任何反力和内力。又如图 3.49（b）所示简支梁，当 B 支座发生沉降时，只会引起刚体位移，而在梁内并不引起反力和内力。事实上，当荷载为零时，零内力状态能够满足静定结构各部分的平衡条件。由静定结构解答的唯一性可知，这就是唯一的、真正的解答。由此可以推断，除荷载外其他任何因素均不会使静定结构产生反力和内力。

In the cantilever beam shown in Figure 3.49(a), when the temperature on the upper side increases by t_1 and the lower side increases by t_2 (set $t_1>t_2$), the beam will undergo free elongation and bending without any reaction or internal forces. Another example is the simply supported beam shown in Figure 3.49(b), when the settlement of the support B occurs, it will only cause the rigid body displacement, but not the reaction and internal forces in the beam. When the load is zero, the zero internal force state can satisfy the equilibrium condition of all parts of the statically determinate structure. It is known from the uniqueness of the static force solution that this is the only and true solution. This can infer that no other factors other than the loads will cause the reaction and internal forces in the statically determinate structure.

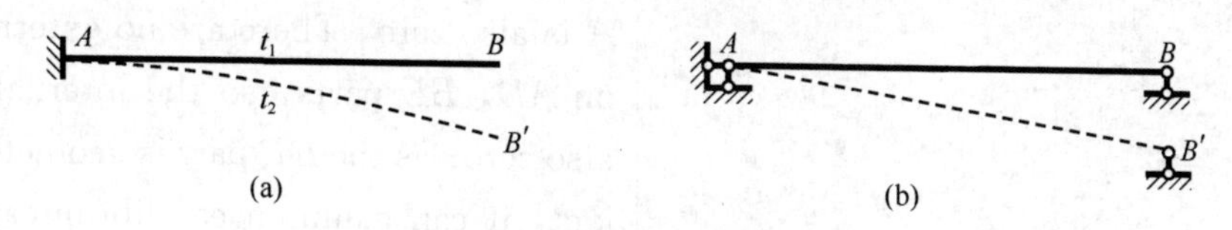

图 3.49 静定结构中的温度变化和支座沉降

Figure 3.49 The temperature changes and support settlements in statically determinate structure

（3）平衡力系的影响。

当平衡力系作用在静定结构的某一内部几何不变部分时，除了该部分受力外，其余部分的反力和内力均为零，如图 3.50 所示。

(3) Effect of equilibrium force system.

When the equilibrium force system acts on an internal geometric invariant part of a statically determinate structure, the reaction and internal forces in the rest of the part are zero except for the forces in that part, as shown in Figure 3.50.

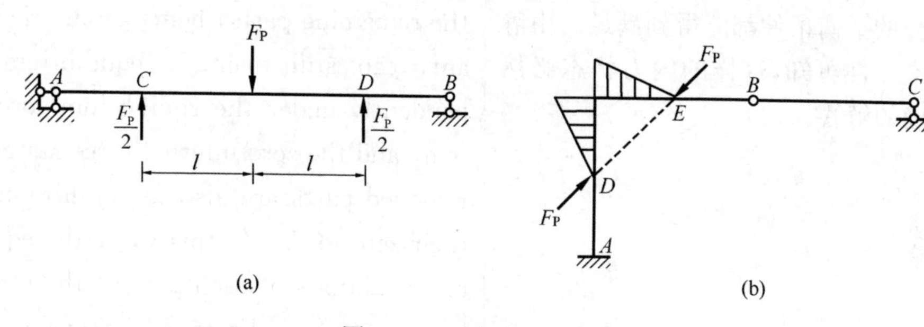

图 3.50 平衡力系的影响

Figure 3.50 Effect of equilibrium force system

如图 3.50（a）所示简支梁，*CD* 段为一内部几何不变部分，作用有平衡力系，则只有 *CD* 段受力，其余的 *AC* 段、*BD* 段均没有反力和内力。又如图 3.50（b）所示刚架，内部几何不变部分 *DE* 上作用有平衡力系，由于附属部分 *BC* 上无荷载，由平衡条件可知其反力和内力均为零；再以 *AB* 为隔离体，可知 *A* 支座反力也为零，*AD*、*BE* 部分均无外力，内力亦全为零；而 *DE* 部分由于本身为几何不变，故在平衡力系下仍能独立地维持平衡。弯矩图如图 3.50（b）所示。

As the simply supported beam shown in Figure 3.50(a), the *CD* segment is an internal geometric invariant part acted a balanced force system, so there are internal forces only in *CD* segment, and the other parts of *AC* and *BD* segments have no reaction force or internal force. Another example is the rigid frame shown in Figure 3.50 (b), there is a balance force system acting on the internal geometric invariant part *DE*. Since there is no load on the secondary part *BC*, the reaction force and internal force are both zero according to the equilibrium condition. Now take *AB* as the isolator, it can be known that the reaction force of support

A is also zero, There are no external forces on *AD*, *BE* parts, so the internal force is also zero; as the *DE* part is geometric invariant, it can maintain equilibrium independently under the equilibrium force system. The bending moment diagram is shown in Figure 3.50(b).

这种情形实际上具有普遍性。因为当平衡力系作用于静定结构的任何几何不变部分上时,设想其余部分不受力而将它们撤除,则所剩部分由于本身是几何不变的,在平衡力系下仍能独立地维持平衡,而所去除部分的零内力状态也与其零荷载相平衡。这样,结构上各部分的平衡条件都能得到满足。由静力解答的唯一性可知,这样的内力状态必然是唯一正确的解答。

This situation is practically universal. Because when the equilibrium force system acts on any geometric invariant part of a statically determinate structure, it is assumed that the remaining parts are not subjected to forces, and they are removed, then the remaining parts, being geometric invariant, can still maintain equilibrium independently under the equilibrium force system, and the zero internal force state of the removed parts are also in equilibrium with their zero loads. In this way, the equilibrium conditions of each part of the structure can be satisfied. From the uniqueness of the static force solution, it is clear that such a state of internal forces must be the only correct solution.

(4) 荷载等效变换的影响。

(4) The effect of load equivalent transformation.

当静定结构的一个内部几何不变部分上的荷载作等效变换时,则只有该部分的内力发生变化,其余部分的内力不变。这里,等效荷载是指荷载分布不同,但其合力彼此相等的荷载。

When the load on an internal geometric invariant part of a statically determinate structure is transformed equivalently, only the internal force of that part changes, while the internal force of the rest remains unchanged. Here, the equivalent load refers to the different load distribution, but the combined forces are equal.

如图 3.51 (a) 所示简支梁在 F_P 作用下,若把 F_P 等效变换为图 3.51 (b) 所示情况,则除了 *CD* 范围内的受力状态有变化

If F_P in Figure 3.51(a) is transformed equivalently in the case shown in Figure 3.51(b), the internal forces in the rest of

外，其余部分的内力均保持不变。

the beam remain unchanged except for the changes in the state of forces in the *CD* segment.

静定结构在等效荷载作用下的这一特性，可用平衡力系的影响来证明。图 3.51（a）的受力情况等价于图 3.51（b）和图 3.51（c）两种情况的叠加。而图 3.51（c）是静定结构受一平衡力系，所以除 *CD* 段以外，其余部分的内力为零。因而有，图 3.51（a）和图 3.51（b）两种情况除 *CD* 段以外，其余部分的内力相同。这就证明了上述结论。

This property of the statically determinate structure under equivalent loads can be demonstrated by the effect of the equilibrium force system. The forces in Figure 3.51 (a) are equivalent to the superposition of the two cases in Figure 3.51(b) and Figure 3.51(c). In Figure 3.51(c), there is a statically determinate structure subjected to an equilibrium force system, so the internal force in the rest of the structure is zero except for the *CD* segment. Thus, the internal forces of the two cases in Figure 3.51 (a) and Figure 3.51(b) are the same except for the *CD* segment. This proves the above conclusion.

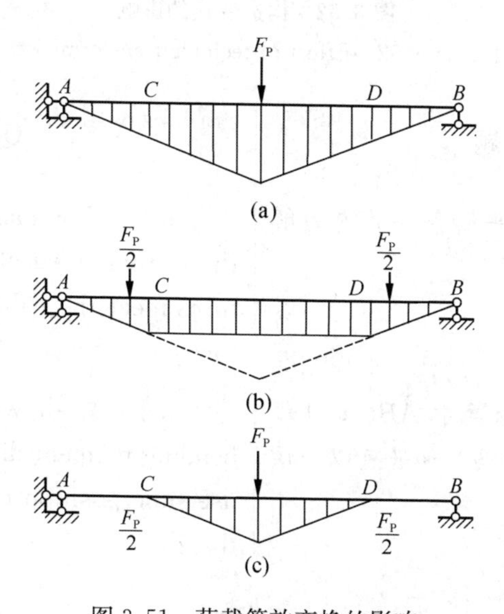

图 3.51 荷载等效变换的影响

Figure 3.51 The effect of load equivalent transformation

（5）构造变换的影响。

(5) Effect of tectonic transformation.

当静定结构的一个内部几何不变部分作构造上的局部改变时，只有在该部分的内力发生变化，其余部分的内力均保持不变。

When an internal geometrically invariant part of a statically determinate structure is partially changed, only the internal forces in that part change, while the internal forces in the rest of the structure remain unchanged.

如图3.52（a）所示桁架，把AB杆换成图3.52（b）所示的小桁架，而作用的荷载和端部A、B的约束性质保持不变，则在作上述组成的局部发生改变后，只有AB部分的内力发生变化，其余部分的内力保持不变。

The rod AB in truss in Figure 3.52(a) is replaced by the small truss, as shown in Figure 3.52(b), with the same loads and restraint properties of ends A and B, then only the internal forces of the AB part will be changed, while the internal forces of the remaining parts will remain unchanged after making the local change.

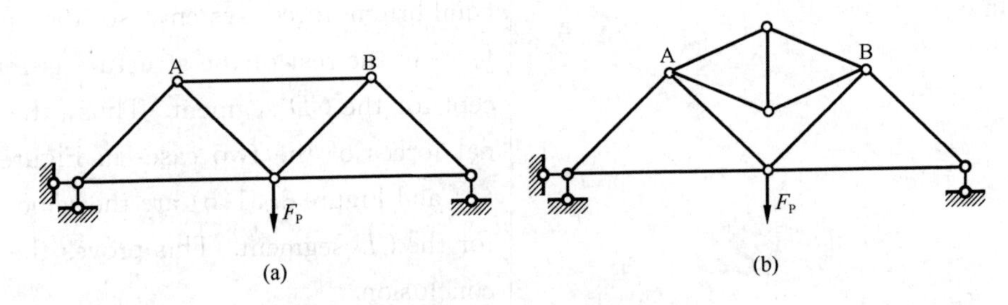

图3.52　构造变换的影响

Figure 3.52　Effect of tectonic transformation

思　考　题

Questions

3.1　静定结构满足平衡条件的内力解答有多少种？

3.1　How many kinds of solutions are there that statically indeterminate structures meet the balance condition of internal force?

3.2　图示3.53所示梁中AB、CD段的弯矩图能用叠加法绘制吗？如果可以，该如何做？

3.2　As shown in Figure 3.53, can the bending moment diagram in beams AB, CD use superposition method? If so, how to do that?

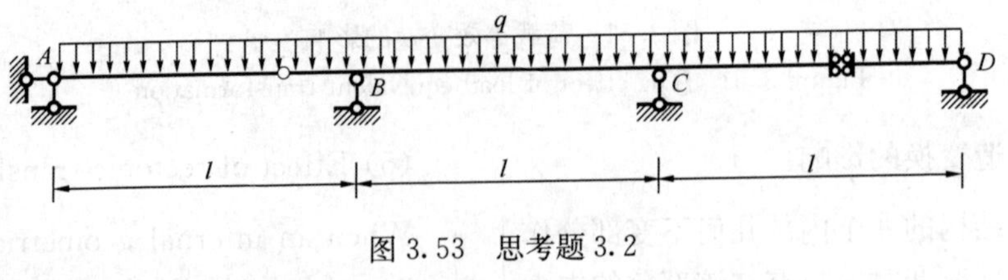

图3.53　思考题3.2

Figure 3.53　Question 3.2

3.3 根据图 3.54 中梁的弯矩图和剪力图的形状，梁上有什么样的荷载？

3.3 According to the shapes of bending moment diagram and shear force diagram of the beam in Figure 3.54, what kinds of loads are there on the beam?

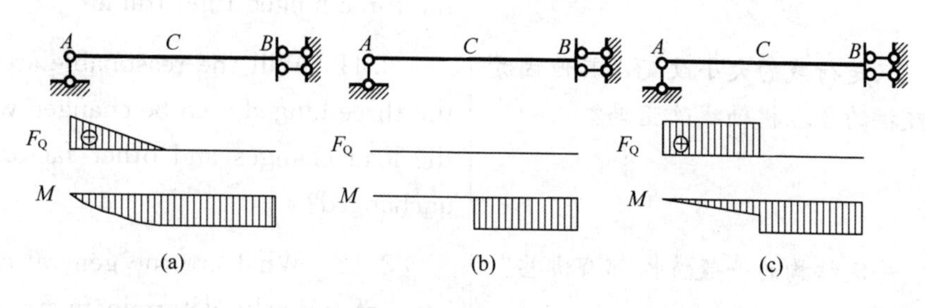

图 3.54 思考题 3.3

Figure 3.54 Question 3.3

3.4 静定多跨梁的几何组成特点和受力特点是什么？

3.4 What are the geometric characteristics and mechanical characteristics of statically determinate multi-span beams?

3.5 对于静定结构，它的某一局部能平衡外力时，其他部分还受力吗？为什么？

3.5 For statically determinate structure, when one part can balance the external force itself, will other parts have internal forces? Why?

3.6 对于静定结构，改变材料的性质，或改变横截面的形状和尺寸，会不会改变其内力分布，会不会改变其变形和位移？

3.6 For statically determinate structure, will changing the nature of the material or changing the shape and size of the cross section change its internal force distribution, or change its deformation and displacement?

3.7 理想桁架的组成特点是什么？桁架中的杆件都有什么内力？

3.7 What are the composition characteristics of the ideal truss? What are the internal forces of the rods in the truss?

3.8 桁架在给定荷载作用下，有些杆件的轴力为零。这些杆件有什么作用？能不能将其去掉？

3.8 The axis forces in some of the rods in the truss are zero. What are the functions of these rods? Can we remove them?

3.9 在静定结构内力计算时，都在哪里用到了几何组成分析？几何组成分析给静定结构的内力计算带来了哪些方便？

3.9 When calculating the internal forces of statically determinate structures, when can we use geometric composition analysis? What convenience does geometric composition analysis bring to the internal force calculation of statically determinate structures?

3.10 三铰拱的反力计算公式（3.6）能计算三铰刚架的反力吗？什么情况下可以？

3.10 Can the reaction force calculation Equation（3.7）of the three-hinged arch calculate the reaction force of three-hinged rigid frame? When can we use that in the three-hinged rigid frame?

3.11 只是荷载的大小改变，其他因素不变，三铰拱的合理拱轴线改变吗？

3.11 Will the reasonable arch axis of the three-hinged arch be changed when only the load changes and other factors remain unchanged?

3.12 静定结构的一般特性都有哪些？

3.12 What are the general characteristics of statically determinate structures?

习　题

Exercises

3.1 试用叠加法绘制图3.55所示梁的弯矩图。

3.1 Try to use the superposition method to draw the bending moment diagrams in Figure 3.55.

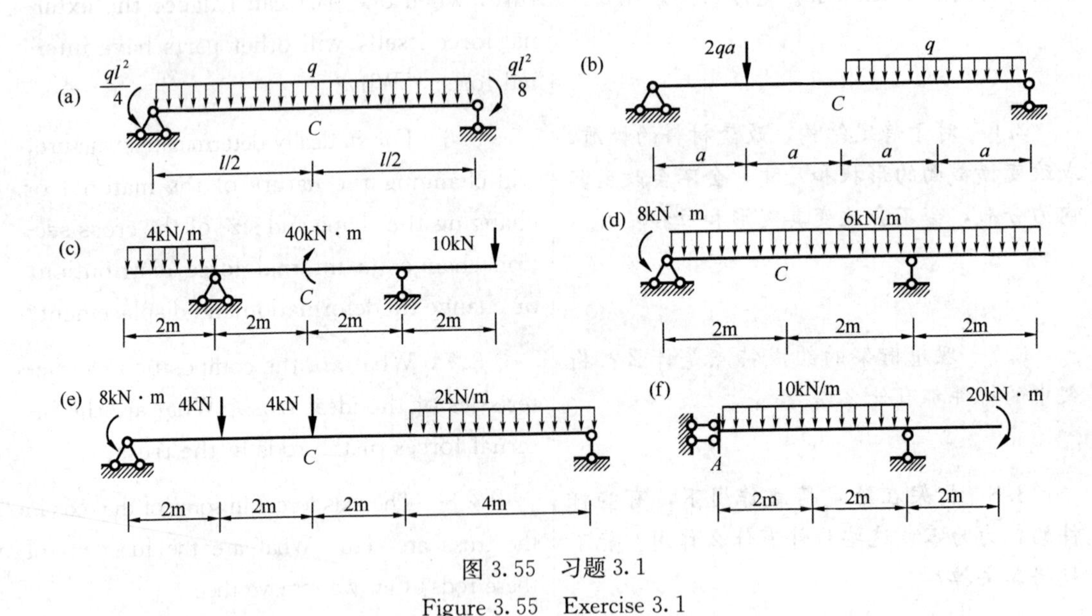

图3.55 习题3.1

Figure 3.55 Exercise 3.1

3.2 判断图3.56所示梁的内力图形状正确与否，并将错误的地方进行改正。

3.2 Try to determine whether internal force diagrams of each beam in Figure 3.56 are correct or not, and correct the wrong diagrams.

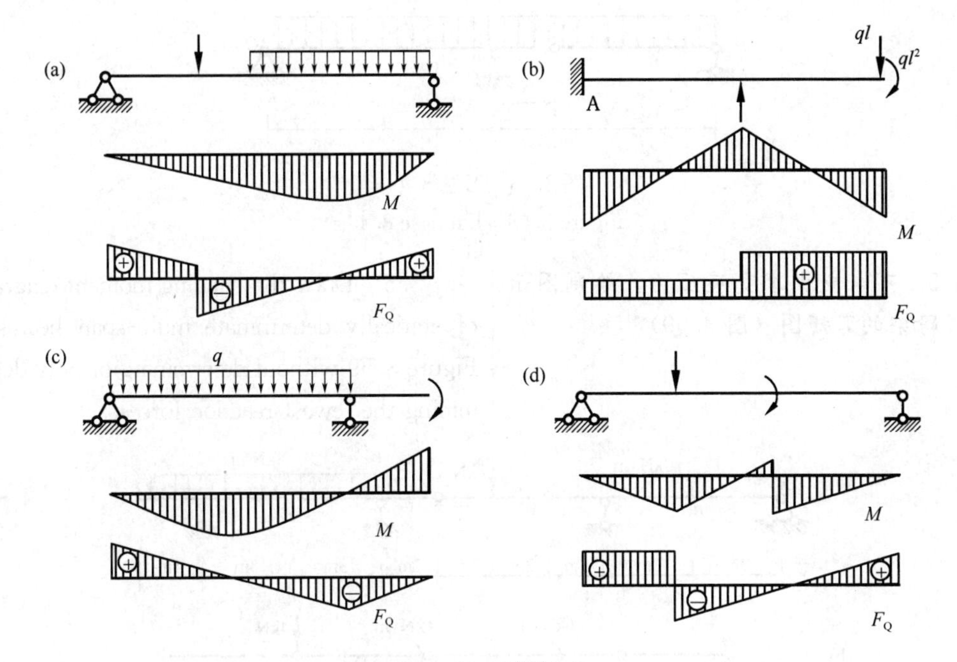

图 3.56　习题 3.2

Figure 3.56　Exercise 3.2

3.3　求图 3.57 所示静定多跨梁的支座反力，并绘制内力图。

3.3　Find out the bearing reaction forces of the statically determinate multi-span beams in Figure 3.57 and draw the internal force diagrams.

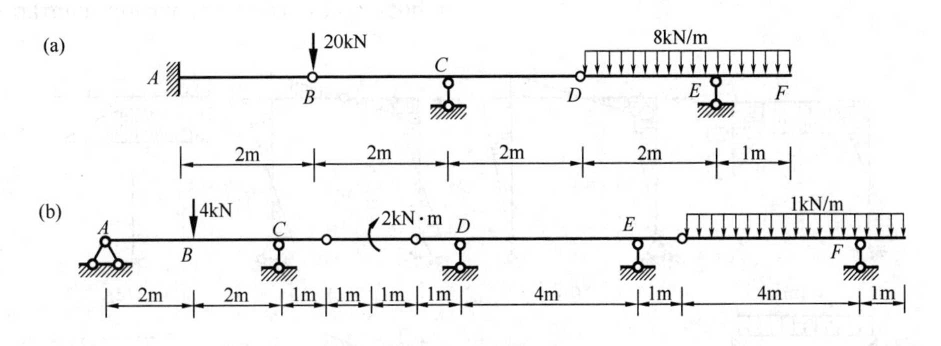

图 3.57　习题 3.3

Figure 3.57　Exercise 3.3

3.4　如使梁中正负弯矩峰值相等，那么铰 D 应放在何处（图 3.58)？

3.4　Where should the hinge D be placed if the peak values of positive and negative bending moments in the beam are equal(Figure 3.58)?

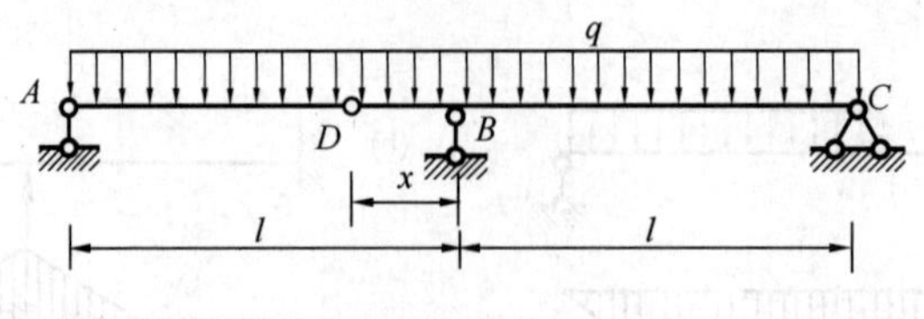

图 3.58　习题 3.4

Figure 3.58　Exercise 3.4

3.5　不求或少求支座反力，绘制图示静定多跨梁的弯矩图（图 3.59）。

3.5　Draw the bending moment diagrams of statically determinate multi-span beams in Figure 3.59 without determining or only determining the fewest reaction forces.

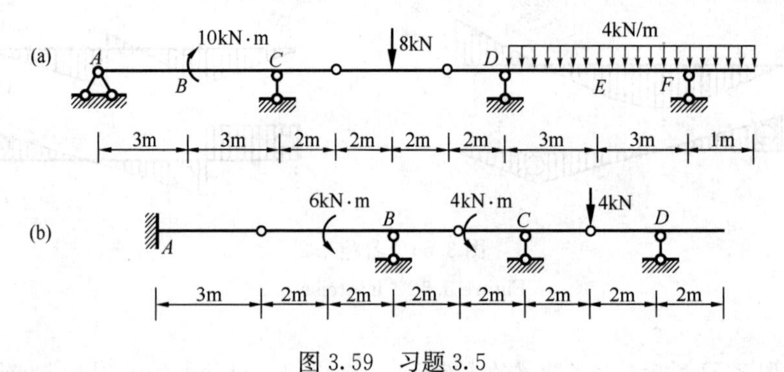

图 3.59　习题 3.5

Figure 3.59　Exercise 3.5

3.6　判断图 3.60 所示刚架弯矩图形状正确与否，并将错误的地方进行改正。

3.6　Determine whether the bending moment diagrams in Figure 3.60 are correct or not, and correct the wrong diagrams.

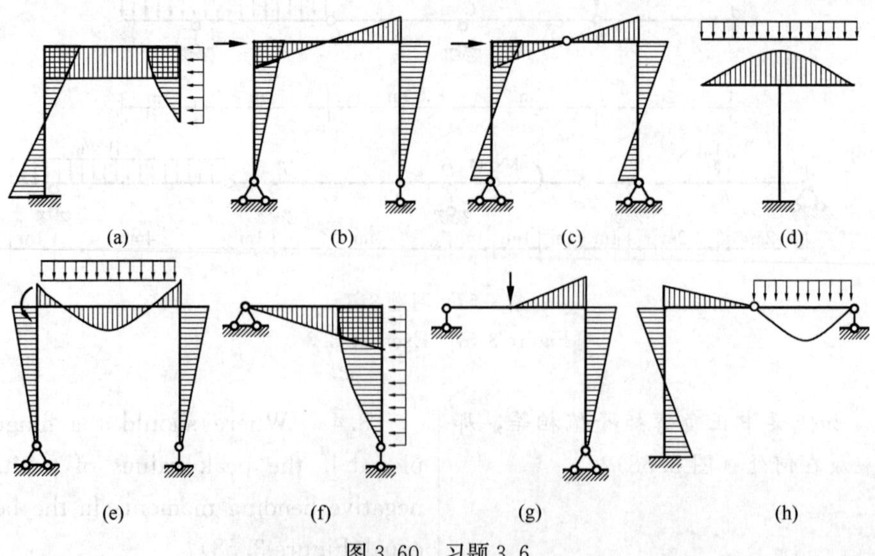

图 3.60　习题 3.6

Figure 3.60　Exercise 3.6

3.7 绘制图 3.61 所示刚架的内力图。

3.7 Draw the internal force diagrams of rigid frames in Figure 3.61.

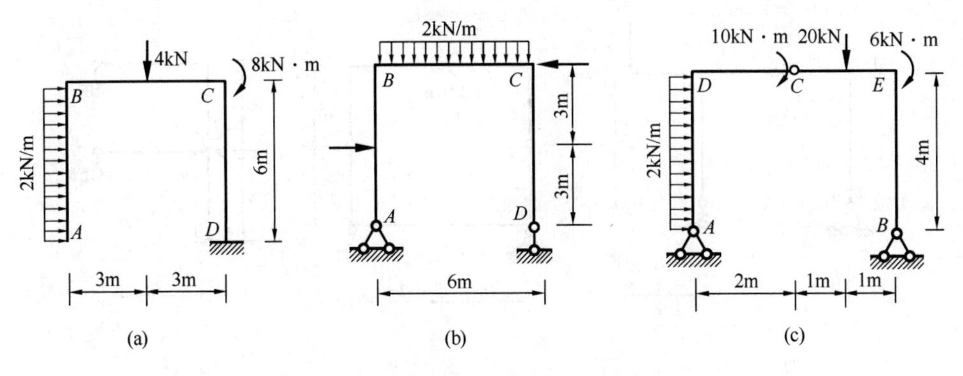

图 3.61 习题 3.7
Figure 3.61 Exercise 3.7

3.8 绘制图 3.62 所示刚架弯矩图。

3.8 Draw the bending moment diagrams of rigid frames in Figure 3.62.

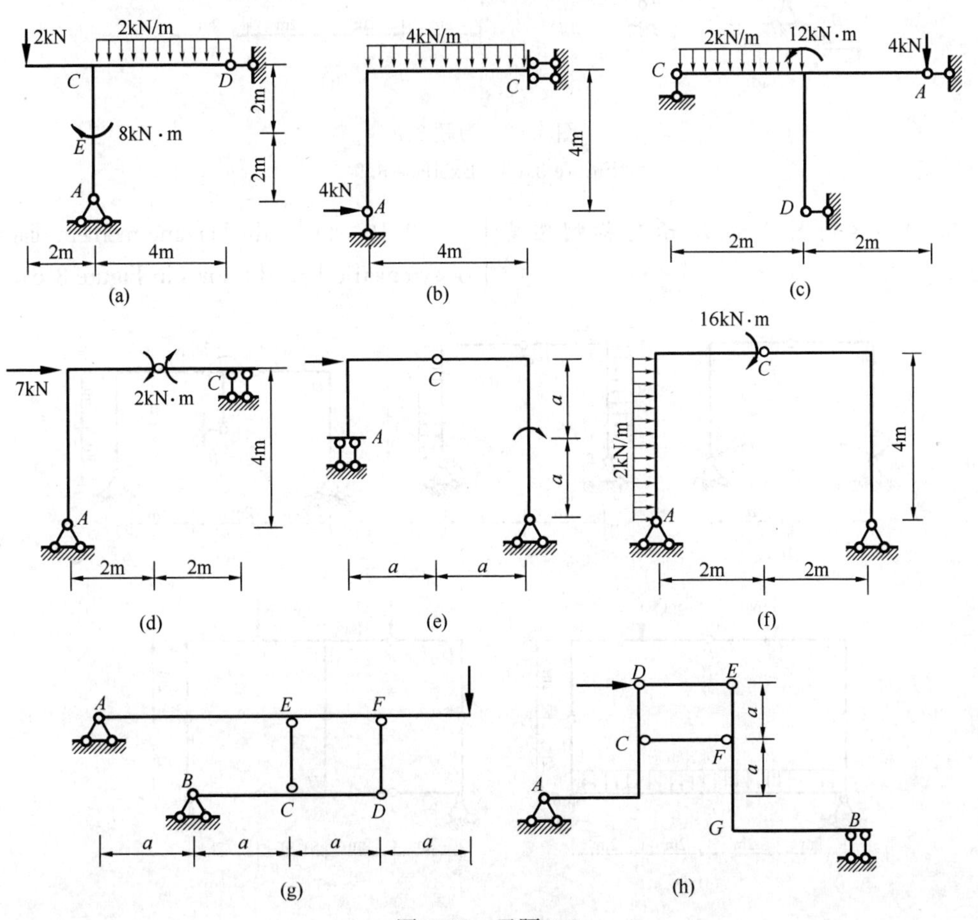

图 3.62 习题 3.8
Figure 3.62 Exercise 3.8

3.9　绘制图3.63所示刚架弯矩图。

3.9　Draw the bending moment diagrams of rigid frames in Figure 3.63.

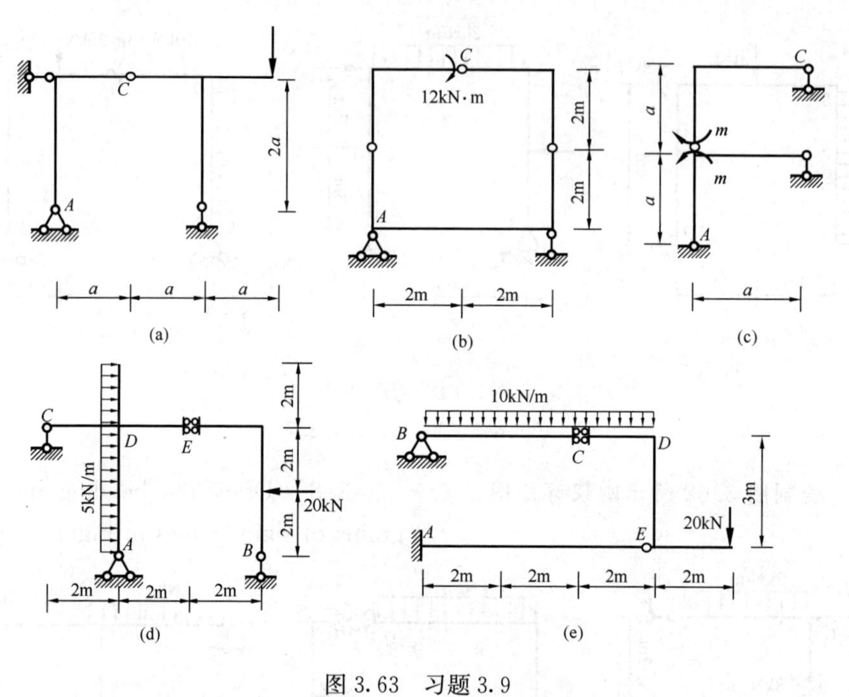

图3.63　习题3.9

Figure 3.63　Exercise 3.9

3.10　绘制图3.64所示对称刚架弯矩图。

3.10　Draw the bending moment diagrams of symmetrical rigid frames in Figure 3.64.

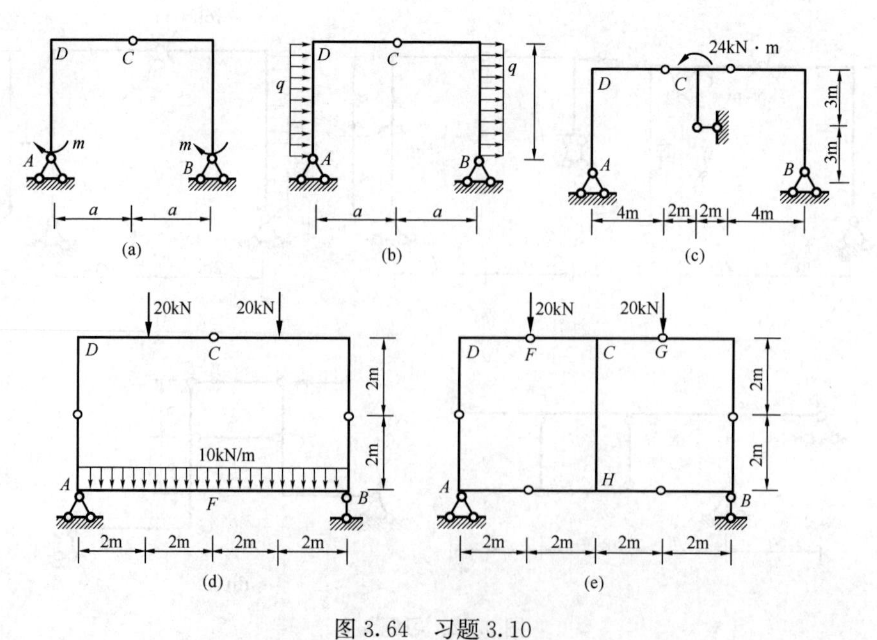

图3.64　习题3.10

Figure 3.64　Exercise 3.10

3.11 试用节点法计算图 3.65 所示桁架各杆轴力。

3.11 Try to use the nodal method to calculate axial forces of each rod in trusses shown in Figure 3.65.

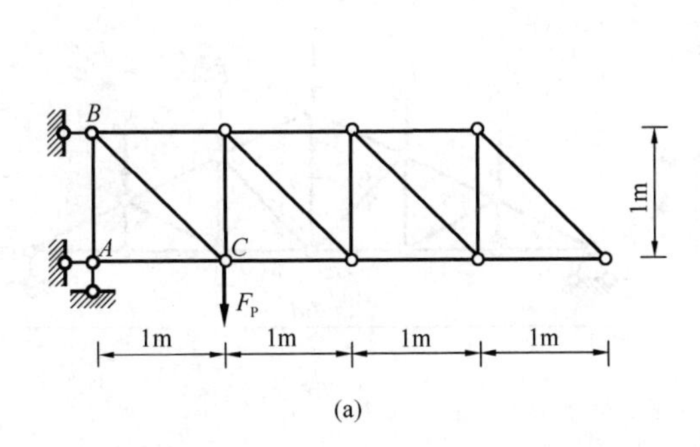

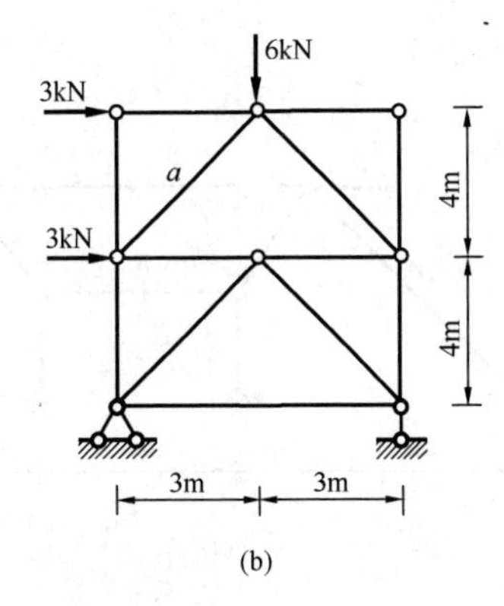

图 3.65 习题 3.11

Figure 3.65 Exercise 3.11

3.12 试判断图 3.66 所示桁架中的零杆。

3.12 Try to determine the zero rods in trusses in Figure 3.66.

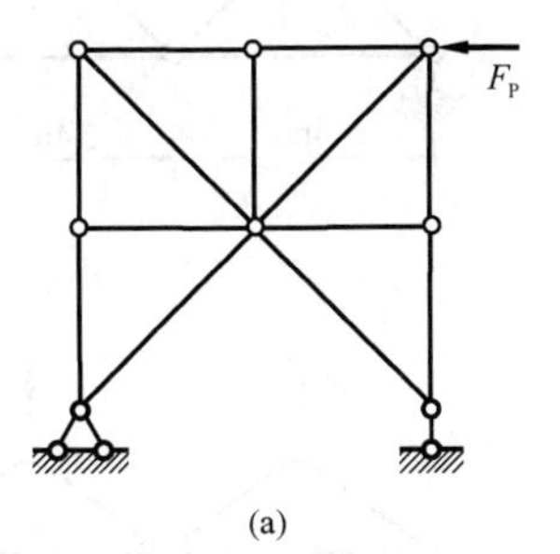

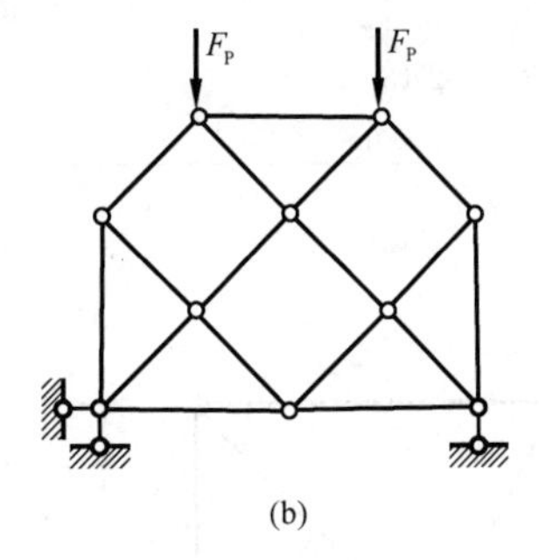

图 3.66 习题 3.12

Figure 3.66 Exercise 3.12

3.13 试用截面法计算图 3.67 所示桁架中指定杆件的内力。

3.13 Try to use method of sections to calculate the internal forces of the specified rods in trusses in Figure 3.67.

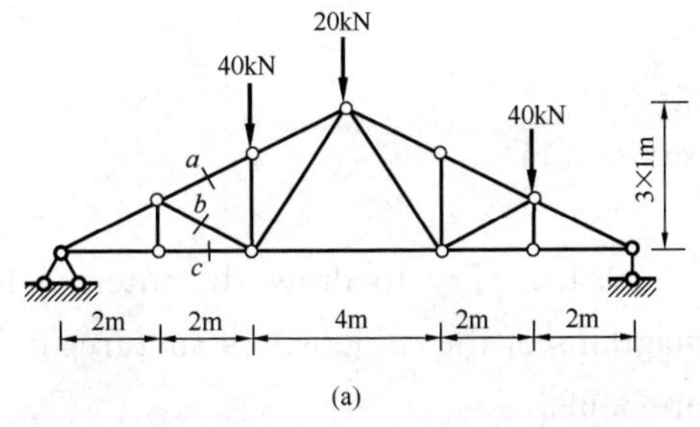

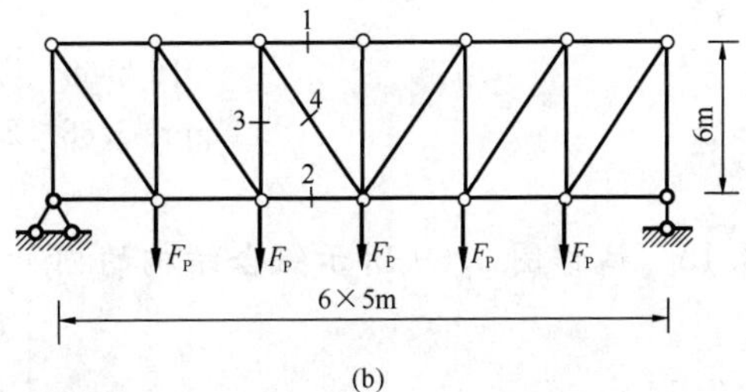

图 3.67 习题 3.13

Figure 3.67 Exercise 3.13

3.14 试用较简便的方法计算图3.68所示桁架中指定杆件的内力。

3.14 Try to use simple methods to calculate the internal forces of the specified rods in trusses in Figure 3.68.

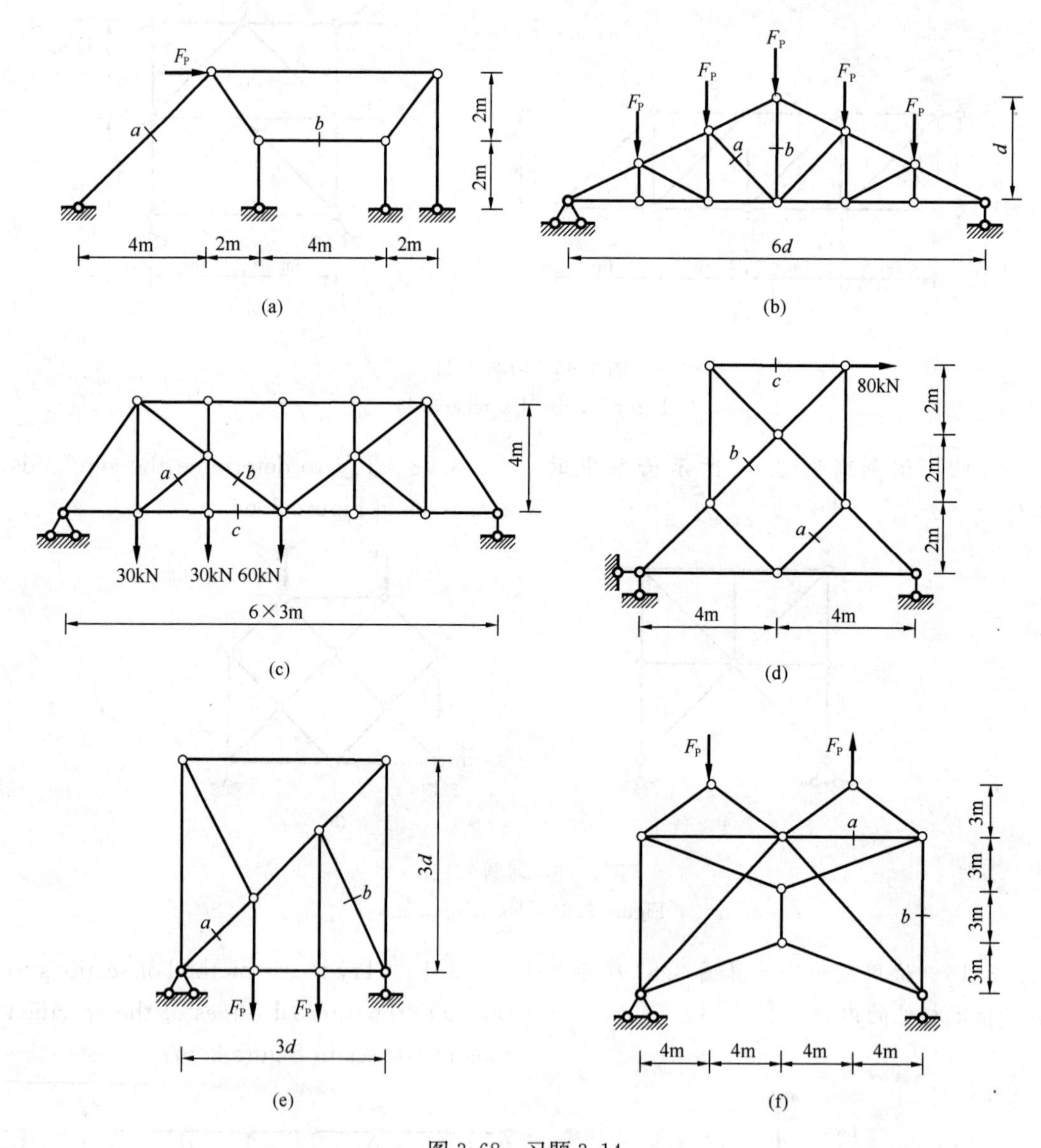

图3.68 习题3.14

Figure 3.68 Exercise 3.14

3.15 试作图3.69所示组合结构的内力图。

3.15 Try to draw the internal force diagrams of the composite structures in Figure 3.69.

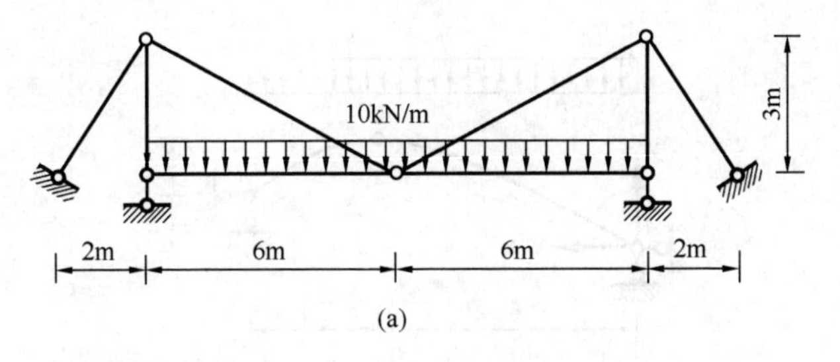

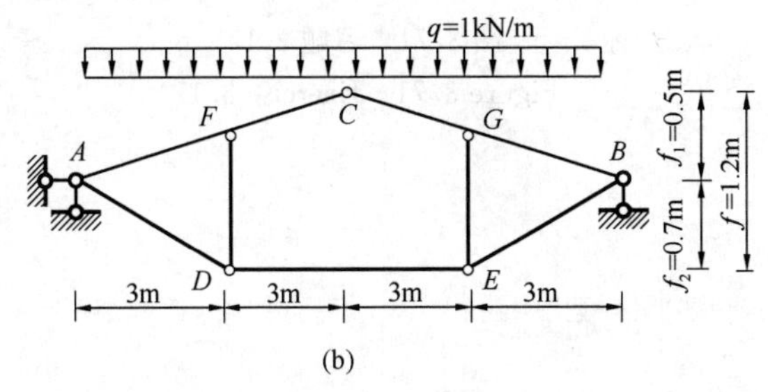

图 3.69 习题 3.15

Figure 3.69 Exercise 3.15

3.16 图 3.70 所示三铰拱轴线方程为抛物线，$y=\frac{4f}{l^2}x(l-x)$，$l=16\text{m}$，$f=4\text{m}$。试求：

(a) 支座反力；

(b) 截面 E 的 M、F_Q、F_N值；

(c) D 点左右两侧截面的 F_Q、F_N值。

3.16 Figure 3.70 shows that the axis in the three-hinged arch obeys a parabolic line with the equation $y=\frac{4f}{l^2}x(l-x)$, $l=16\text{m}$, $f=4\text{m}$. Try to calculate:

(a) Bearing reactions;

(b) M, F_Q and F_N of section E;

(c) F_Q and F_N of the left and right cross sections at point D.

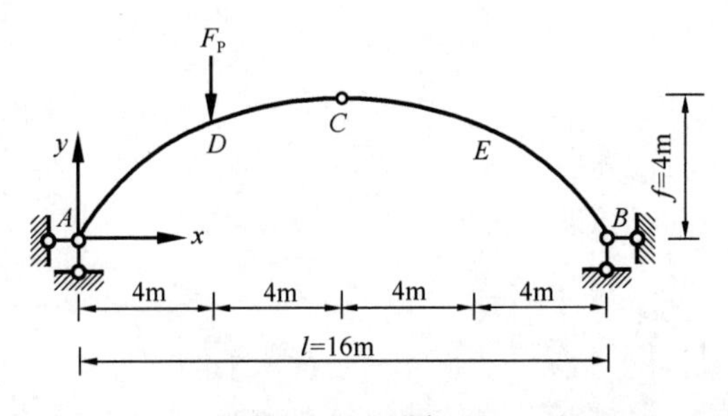

图 3.70 习题 3.16

Figure 3.70 Exercise 3.16

3.17 试求图 3.71 所示均布荷载作用下三铰拱的合理拱轴线。

3.17 Try to find out the reasonable arch axis of three-hinged arch under uniformly distributed load in Figure 3.71.

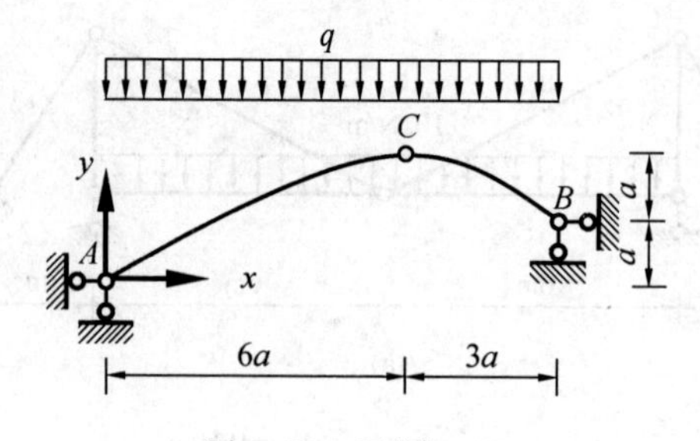

图 3.71 习题 3.17
Figure 3.71 Exercise 3.17

第4章 Chapter 4

静定结构位移计算 Displacement Calculations for Statically Determinate Structures

要点

➢ 位移计算的基本概念
➢ 广义力与广义位移
➢ 虚功原理
➢ 荷载作用下的位移计算
➢ 图乘法
➢ 非荷载因素引起的位移计算
➢ 互等定理

Keys

➢ Basic concepts of displacement calculation
➢ Generalized forces and generalized displacements
➢ Principle of virtual work
➢ Displacement calculations under loads
➢ Graph multiplication
➢ Calculation of displacements due to non-load factors
➢ Reciprocity theorem

4.1 结构位移计算概述

4.1 Overview of Structural Displacement Calculation

结构的位移计算，在结构力学中是一个重要内容，其目的有以下三个方面：

Displacement calculation for structures is an essential element in structural mechanics for three purposes:

1. 验算结构的刚度。在结构设计时，不仅要考虑其强度要求，还需保证其刚度条件，即结构的变形不能超过规范规定的限值。如钢筋混凝土屋盖和楼盖梁的挠度限值是其跨度的 1/400～1/200，吊车梁的挠度限值是其跨度的 1/600～1/500。

1. Checking the stiffness of the structure. In the design of the structure, it is necessary not only to consider the structure's strength requirements, but also to ensure its stiffness conditions, i. e. , the deformation of the structure cannot exceed the specified limits in the code. For example, the deflection limit of reinforced concrete roof and floor cover beam is the span of 1/400—1/200,

the deflection limit of crane beam is the limit of the span of 1/600—1/500.

2. 为超静定结构的内力分析打下基础。在计算超静定结构内力时，除利用静力平衡条件外，还需考虑结构的变形协调条件，这就要以位移计算作为基础。

2. To lay the foundation for the internal force analysis of a statically indeterminate structure. When calculating the internal forces of a statically indeterminate structure, except for using the static equilibrium conditions, it is necessary to consider the deformation coordination conditions of the structure, which takes the displacement calculation as a basis.

3. 在结构的制作、架设、养护等过程中，常需预先知道结构的位移，以便采取一定的施工措施，如桥梁工程中的预拱度计算。

3. During the fabrication, erection, and maintenance of a structure, it is often necessary to know the displacement of the structure in advance so that specific construction measures can be taken, such as the calculation of camber in bridge engineering.

4.1.1　位移计算的基本概念

4.1.1　Basic Concepts of Displacement Calculation

变形：结构原有形状发生的改变。

Deformation: changes in the original shape of a structure.

位移：由于结构的变形，导致其上各点发生移动。截面发生移动和转动，统称为位移。

Displacement: the movement of a structure leading to deformation of its points. The movement and rotation of a section is collectively referred to as displacement.

注意，变形一定会导致位移，但发生位移不一定导致变形，比如静定结构的支座移动等。

It should be noted that deformation must lead to displacement, but the occurrence of displacement does not necessarily lead to deformation, such as the support movement of statically determinate structure.

产生位移的原因有：荷载、温度改变、材料收缩徐变、支座移动、制造误差等。

The causes of displacement include load, temperature change, material shrinkage and creep, support movement, manufacturing error, etc.

结构的位移分为两大类：

The displacement of structures is divided into two main categories:

第一类是线位移：指结构上某点或某截面沿直线方向移动的距离，如图 4.1（a）中 Δ。

The first one is linear displacement. It means the distance that a point or section on a structure moves in a linear direction, as Δ shown in Figure 4.1(a).

第二类是角位移：也称转角，指截面转动的角度，如图 4.1（b）中 θ。

The second one is angular displacement, which is also known as the angle of rotation, it refers to the angle at which the section rotates, as θ shown in Figure 4.1(b).

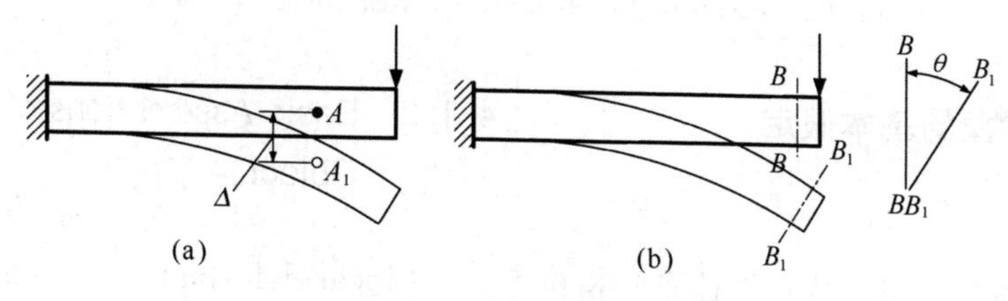

图 4.1　位移的分类

（a）线位移；（b）角位移

Figure 4.1　The classification of displacement

(a) Linear displacement; (b) Angular displacement

线位移通常可分解为水平和竖直分量，如图 4.2（a）所示。

The linear displacement can usually be decomposed into horizontal and vertical components, as shown in Figure 4.2(a).

两个截面之间转角之差称之为两截面之间的**相对转角**或**相对角位移**；两点之间线位移之差称为两点间的**相对线位移**。如图 4.2（b）所示，此处需注意，位移是有方向的，方向相反的位移正负号也相反。

The difference in the rotation angle between two sections is called the **relative angle of rotation** or **relative angular displacement** of the two sections; the difference in the linear displacement between two points is called **the relative linear displacement** between the two points. As shown in Figure 4.2(b), it should be noted that the displacements are directional, and the positive and negative signs of displacements means opposite directions of displacements.

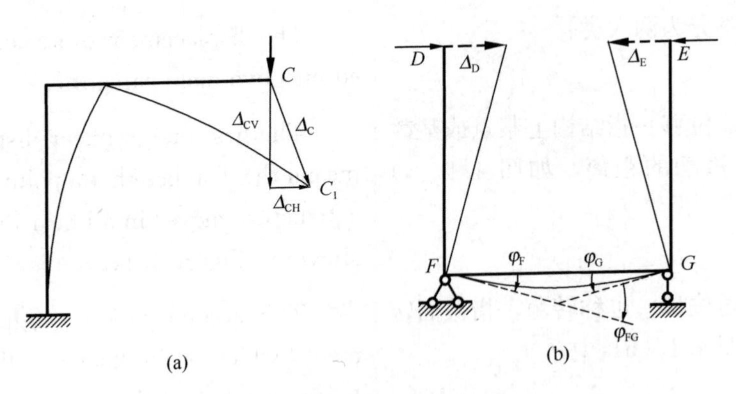

图4.2 位移的分解

(a) 位移的分量；(b) 相对位移

Figure 4.2 The decomposition of linear displacement

(a) Components of displacement; (b) Relative displacement

4.1.2 研究对象的基本假定

线性变形体系（也称线弹性体系）指位移与荷载呈线性关系的体系，而且当荷载全部撤除后，位移完全消失。

线弹性体系需要满足两个条件：

1. 线弹性材料：应力应变关系满足胡克定律；

2. 小变形假设：变形前后荷载作用位置不变，不考虑由于杆弯曲所引起的杆端轴力对弯矩及弯曲变形的影响，如图4.3所示。

4.1.2 Basic Assumptions of Research Object

Linear deformation system (also called linear elastic system) means the one in which the displacement is linearly related to the loads and where the displacement disappears completely when the load is fully withdrawn.

The linear deformation system should satisfy the following conditions:

1. Linearly elastic materials: The stress-strain relationship satisfies Hooke's Law;

2. Small deformation assumption: The position of the load is constant before and after deformation, and the effect of the axial force at the rod end on the bending moment and bending deformation due to rod bending is not considered, as shown in Figure 4.3.

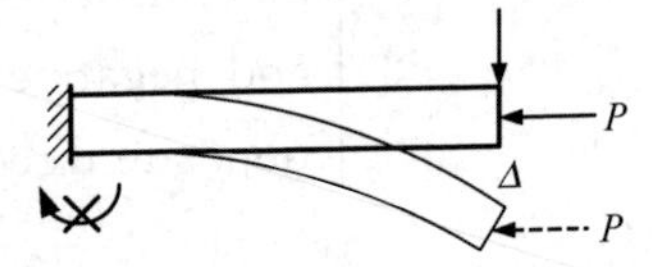

图4.3 忽略轴向力对弯矩的影响

Figure 4.3 Neglecting the effect of axial forces on bending moments

此外，在位移计算中，假定结构各部分之间不计摩擦阻力（理想约束），线性变形体系的位移计算可以使用叠加原理。

Furthermore, in displacement calculations, frictional resistance between different parts of the structure are ignored (ideal restraint), the superposition principle can be used for displacement calculations in linear deformation systems.

4.1.3 实功与虚功

4.1.3 Real Work and Virtual Work

力的实功：力在其本身引起的位移上所做的功；

Real work of force: the work done by a force on the displacement caused by the force itself;

力的虚功：力在其他原因引起的位移上所做的功。

Virtual work of a force: work done by a force on a displacement caused by other causes.

图 4.4（a）所示悬臂梁先后受到 F_{P1} 和 F_{P2} 作用，将加载过程分为Ⅰ、Ⅱ两个阶段，如图 4.4（b）和图 4.4（c）所示（图中符号第一个脚标表示位置，第二个脚标表示原因，如：Δ_{21} 表示②号位置由 F_{P1} 引起的位移）。

The cantilever beam shown in Figure 4.4(a) is successively subjected to F_{P1} and F_{P2}. The loading process is divided into two stages, Ⅰ and Ⅱ, as shown in Figure 4.4 (b) and Figure 4.4 (c) (The first foot marker of the symbols in the figure indicates the position, and the second foot marker indicates the cause, e.g., Δ_{21} indicates the displacement caused by F_{P1} at position ②).

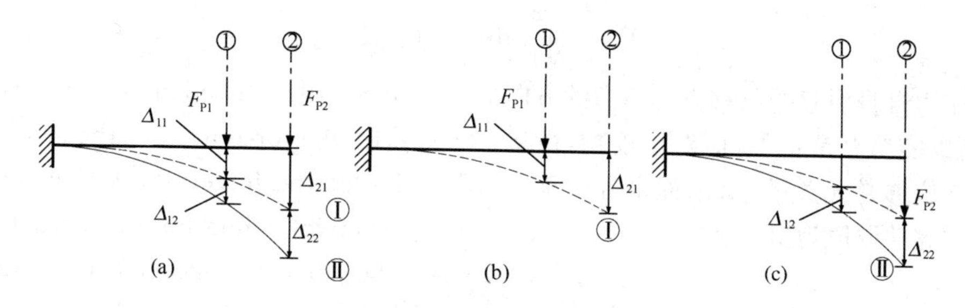

图 4.4 力的加载步骤

Figure 4.4 Force loading steps

阶段Ⅰ（图 4.4b）中：F_{P1} 对应①号的位移 Δ_{11}，因 Δ_{11} 由 F_{P1} 所引起，所以这一个过程所做的就是实功；

In stage Ⅰ (Figure 4.4b), F_{P1} corresponds to the displacement Δ_{11} of position ①, because Δ_{11} is caused by F_{P1}, this process is real work;

阶段Ⅱ（图4.4c）中：F_{P1}对应①号的位移Δ_{12}，因Δ_{12}不是由F_{P1}所引起，这一个过程所做的就是虚功。

In stage Ⅱ (Figure 4.4c): Displacement Δ_{12} of position ① corresponds to F_{P1}, because Δ_{12} is not caused by F_{P1}, this process is virtual work.

考虑到荷载值从零增大的过程，实功和虚功的力与位移关系可表示为图4.5。

Considering that load increases from zero, the force-displacement relationship of real work and virtual work can be shown in Figure 4.5.

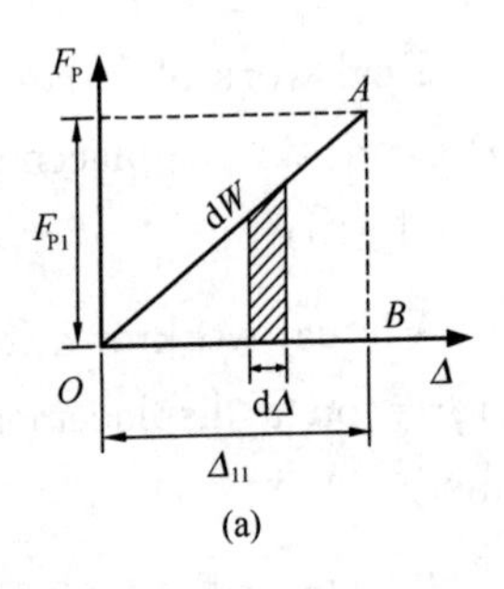

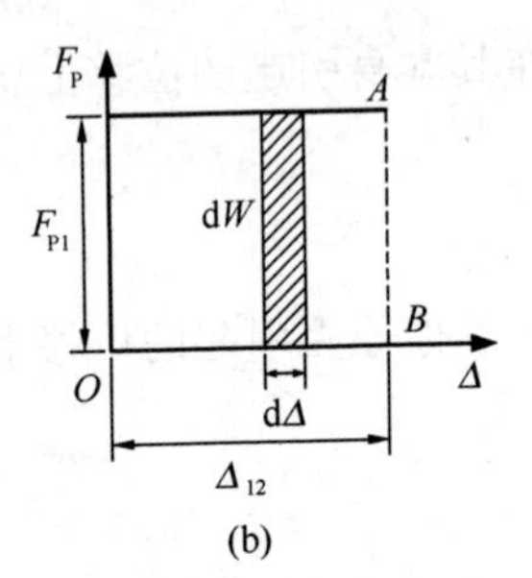

图4.5 力-位移关系

（a）实功；（b）虚功

Figure 4.5 Force-displacement relationship

(a) Real work; (b) Virtual work

对于实功（图4.5a），F_{P1}所做的总功W_{11}为：

For real work (Figure 4.5a), the total work W_{11} made by F_{P1} is:

$$W_{11} = \int_0^{\Delta_{11}} \mathrm{d}W = \frac{1}{2} F_{P1} \Delta_{11}$$

对于虚功（图4.5b），F_{P1}所做的总功W_{12}为：

For virtual work (Figure 4.5b), the total work W_{12} made by F_{P1} is:

$$W_{12} = \int_0^{\Delta_{12}} \mathrm{d}W = F_{P1} \Delta_{12}$$

由于力自身引起的位移方向与力的方向一致，所以实功总是正的，而对于虚功，由于其中力与位移是无关的，所以如果方向不一致的时候虚功取值为负。

Since the direction of displacement caused by the force itself is the same as the direction of the force, the real work is always positive, while for the virtual work, since the force is independent of the displacement, the virtual work takes a negative value if the direction is not the same.

4.1.4 广义力和广义位移

4.1.4 Generalized Forces and Generalized Displacements

在虚功表达式（4.1）中，包含了两个

In Equation (4.1) for virtual work,

方面的因素：

two elements are included:

$$W = P\Delta \tag{4.1}$$

一个是力相关的，可以是单个或一组集中力，单个或一组力偶等，甚至还可能是一个力系，这些与力相关的因素称为广义力 P。

One is related to forces, which can either be a single or a set of concentrated forces, a single or a set of force couples, etc., and possibly even a force system, and these force-related factors are called generalized forces P.

一个是和位移相关的，与广义力相对应，这些和位移有关的因素称为广义位移 Δ。

The other is related to displacements, which corresponds to generalized forces, and these displacement-related factors are called generalized displacements Δ.

例如，如果广义力为单个集中力，则其对应的广义位移为该力作用点沿力的方向上的线位移。常见的广义力对应的广义位移如表 4.1 所示。

For example, if the generalized force is a single concentrated force, its corresponding generalized displacement is the linear displacement of the action point along the direction of the force. The generalized displacements corresponding to common generalized forces are shown in Table 4.1.

广义力和广义位移 **表 4.1**

Generalized forces and generalized displacements **Table 4.1**

广义力 Generalized forces	广义位移 Generalized displacements
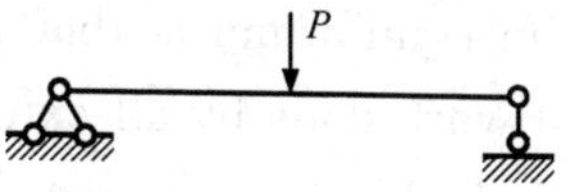 P	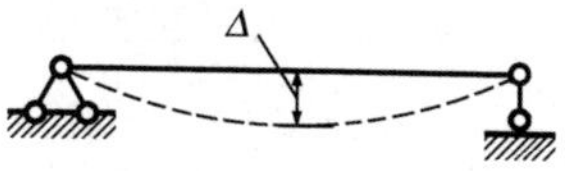 Δ
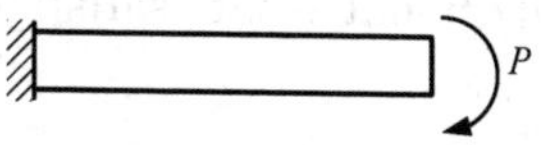 P	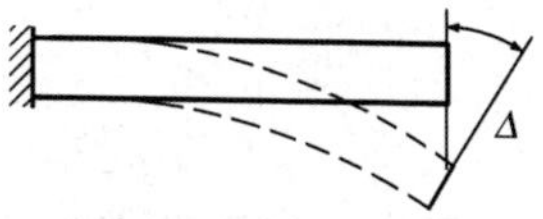 Δ
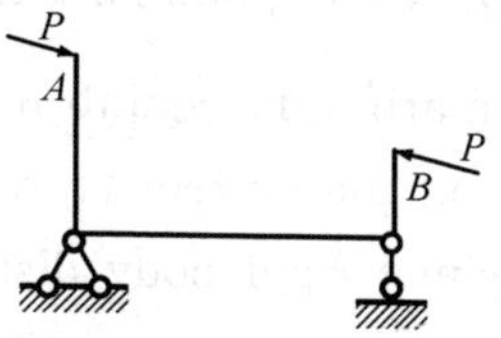 P, A, P, B	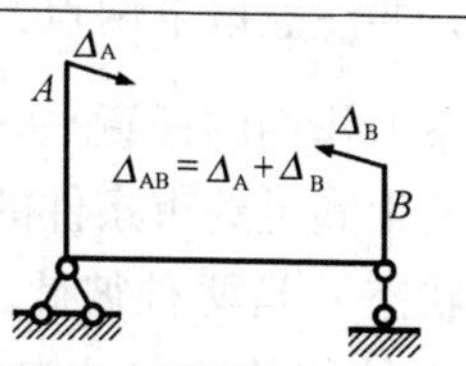 Δ_A, A, Δ_B, $\Delta_{AB}=\Delta_A+\Delta_B$, B
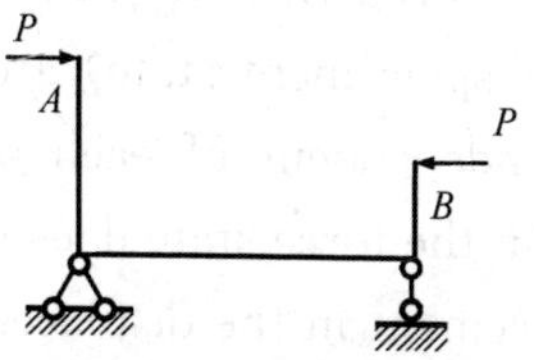 P, A, P, B	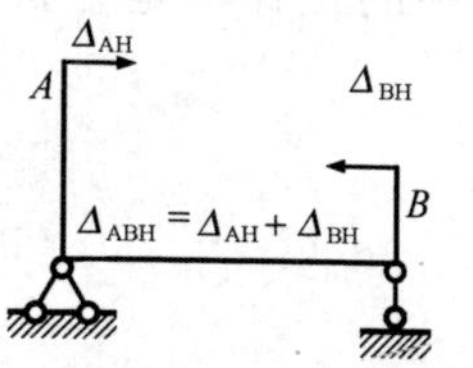 Δ_{AH}, A, Δ_{BH}, $\Delta_{ABH}=\Delta_{AH}+\Delta_{BH}$, B

续表
Continued

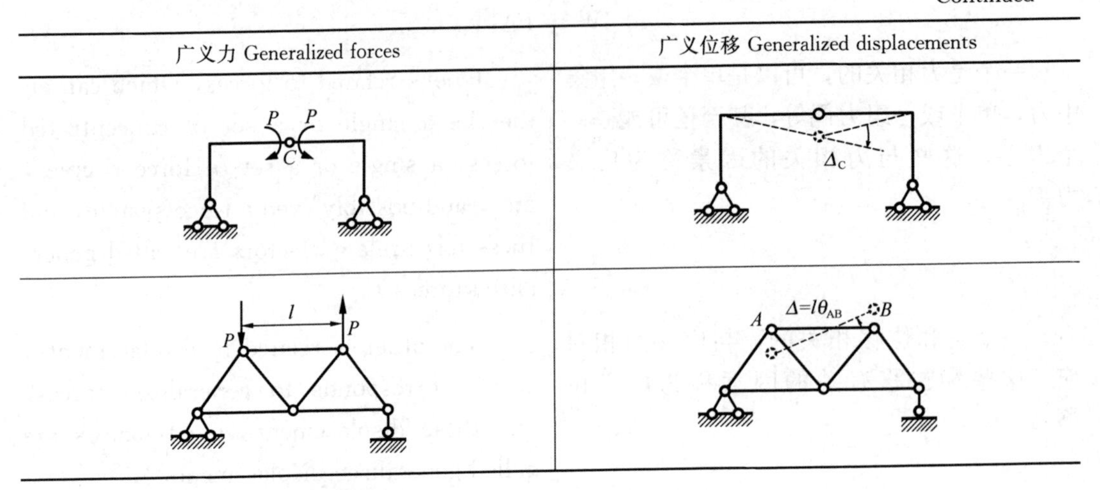

4.2　虚功原理

4.2 Principle of Virtual Work

4.2.1　刚体体系虚功原理

4.2.1 Principle of Virtual Work for Rigid Body Systems

理论力学中，学习了刚体体系的虚功原理：对于具有理想约束的刚体体系，处于平衡的充分必要条件是，对于符合约束条件的任意虚位移，刚体体系所有外力所做虚功之和为零。

In theoretical mechanics, the principle of virtual work for rigid body systems is studied: for a rigid body system with ideal constraints, a sufficient necessary condition for being in equilibrium is that the sum of the virtual work done by all external forces on the rigid body system is zero for any virtual displacement that satisfies the constraints.

在这里，要注意以下两点：

Following two points are to be noted:

(1) 体系上作用的任意平衡力系（力状态）与体系发生满足约束条件的无限小刚体位移（位移状态）是两种彼此无关的状态，即位移状态中的位移不是力状态中的力产生的。

(1) An arbitrary equilibrium force system acting on the system (force state) and an infinitesimal rigid body displacement of the system occurring to satisfy the constraints (displacement state) are two states that are independent of each other, i. e., the force in the force state does not generate the displacement in the displacement state.

(2) 体系上的主动力除荷载外，还包括撤除约束处与约束相应的约束力。因此对一般情况，刚体体系虚功原理的表达式为：

(2) The main active force on the system includes not only the loads, but also the restraining forces corresponding to the restraint at the withdrawal of the restraint. Thus, for general cases, the expression for the principle of virtual work on a rigid body system is:

$$\Sigma F_{Pi}\Delta_i + \Sigma F_{Rk}c_k = 0 \tag{4.2}$$

式中 F_{Pi}——体系所受荷载；

F_{Rk}——体系的约束力；

Δ_i——与力 F_{Pi} 相应的位移，与力 F_{Pi} 同向取正，否则取负；

c_k——与约束力 F_{Rk} 相应的位移，与力 F_{Rk} 同向取正，否则取负。

Where F_{Pi}——the load on the system;

F_{Rk}——the restraining force of the system;

Δ_i——displacement corresponding to the force F_{Pi}, taken as positive in the same direction as the force F_{Pi}, otherwise as negative;

c_k——the displacement corresponding to the restraining force F_{Rk}, taken as positive in the same direction as the force F_{Rk}, otherwise as negative.

由于在虚功原理中所涉及的力状态和位移状态是彼此独立无关的两种状态，因此，不仅可以把位移状态看作是虚设的，也可以把力状态看作是虚设的。相应地，虚功原理的应用有两种形式：

Since the force and displacement states involved in the principle of virtual work are independent, it is possible to consider either the displacement state as virtual state or the force state as virtual state. Accordingly, the application of the principle of virtual work has two forms:

一是虚位移原理，虚设位移状态，求实际力状态中的未知力。

The first is the principle of virtual displacement, where a virtual displacement state is supposed to find the unknown force in the actual force state.

二是虚力原理，虚设力状态，求实际位移状态中的未知位移。

The second is the principle of virtual force, where a virtual force state is set up to find the unknown displacement in the actual displacement state.

4.2.1.1　虚位移原理

虚设位移状态的目的是求未知力：

将与拟求未知力 X 相应的约束撤除，代以相应的力 X（这时 X 已是主动力），再让该可变体系发生满足约束条件的刚体虚位移，得到一虚设的可能位移状态，结合初始力状态，建立体系的虚功方程，进而求解得到未知力 X。下面举例说明。

【例 4.1】 求图 4.6（a）所示简支梁 B 支座反力。

4.2.1.1 Principle of Virtual Displacement

The purpose of virtual displacement state is to find the unknown force:

The constraint corresponding to the proposed unknown force X is removed and replaced by the corresponding force X (where X is already the active force), then the variable system is allowed to undergo a rigid body virtual displacement satisfying the constraints to obtain a fictitious possible displacement state, which is combined with the initial force state and the system's virtual work equation is established, which is then solved to obtain the unknown force X. The following examples are given.

[**Example 4.1**] Find the reaction force on the support B of the simply supported beam shown in Figure 4.6(a).

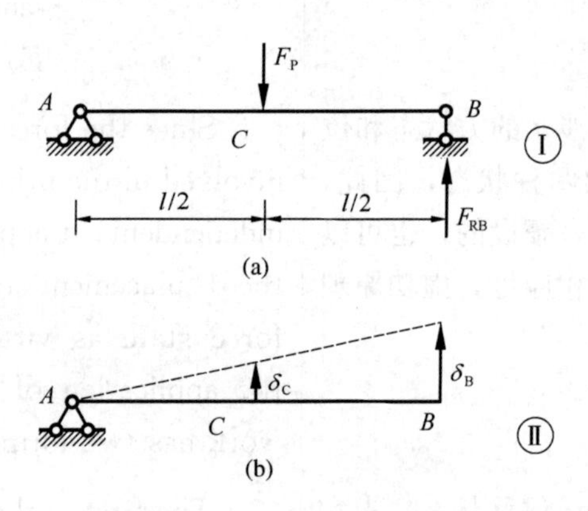

图 4.6　刚体虚位移原理

Figure 4.6　Principle of virtual displacement for rigid body

【解】（1）图 4.6（b）显示了初始的、实际的力状态 Ⅰ。用待求支座反力 F_{RB} 代替 B 支座。

（2）虚设 B 端一个位移 δ_B，得到虚拟的位移状态Ⅱ，如图 4.6（b）所示。根据

[**Solution**] (1) Figure 4.6(b) shows the initial, actual force state Ⅰ. Substitute support B with the unknown support reaction force F_{RB}.

(2) Suppose a virtual displacement δ_B at end B to obtain a virtual displacement

几何关系可知：

state Ⅱ, as shown in Figure 4.6(b). From the geometric relations:

$$\delta_C = \frac{\delta_B}{2}$$

（3）建立虚功方程：

(3) Establish the virtual work equation:

$$F_{RB}\delta_B - F_P\delta_C = 0$$

（4）解得：

(4) The solution is:

$$F_{RB} = \frac{F_P}{2}$$

上述计算是在给定的平衡力系与虚设的可能位移状态之间应用虚功原理，这种形式的应用即为虚位移原理。根据虚位移原理建立的虚功方程，实质上是拟求未知力与荷载之间的静力平衡方程。

The above calculations apply the principle of virtual work between a given system of equilibrium forces and a fictitious state of possible displacements, and this form of application is known as the principle of virtual displacement. The virtual work equation established from the principle of virtual displacement is a proposed equation for the static equilibrium between the unknown forces and the loads.

由于 δ_B 与 δ_C 的比值是不变的，所以通常虚设单位位移 1 以简化计算，这种方法称为**单位位移法。**

Due to the ratio of δ_B to δ_C is constant, the unit displacement 1 is usually supposed to simplify the calculation, and this method is called the **unit displacement method.**

单位位移法求解静定结构某一未知约束力 X（包括支座反力或任一截面内力）的计算步骤是：

The unit displacement method for solving a specific unknown constraint force X in statically determinate structure (including the support reaction force or the internal force in any section) is calculated as:

（1）撤除与 X 相应的约束，代以相应的未知力 X，使原静定结构变为具有一个自由度的机构，约束力 X 变为主动力，X 与原力系维持平衡。

(1) The corresponding constraint with X is removed, and the corresponding unknown force X is substituted so that the original static structure becomes a body with one degree of freedom, the constraint force X becomes the active force, and X is maintained in equilibrium with the original force system.

（2）令机构沿 X 正向发生单位虚位移，作出机构满足约束条件的刚体虚位移图，并由几何关系求出与荷载 F_P 相应的位移 δ_P。

(2) Let the mechanism undergo a unit virtual displacement along X's positive direction, make the rigid body virtual displacement diagram for the mechanism satisfying the constraints, and find the displacement δ_P corresponding to the load F_P from the geometric relationship.

（3）根据虚功原理，在平衡力系和虚设位移之间建立虚功方程。

(3) Based on the principle of virtual work, the virtual work equation is established between the equilibrium force system and the virtual displacement.

（4）代入几何关系，解出约束力 X。

(4) Substitute the geometric relations and solve for the constraint force X.

【例 4.2】 求图 4.7（a）中连续梁 B 截面弯矩、剪力。

[Example 4.2] Find out the bending moment and shear force of section B of the continuous beam in Figure 4.7 (a).

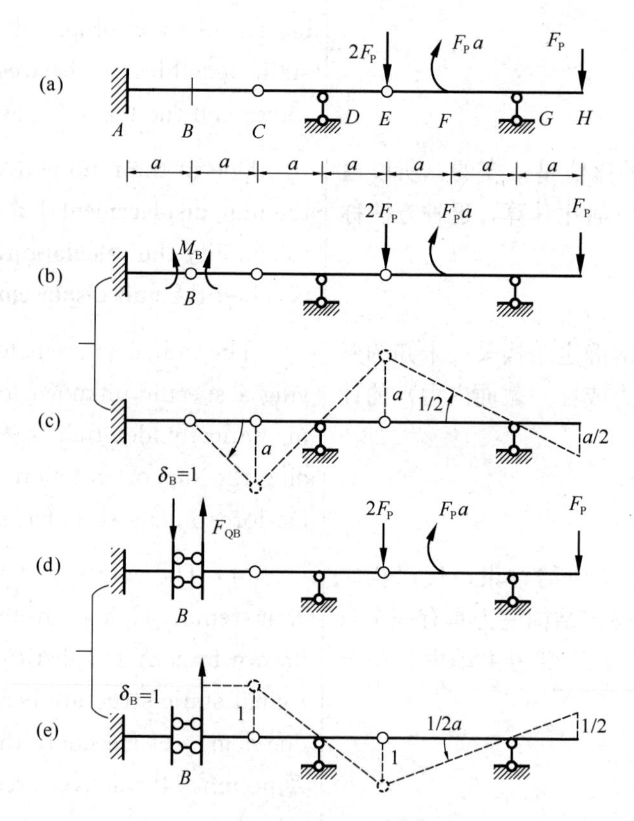

图 4.7　利用虚位移原理求内力

Figure 4.7　Using the principle of virtual displacement to find out internal forces

【解】弯矩求解见图 4.7（b）、图 4.7（c）：

[Solution] The solution of bending moment is shown in Figure 4.7(b) and Figure 4.7(c):

$$M_B \times 1 - 2F_P a + F_P a \times \frac{1}{2} + F_P \frac{a}{2} = 0, \ M_B = F_P a \text{ (lateral tension)}$$

剪力求解见图 4.7（d）、图 4.7（e）：

Shear force solving is shown in Figure 4.7(d) and Figure 4.7(e):

$$F_{QB} \times 1 + 2F_P \times 1 - F_P a \frac{1}{2a} - F_P \frac{a}{2} = 0, \ F_{QB} = -F_P$$

连续梁的虚位移图特点如表 4.2 所示。

The characteristics of virtual displacement diagram of continuous beam are shown in Table 4.2.

连续梁虚位移图的特点 **表 4.2**

Characteristics of the virtual displacement diagram for continuous beam **Table 4.2**

界（边界）Boundaries 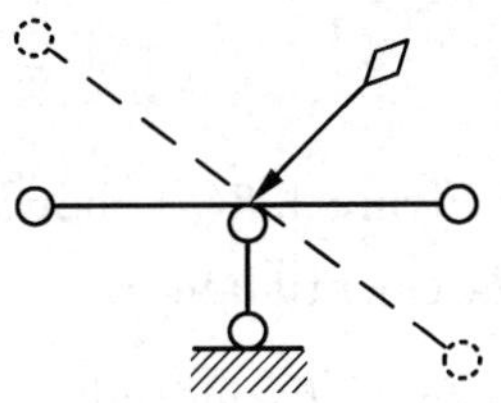	拐（拐折）Abduction fold 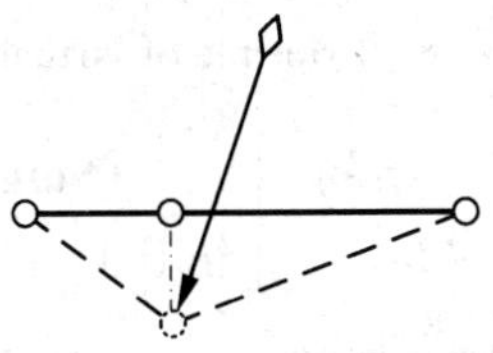	平（平行）Parallel 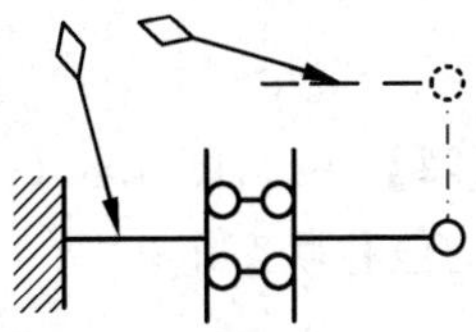

4.2.1.2 虚力原理

4.2.1.2 Principle of Virtual Force

虚设力状态，求未知位移：

The purpose of virtual force states is to find unknown displacements:

虚设一平衡力系。为了在虚功方程中只包含拟求位移，而不再含有其他未知位移，应只在拟求位移点沿拟求位移方向虚设力 F_P，此力与相应的支座反力组成一平衡力系，根据平衡条件，可求出支座反力，建立虚功方程，求解方程得到位移未知量。下面举例说明。

Set up a system of equilibrium forces fictitiously. In order to include only the proposed displacements in the virtual work equation and no other unknown displacements, it should only set force F_P in the proposed displacement point along the proposed virtual displacement direction, this force and the corresponding support reaction force form an equilibrium force system, according to the equilibrium conditions, the support reaction force can be found, the virtual work equation is established and the equation to get the unknown displacement is solved. The following example is given.

【例 4.3】 图 4.8（a）所示结构 B 支座产生沉降 c，求 A 端竖向位移。

［**Example 4.3**］Find the vertical displacement at end A for the settlement c of support B of the structure shown in Figure 4.8(a).

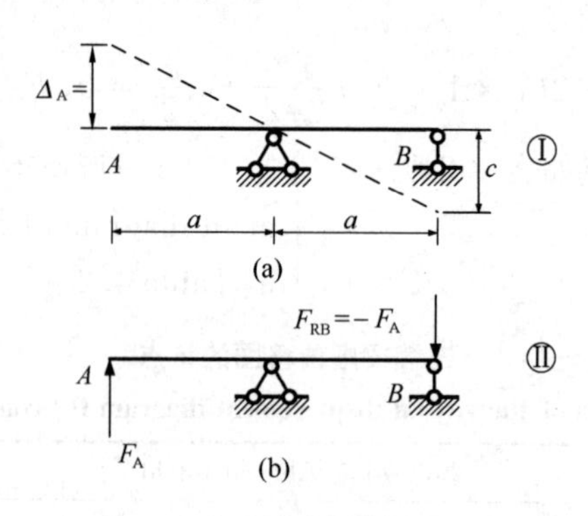

图 4.8 虚力原理

Figure 4.8 Principle of virtual force

【解】（1）图 4.8（a）显示了初始的、实际的位移状态Ⅰ。

［**Solution**］(1) Figure 4.8(a) shows the initial, actual displacement state Ⅰ.

（2）虚设 A 端一个力 F_A，得到虚拟的力状态Ⅱ，如图 4.8（b）所示。根据力的平衡可知：

(2) Suppose a force F_A at end A, the virtual force state Ⅱ is obtained as shown in Figure 4.8(b). According to the equilibrium of the force:

$$F_{RB}=-F_A$$

（3）建立虚功方程：

(3) Establish the virtual work equation:

$$F_A\Delta_A+F_{RB}c=0$$

（4）解得：

(4) The solution is:

$$\Delta_A=c$$

可以看出，Δ_A 与 F_A 无关。因此，可虚设单位荷载以简便计算，这种方法称为**单位荷载法**。

It can be seen that Δ_A is independent of F_A. Therefore, a virtual unit load can be created to simplify the calculation, and this method is called the **unit load method.**

上述计算是在虚设的平衡力系与给定的位移状态之间应用虚功原理，这种形式的应

The above calculations apply the principle of virtual work between a fictitious sys-

用即为虚力原理。根据虚力原理建立的虚功方程，实质上是拟求的未知位移与已知位移之间的几何方程。

tem of equilibrium forces and a given state of displacement, and this form of application is known as the principle of virtual force. The virtual work equation established from the principle of virtual force is the geometric equation between the proposed unknown displacements and the known displacements.

4.2.2 变形体系虚功原理

4.2.2 Principle of Virtual Work of Deformed Body

对于任何一个变形体系，处于两种彼此无关的状态，如图 4.9 所示：

Figure 4.9 shows any deformed body in two mutually independent states：

力状态Ⅰ：力的平衡状态。

Force state Ⅰ：The state of equilibrium of forces.

位移状态Ⅱ：变形连续。

Displacement state Ⅱ：Deformation is continuous.

下面从力和位移两个切入点来计算虚功。

The following shows calculation of the virtual work from the perspective of force and displacement.

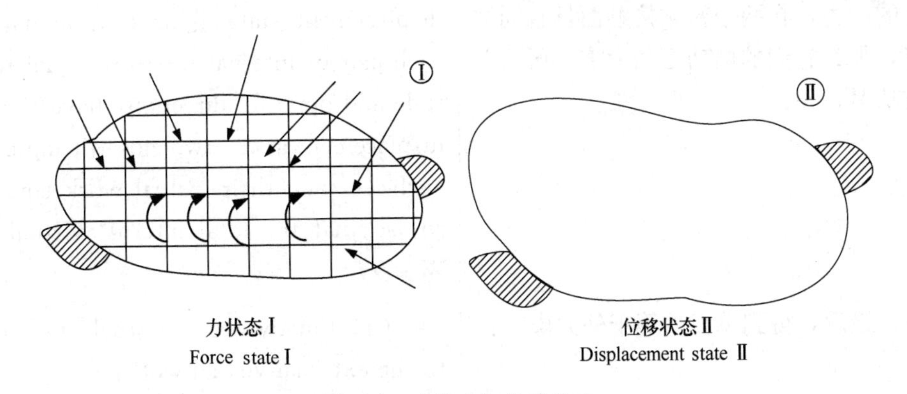

图 4.9 力状态与位移状态

Figure 4.9 Force state and displacement state

4.2.2.1 力途径计算虚功

4.2.2.1 Calculation of Virtual Work by the Force Path

(1) 首先，将变形体进行切割，变形体所受外荷载称为**外力**，如图 4.9；切割面内

(1) First, the deformed body is cut, and the external loads on the deformed body

力称为**内力**，如图4.10所示。

are called the **external forces**, as shown in Figure 4.9; the forces in the cutting surface are called the **internal forces**, as shown in Figure 4.10.

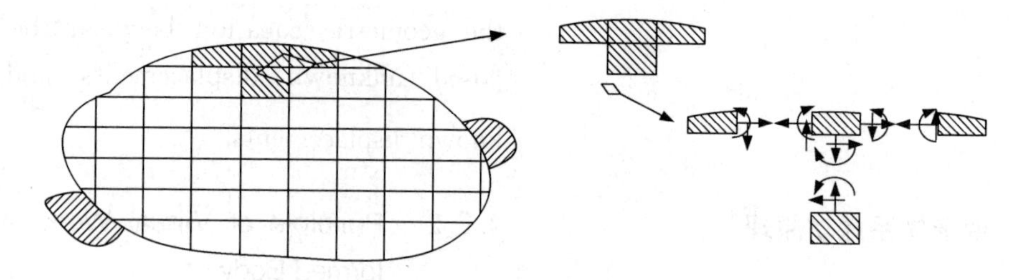

图4.10 内力

Figure 4.10 Internal forces

（2）其次，结构总虚功 W 可以为外力虚功 W_e 和内力虚功 W_i 之和，即：

(2) Second, the total virtual work W of the structure can be the sum of the external virtual work W_e and the internal virtual work W_i, namely:

$$W = W_e + W_i$$

（3）然后，如图4.10所示，内力总是成对出现的，大小相等方向相反，又因为位移状态Ⅱ是连续的，那么每一对大小相等方向相反的内力，在两个密贴接触面具有同样的位移，那么它们的虚功之和为零，所以总内力虚功 $W_i=0$。

(3) Third, as in Figure 4.10, internal forces always occur in pairs of equal magnitude and opposite direction, and because displacement state Ⅱ is continuous, then each pair of internal forces of equal magnitude and opposite direction, have the same displacement at the two close-fitting contact surfaces, and their virtual work equals zero, so that the total internal virtual work $W_i=0$.

（4）最后，得到总虚功等于外力虚功：

(4) Finally, total virtual work equals to the external virtual work:

$$W = W_e$$

4.2.2.2 位移途径计算虚功

4.2.2.2 Calculation of Virtual Work by the Displacement Path

将微元的虚位移进行分解：先发生刚体虚位移，再发生变形虚位移。如图4.11所示。

The virtual displacement of the micro segment is decomposed: The virtual displacement of rigid body occurs first, then the virtual displacement of deformation occurs. This is shown in Figure 4.11.

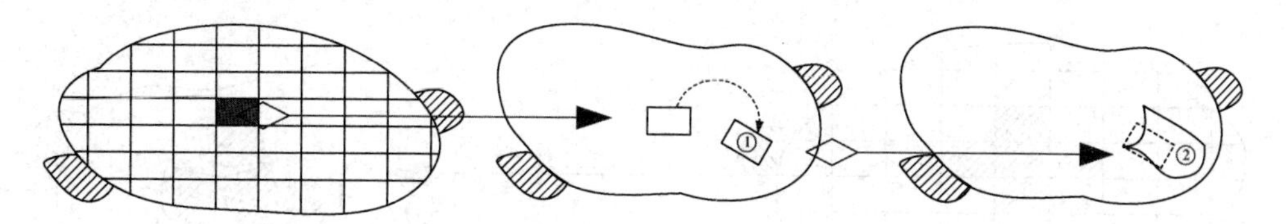

图 4.11 虚位移的分解
① 刚体虚位移；② 变形虚位移

Figure 4.11 Decomposition of virtual displacements
① Virtual displacement of rigid body; ② Virtual displacement of deformation

则微元上总虚功 dW 为所有力在刚体虚位移上所做虚功 dW_s 和微元上所有力在变形虚位移上所做虚功 dW_v 之和，即：

Then the total virtual work dW on the micro segment is the sum of the virtual work dW_s done by all forces on the rigid body and the virtual work dW_v done by all forces on the virtual displacements of deformation, i. e. :

$$dW = dW_s + dW_v$$

由刚体虚功原理，微段处于平衡状态 $\Leftrightarrow dW_s = 0$，所以：

Based on the principle of virtual work of rigid body, the micro segments are in balance $\Leftrightarrow dW_s = 0$, so:

$$dW = dW_v$$

对于整个结构，有：

For the entire structure, there are:

$$\Sigma\int dW = \Sigma\int dW_v$$

即：

Namely:

$$W = W_v$$

与上节结论 $W=W_e$ 联立，得到：

Simultaneous with the above conclusion $W=W_e$, we can get:

$$W_e = W_v \tag{4.3}$$

此处，有一个问题，“变形虚位移对应的广义力是内力，内力互为反作用力，为什么 $W_v \neq 0$?”，其原因在于 W_v 只是变形虚位移，而变形虚位移不是连续的，如图 4.12、图 4.13 所示。

There is one question, “The generalized forces corresponding to the virtual displacements of deformation are internal forces, and the internal forces are mutually reactive, so why is $W_v \neq 0$?”, the reason for this is that

W_v is only the virtual displacement of deformation, which is not continuous, as shown in Figure 4.12 and Figure 4.13.

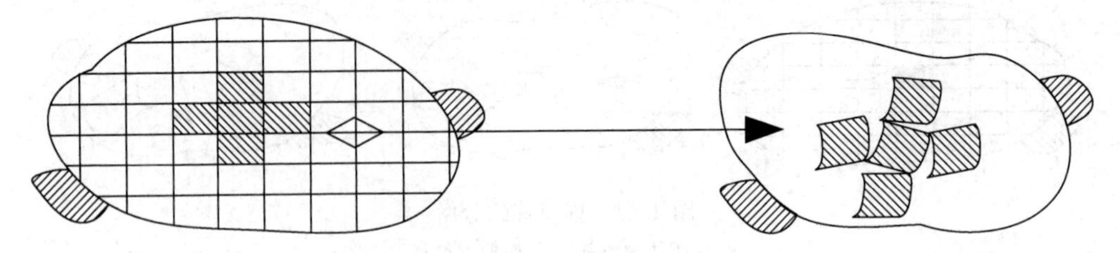

图 4.12 变形虚位移的不连续性

Figure 4.12 Discontinuity of virtual displacement of deformation

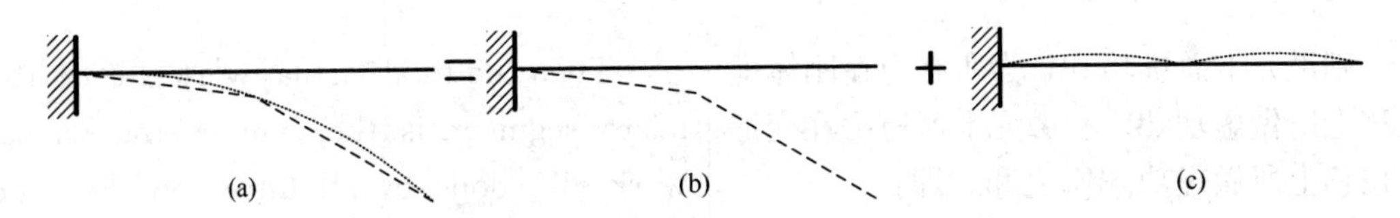

图 4.13 变形虚位移不连续性的宏观举例

(a) 总虚位移；(b) 刚体虚位移；(c) 变形虚位移

Figure 4.13 Macroscopic examples of discontinuities of virtual displacement of deformation

(a) Total virtual displacement; (b) Virtual displacement of rigid body; (c) Virtual displacement of deformation

4.2.2.3 变形体虚功原理的完整表述

虚功原理可表述为：设变形体系在力系作用下处于平衡状态，又设变形体系由于其他原因产生满足变形连续条件的微小位移和变形，则体系上所有外力在相应位移上所做外力虚功之和恒等于体系各微段截面上的内力在相应微段变形上所做虚功之和，便于叙述，简称外力虚功等于变形虚功。

4.2.2.3 Complete Formulation of Principle of the Virtual Work of Deformed Body

The principle of virtual work can be expressed as follows: Suppose that the deformation system is in equilibrium under the action of the force system, and the deformation system produces small displacements and deformations satisfying the conditions of deformation continuity due to other reasons, then the sum of virtual work done by all external forces on the system at the corresponding displacements is constantly equal to the sum of virtual work done by the internal forces on each micro section of the system at the corresponding micro section deformation, which is convenient to describe and referred to as external virtual work equals virtual deformation work.

几点说明：

Several points are needed to be mentioned:

（1）虚功原理的证明过程没有用到“变形体”的力学性质，因此本原理对任意力-变形关系（力学中常称为本构关系）的可变形物体都适用。

(1) The proof process of the principle of virtual work does not use the mechanical properties of the “deformed body”, so the principle applies to deformable objects with any force-deformation relationship (often called intrinsic relations in mechanics).

（2）证明过程也没有限制变形体的形状、组成，因此本原理对杆件体系结构、平面和空间结构、板壳结构及各种组合形式的结构等都适用。

(2) The proof process also does not restrict the shape and composition of the deformed bodies, so the principle is applicable to rod system structures, plane and space structures, plate and shell structures and structures in various combinations, etc.

（3）本小节内容仅证明了虚功原理是一个必要性命题。

(3) The content of this chapter proves only that the principle of virtual work is a necessity proposition.

4.2.3 平面杆系结构位移计算的一般公式

4.2.3 General Formulae for the Calculation of Displacements in Planar Rod System Structures

对于平面杆系结构，微段在位移状态的虚变形包括：弯曲变形、剪切变形、轴向变形，如图4.14所示。

For a planar rod system structure, the virtual deformation of the micro segment in the displacement state includes bending deformation, shear deformation and axial deformation, as shown in Figure 4.14.

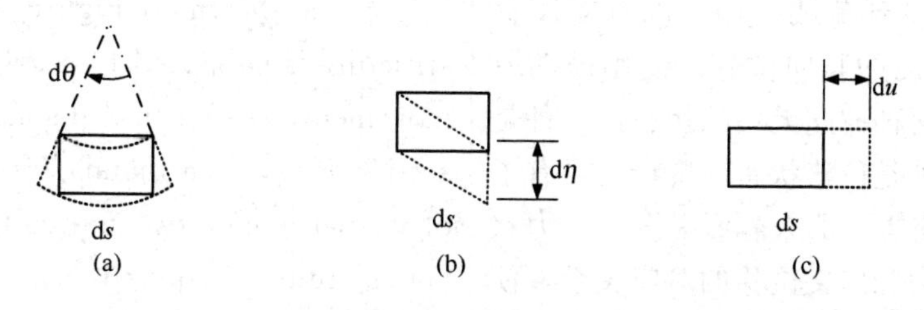

图4.14 微段变形

(a)弯曲变形;(b)剪切变形;(c)轴向变形

Figure 4.14 Micro segment deformation

(a) Bending deformation; (b) Shear deformation; (c) Axial deformation

变形虚功可写为：

The virtual deformation work can be written as:

$$dW_v = Md\theta + F_Q d\eta + F_N du \tag{4.4}$$

式中 M、F_Q、F_N为对应于变形微段 ds 的力状态中的内力。

Where M, F_Q and F_N are the internal forces in the force state corresponding to the deformed micro segment ds.

此处说明：

(1) dM、dF_Q、dF_N、分布荷载所做虚功作为高阶微量略去不计；

(2) 如果对应于变形微段 ds 的力状态中有集中荷载或集中力偶，可以认为它们作用在截面而非微段上，因而当 ds 变形时，无对应虚功。

Note:

(1) Virtual work done by dM, dF_Q, dF_N and distributed load is omitted as higher-order trace;

(2) If there are concentrated loads or concentrated force couples in the force states corresponding to the deformed micro segment ds, they can be considered to act on the section rather than on the micro segment and thus there are no corresponding virtual work when ds is deformed.

整个结构的变形虚功为：

The virtual work of deformation of the whole structure is:

$$W_v = \Sigma\int Md\theta + \Sigma\int F_Q d\eta + \Sigma\int F_N du \tag{4.5}$$

结合虚功原理（式 4.3），有：

Combined with the principle of virtual work (Equation 4.3):

$$W_e = W_v = \Sigma\int Md\theta + \Sigma\int F_Q d\eta + \Sigma\int F_N du \tag{4.6}$$

下面举例说明其应用，如图 4.15（a）所示，结构既受到荷载作用，也有支座移动，要求 K 截面的竖向位移。使用虚功原理，首先需要两种彼此无关的状态：一种是力状态、另一种是位移状态。现在已经有了实际的位移状态①，那么需要虚设一个力状态，根据广义力和广义位移的对应关系，使用单位荷载法，在 K 截面设置一个竖直向下的虚拟单位荷载 $F_P=1$，得到状态Ⅱ，如图 4.15（b）所示。

An example of its application is given below, as shown in Figure 4.15(a), the structure is subjected to a load and support movement, the vertical displacement of the section K is to be found. Use the principle of virtual work, two mutually independent states are first required: a force state and a displacement state. The actual displacement state ① is available, then a fictitious force state is required, and according to the correspondence between generalized force and generalized displacement, using the unit load method, a virtual unit load $F_P=1$ is set vertically downward at section K to obtain state Ⅱ as shown in Figure 4.15(b).

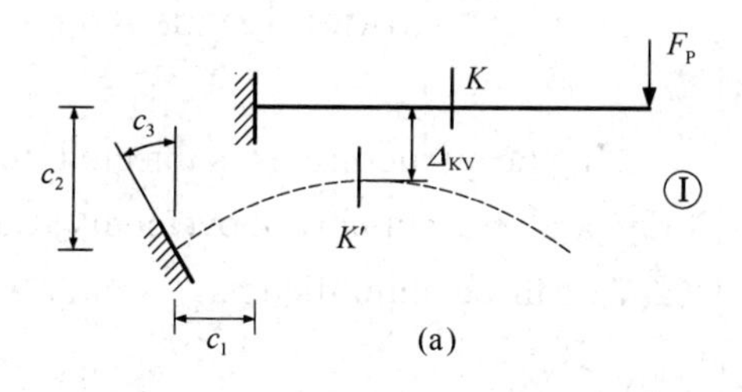

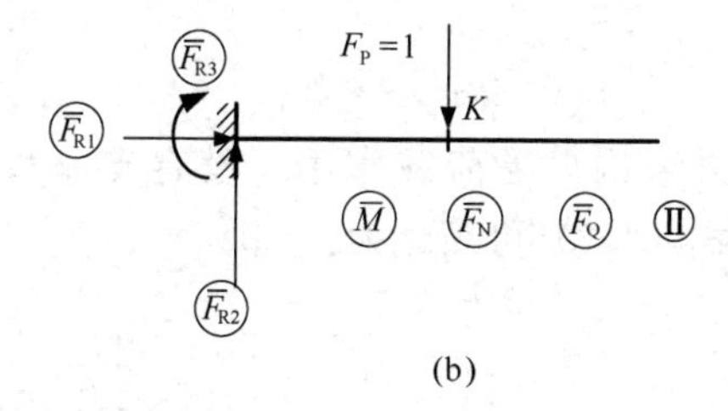

图 4.15 示例

(a) 虚拟力状态；(b) 虚拟位移状态

Figure 4.15 Example

(a) Virtual displacement state; (b) Virtual force state

力状态中的内力、支座反力都可以根据静定结构的受力平衡方程计算得出，可写出外力虚功表达式为：

The internal forces and support reactions in the force state can be calculated from the equations of force equilibrium of a statically determinate structure, and the expression for the virtual work of the external force can be written as:

$$W_e = 1 \times \Delta_{KV} + \Sigma \overline{F}_{Ri} c_i$$

根据虚功原理，有：

According to the principle of virtual work, there are:

$$1 \times \Delta_{KV} + \Sigma \overline{F}_{Ri} c_i = \Sigma \int \overline{M} d\theta + \Sigma \int \overline{F}_Q d\eta + \Sigma \int \overline{F}_N du$$

将该式进行求解，即可得到 Δ_{KV}。

By solving this equation, Δ_{KV} can be obtained.

对于任意待求位移 Δ，可总结平面杆系结构的一般公式如下：

For any displacement Δ to be found, the general formulae for a planar rod system structure can be summarized:

$$1 \times \Delta + \Sigma \overline{F}_{Ri} c_i = \Sigma \int \overline{M} d\theta + \Sigma \int \overline{F}_Q d\eta + \Sigma \int \overline{F}_N du$$

$$\Delta = \Sigma \int \overline{M} d\theta + \Sigma \int \overline{F}_Q d\eta + \Sigma \int \overline{F}_N du - \Sigma \overline{F}_{Ri} c_i \tag{4.7}$$

式 (4.7) 中，所有乘式的第一项均来自于虚拟力状态，第二项均来自于实际位移状态。

In Equation (4.7), the first term of all multiplications is from the virtual force state, and the second term is from the actual displacement state.

4.3　荷载作用下的位移计算

4.3　Displacement Calculation under Loads

如果结构只受荷载作用，没有支座移动，则式（4.7）可简化为：

If the structure is subjected to loads only and no support movement, Equation (4.7) can be simplified as:

$$\Delta=\Sigma\int\bar{M}\mathrm{d}\theta+\Sigma\int\bar{F}_{\mathrm{Q}}\mathrm{d}\eta+\Sigma\int\bar{F}_{\mathrm{N}}\mathrm{d}u \tag{4.8}$$

在线弹性范围内，由材料力学知识可得：

In the range of linear elasticity, it follows from knowledge of the mechanics of materials:

$$\begin{aligned}\mathrm{d}\theta&=\frac{M_{\mathrm{P}}}{EI}\mathrm{d}s\\ \mathrm{d}\eta&=\frac{\mu F_{\mathrm{QP}}}{GA}\mathrm{d}s\\ \mathrm{d}u&=\frac{F_{\mathrm{NP}}}{EA}\mathrm{d}s\end{aligned} \tag{4.9}$$

式中，E 为材料弹性模量；G 为剪切模量；I 为截面惯性矩；A 为截面面积；μ 为考虑切应力沿截面高度分布不均匀的修正系数，对于矩形截面取 1.2，圆形截面取 10/9，薄壁圆环截面取 2，工字形截面按 A/A_{f} 计算，A_{f} 为腹板截面积。

Where, E is the material modulus of elasticity; G is the shear modulus; I is the moment of inertia of the section; A is the cross-sectional area; μ is the correction factor considering the uneven distribution of shear stress along the height of the section, for rectangular sections it takes 1.2, for circular sections it takes 10/9, for thin-walled circular sections it takes 2, for I-beam sections it is calculated according to A/A_{f}, A_{f} is the cross-sectional area of the web.

将式（4.9）代入式（4.8）可得：

Substituting Equation (4.9) into Equation (4.8):

$$\Delta=\Sigma\int\frac{\bar{M}M_{\mathrm{P}}}{EI}\mathrm{d}s+\Sigma\int\mu\frac{\bar{F}_{\mathrm{Q}}F_{\mathrm{V}}}{GA}\mathrm{d}s+\Sigma\int\frac{\bar{F}_{\mathrm{N}}F_{\mathrm{NP}}}{EA}\mathrm{d}s \tag{4.10}$$

4.3.1 梁杆结构

4.3.1 Beam-bar Structure

此类结构以弯曲变形为主，对于由细长杆件组成的梁杆结构，可忽略剪切变形的影响（有特别说明的除外，如高跨比较大的深梁），另外，如果轴力较小，还可忽略轴向变形的影响（有特别说明的除外，如高层建筑中的柱）。此时，式（4.10）可简化为：

Such structures are dominated by bending deformation, and the effect of shear deformation can be neglected for beam-bar structures consisting of slender members (except for special conditions such as deep beams with relatively high spans), and in addition, the effect of axial deformation can be neglected if the axial forces are small (except for special conditions such as columns in high buildings), At this time, Equation (4.10) can be simplified as:

$$\Delta = \Sigma\int \frac{\bar{M} M_P}{EI} ds \tag{4.11}$$

【例 4.4】试求图 4.16（a）所示矩形截面简支梁中点 C 的竖向位移 Δ_{CV}。

[Example 4.4] Try to find out the vertical displacement Δ_{CV} at midpoint C in the simply supported beam of the rectangular section shown in Figure 4.16(a).

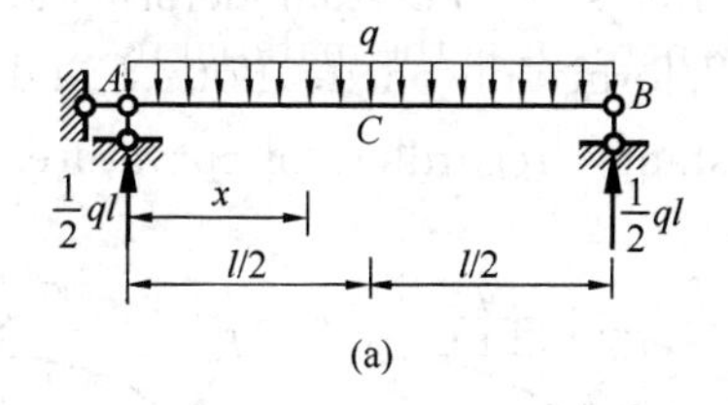

(a)

(b)

图 4.16 例 4.4

Figure 4.16 Example 4.4

【解】（1）虚设力状态：在梁中点 C 加一竖向单位力 $F_P=1$，如图 4.16（b）所示。

[Solution] (1) Virtual force state: add a vertical unit force $F_P=1$ at the midpoint C of the beam, as shown in Figure 4.16 (b).

（2）分别计算梁在实际荷载和虚单位荷载作用下的内力。取 A 为原点，当 $0\leqslant x\leqslant \frac{l}{2}$ 时，由平衡条件得任意截面 x 的内力表达式为：

(2) Respectively calculate internal forces in the beam under the actions of the actual load and the virtual unit load. Take A as the original point, when $0\leqslant x\leqslant \frac{l}{2}$, by the balance condition, internal force expression at any section x is:

实际荷载：

Actual load:

$$M_P = \frac{q}{2}(lx - x^2)$$

虚单位荷载：

Virtual unit load:

$$\bar{M} = \frac{1}{2}x$$

（3）计算 Δ_{CV}。由于梁及荷载对称，故积分限取长度的一半，再把计算结果乘以2倍。由式（4.11）得弯曲变形引起的位移为：

(3) Calculate Δ_{CV}. Due to the beam and load symmetry, take half of the length as the integral limit, and then get the calculation results multiplied by 2 times. The displacement caused by the bending deformation by Equation (4.11) is:

$$\Delta_{CV}^{M} = \int \frac{\bar{M} M_P}{EI} ds = 2\int_0^{l/2} \frac{\frac{1}{2}x \cdot \frac{q}{2}(lx - x^2)}{EI} dx = \frac{5ql^4}{384EI} (\downarrow)$$

【例4.5】 试求图4.17（a）所示半径为 R 的圆弧曲杆（1/4圆周）B 点的竖向位移 Δ_{BV}，并分析轴向变形与剪切变形对其影响。I 和 A 均为常数，且不考虑曲率的影响。

[Example 4.5] Try to calculate the vertical displacement Δ_{BV} of point B in the circular arc curved bar (1/4 circle) with the radius R in Figure 4.17(a), and analyze the effects of the axial deformation and shear deformation on it. Both I and A are constants, regardless of curvature.

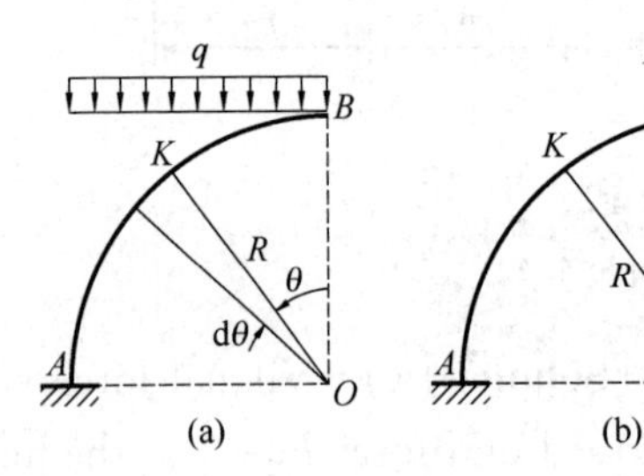

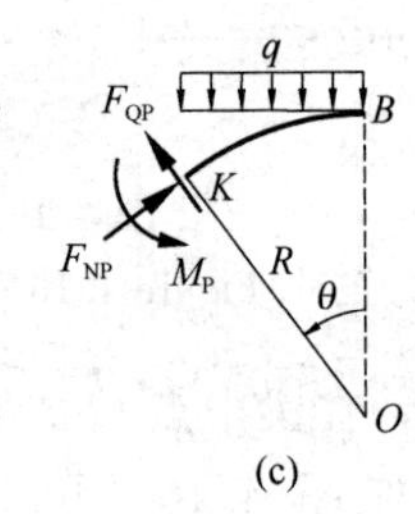

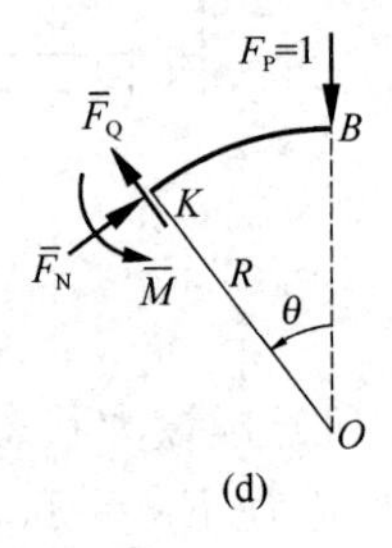

图4.17　例4.5

Figure 4.17　Example 4.5

【解】（1）虚设力状态：在 B 点加一竖向单位力 $F_P=1$，如图4.17（b）所示。

[Solution] (1) Virtual force state: add a vertical unit force $F_P=1$ at point B, as shown in Figure 4.17 (b).

（2）分别计算在实际荷载和虚单位荷载作用下的内力。在与 OB 呈 θ 角的截面 K 上，各内力分量如图4.17（c）和（d）所

(2) Respectively calculate the internal forces under the actions of the actual load and the virtual unit load. At the section K

示，其表达式为：

of an angle θ to OB, the internal force components are shown in Figures 4.17(c) and (d), the expression is:

实际荷载：

Actual load:

$$M_{\mathrm{P}}=\frac{q}{2}R^2\sin^2\theta,\ F_{\mathrm{QP}}=qR\sin\theta\cos\theta,\ F_{\mathrm{NP}}=qR\sin^2\theta$$

虚单位荷载：

Virtual unit load:

$$\bar{M}=R\sin\theta,\ \bar{F}_{\mathrm{Q}}=\cos\theta,\ \bar{F}_{\mathrm{N}}=\sin\theta$$

(3) 计算 Δ_{BV}。将以上各项代入式 (4.10)，并注意到 $\mathrm{d}s=R\mathrm{d}\theta$，得：

(3) Calculate Δ_{BV}. Substitute the above items into Equation (4.10) and note $\mathrm{d}s=R\mathrm{d}\theta$, then:

$$\begin{aligned}\Delta_{\mathrm{BV}}&=\sum\int_B^A\frac{\bar{F}_{\mathrm{N}}F_{\mathrm{NP}}}{EA}\mathrm{d}s+\sum\int_B^A\mu\frac{\bar{F}_{\mathrm{Q}}F_{\mathrm{QP}}}{GA}\mathrm{d}s+\sum\int_B^A\frac{\bar{M}M_{\mathrm{P}}}{EI}\mathrm{d}s\\&=\int_0^{\pi/2}\frac{qR^2\sin^3\theta}{EA}\mathrm{d}\theta+\int_0^{\pi/2}\mu\frac{qR^2\sin\theta\cos^2\theta}{GA}\mathrm{d}\theta+\int_0^{\pi/2}\frac{qR^4\sin^3\theta}{2EI}\mathrm{d}\theta\\&=\frac{2qR^2}{3EA}+\frac{\mu qR^2}{3GA}+\frac{qR^4}{3EI}(\downarrow)\end{aligned}$$

式中三项分别表示曲杆轴向变形、剪切变形和弯曲变形引起的 B 点竖向位移，即：

Where the three items in the equation are respectively axial deformation, shear deformation and bending deformation of the curved bar causing the vertical displacement of point B, namely:

$$\Delta_{\mathrm{BV}}^{\mathrm{N}}=\frac{2qR^2}{3EA}$$

$$\Delta_{\mathrm{BV}}^{\mathrm{Q}}=\frac{kqR^2}{3GA}$$

$$\Delta_{\mathrm{BV}}^{\mathrm{M}}=\frac{qR^4}{3EI}$$

(4) 分析轴向变形与剪切变形对 Δ_{BV} 的影响。

(4) The influence of axial deformation and shear deformation on Δ_{BV} is analyzed.

若曲杆截面为矩形，截面尺寸为 $b\times h$，则有：

If the curved bar section is rectangular with the sectional size $b\times h$, then:

$$\mu=1.2,\ A=\frac{12}{h^2}I$$

此外，设 $G=0.4E$，于是得轴向变形、剪切变形对位移的影响与弯曲变形对位移的

In addition, if $G=0.4E$, the ratios of the effects of axial deformation and shear

影响的比值分别为：

deformation on displacement to the effect of bending deformation on displacement are:

$$\frac{\Delta_{BV}^{N}}{\Delta_{BV}^{M}}=\frac{\dfrac{2qR^2}{3EA}}{\dfrac{qR^4}{3EI}}=\frac{1}{6}\left(\frac{h}{R}\right)^2$$

$$\frac{\Delta_{BV}^{Q}}{\Delta_{BV}^{M}}=\frac{\dfrac{kqR^2}{3GA}}{\dfrac{qR^4}{3EI}}=\frac{1}{4}\left(\frac{h}{R}\right)^2$$

设

Suppose that:

$$\frac{h}{R}=\frac{1}{10}$$

则

Then:

$$\frac{\Delta_{BV}^{N}}{\Delta_{BV}^{M}}=\frac{1}{600}$$

$$\frac{\Delta_{BV}^{Q}}{\Delta_{BV}^{M}}=\frac{1}{400}$$

上述计算结果表明，轴向变形和剪切变形引起的位移很小，可忽略不计，因而只计算弯曲变形一项引起的位移即可。

The above calculation results show that the displacement caused by axial deformation and shear deformation is minimal and negligible, so only the displacement caused by bending deformation can be calculated.

4.3.2 桁架结构

4.3.2 Truss Structure

在节点荷载作用下，桁架中各杆只有轴力，且同一杆件的 F_{NP}、$\overline{F}_N$ 及 EA 沿杆长均为常数，故式（4.10）可简化为：

Under the nodal loads, each rod in the truss only has axial force and F_{NP}, $\overline{F}_N$, EA is constant along the rod length of each rod, so Equation (4.10) can be simplified as:

$$\Delta=\Sigma\int\frac{\overline{F}_N F_{NP}}{EA}ds=\Sigma\frac{\overline{F}_N F_{NP}}{EA}l \tag{4.12}$$

由于每根桁架杆的内力都需要计算两次，故建议列表计算。

Since the internal force of each rod needs to be calculated twice, it is recommended to tabulate the calculation.

【例 4.6】试计算图 4.18（a）所示桁架节点 C 的竖向位移 Δ_{CV}。各杆截面面积 A 分别注于杆旁（单位：cm^2），弹性模量 $E=$

[**Example 4.6**] Try to calculate the vertical displacement Δ_{CV} of node C in the truss shown in Figure 4.18 (a). The cross-

2.0×10⁸kPa。

sectional area A of each rod is separately marked beside the rod (unit: cm^2), and the elastic modulus $E=2.0\times10^8$kPa.

【解】（1）虚设力状态：在节点 C 加一竖向单位力 $F_P=1$，如图 4.18（c）所示。

[Solution]（1）Fictitious force state: add a vertical unit force $F_P=1$ at node C, as shown in Figure 4.18 (c).

（2）分别计算桁架在实际荷载和虚单位荷载作用下的各杆内力。结果分别列于图 4.18（b）、图 4.18（c）中。

（2）Respectively calculate the internal forces of each rod under the actions of the actual load and the virtual unit load. The results are presented in Figure 4.18 (b) and Figure 4.18 (c), respectively.

（3）计算 Δ_{CV}。计算过程列于表 4.3，由此得：

（3）Calculate Δ_{CV}. The calculation is presented in Table 4.3. Then:

$$\Delta_{CV}=\Sigma\frac{\overline{F}_N F_{NP}}{EA}l=1.04\text{cm}(\downarrow)$$

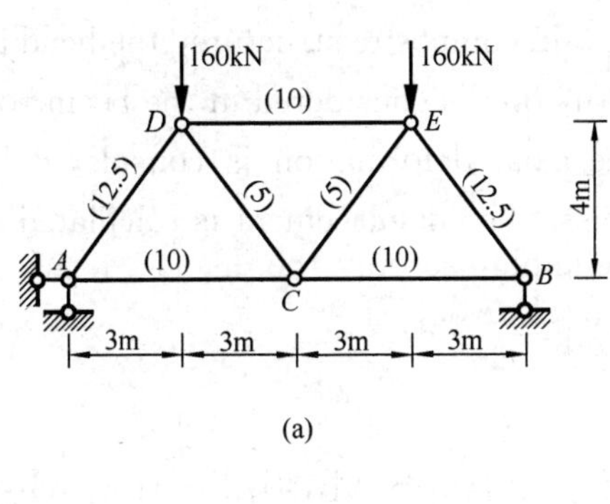

(a)

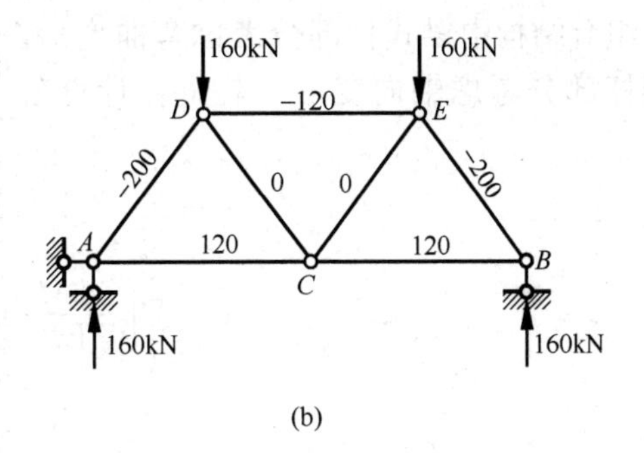

(b)

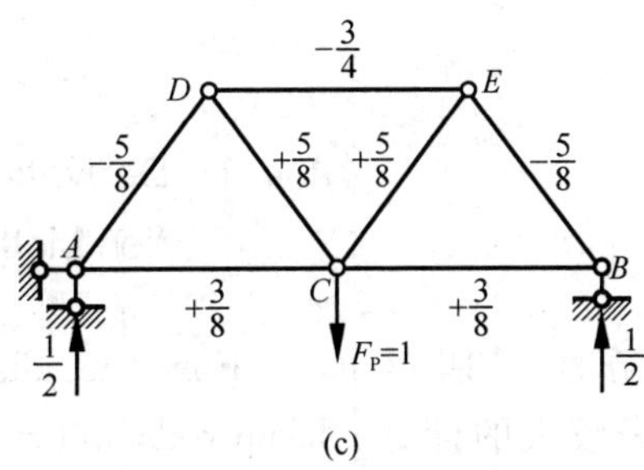

(c)

图 4.18 例 4.6

（a）桁架；（b）实际荷载及 F_{NP}（单位：kN）；（c）虚单位力及 $\overline{F}_N$

Figure 4.18 Example 4.6

(a) Truss; (b) Actual load and F_{NP} (Unit: kN); (c) Virtual unit force and $\overline{F}_N$

Δ_{CV}计算表 表4.3

Calculation table of Δ_{CV} Table 4.3

杆件 Rod	$\overline{F}_N$	F_{NP} (kN)	l (cm)	E (kN/cm²)	A (cm²)	$\frac{\overline{F}_N F_{NP}}{EA}$ (cm)
AC	3/8	120	600	2.0×10^4	10	0.135
BC	3/8	120	600	2.0×10^4	10	0.135
AD	−5/8	−200	500	2.0×10^4	12.5	0.25
BE	−5/8	−200	500	2.0×10^4	12.5	0.25
CD	5/8	0	500	2.0×10^4	5	0
CE	5/8	0	500	2.0×10^4	5	0
DE	−3/4	−120	600	2.0×10^4	10	0.27
Σ						1.04

4.3.3 组合结构

组合结构中梁式杆部分考虑弯曲变形，桁架杆部分考虑轴向变形。其位移计算公式为：

4.3.3 Composite Structure

In composite structure, the bending deformation is considered in the beam rod and the axial deformation is considered in the truss. The displacement is calculated as:

$$\Delta = \Sigma\int \frac{\overline{M}M_P}{EI}ds + \Sigma\int \frac{\overline{F}_N F_{NP}}{EA}ds \tag{4.13}$$

4.4 图乘法

4.4 Graph Multiplication Method

4.4.1 图乘法的推导

对于梁杆、刚架的位移计算时，如果杆件数量较多，则积分运算带来较大的计算量，因此，如果能简化计算过程，对手算有较大意义。下面作几个假设：

4.4.1 Derivation of the Graph Multiplication Method

For the displacement calculation of beam rods and rigid frames, the integral operation brings significant amount of calculation if the number of rods is large. Therefore, if the calculation process can be simplified, the hand calculation has greater significance. Several assumptions are made below:

假定 1：杆件为等截面，即 EI 为常数。则：

Assumption 1: The rod is of equal cross-section, i.e., EI is constant:

$$\Delta=\Sigma\int\frac{\bar{M}M_{P}}{EI}ds=\Sigma\frac{1}{EI}\int\bar{M}M_{P}ds$$

假定 2：杆件为直杆，即 $ds\to dx$。则：

Assumption 2: The rods are straight rods, i.e., $ds\to dx$:

$$\Delta=\Sigma\frac{1}{EI}\int\bar{M}M_{P}dx$$

假定 3：$\bar{M}$、M_P 至少有一个为直线图形，不妨设 $\bar{M}$ 为直线图形（因其是单位荷载作用下弯矩图，多为折线图形），如图 4.19 所示。

Assumption 3: At least one of $\bar{M}$、M_P is a straight-line graph. Suppose that $\bar{M}$ is a straight-line graph (since it is a bending moment graph under unit load, it is mostly a broken line graph), as shown in Figure 4.19.

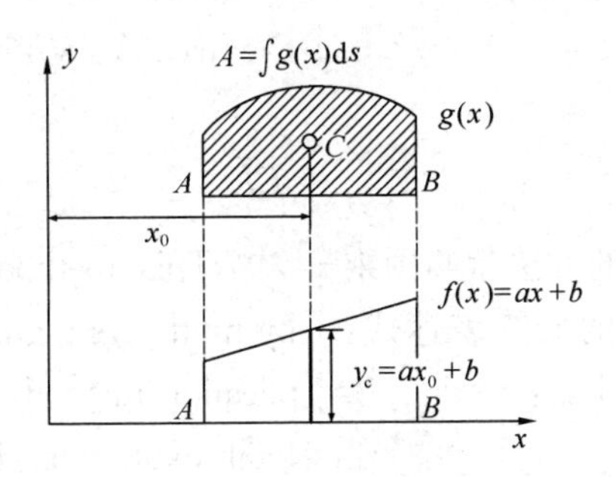

图 4.19 图乘法原理

Figure 4.19 Principle of graph multiplication method

直线的函数表达式可写为 $f(x)=ax+b$，另一个弯矩图的函数表达式记作 $g(x)$，则：

The functional expression for the straight line can be written as $f(x)=ax+b$. The functional expression for the other bending moment diagram is written as $g(x)$:

$$\Delta=\Sigma\frac{1}{EI}\int_A^B\bar{M}M_{P}dx=\Sigma\frac{1}{EI}\int_A^B f(x)g(x)dx=\Sigma\frac{1}{EI}\int_A^B(ax+b)g(x)dx$$

$$=\Sigma\frac{1}{EI}\left[\int_A^B axg(x)dx+\int_A^B bg(x)dx\right]=\Sigma\frac{1}{EI}\left[a\int_A^B xg(x)dx+b\int_A^B g(x)dx\right]$$

由于图 4.19 中 $g(x)$ 函数图形的面积 A 为：

Since the area A of the graph of $g(x)$ in Figure 4.19 is:

$$A=\int_A^B g(x)dx$$

而 $\int_A^B xg(x)\mathrm{d}x$ 则表示 $g(x)$ 面积对 y 轴的静矩，根据面积矩定理，它等于 $g(x)$ 面积 A 乘以其形心到 y 轴的距离 x_0，即：

$\int_A^B xg(x)\mathrm{d}x$ is the static moment of the area of $g(x)$ to y axis, which according to the area moment theorem is equal to the area A of $g(x)$ multiplied by the distance x_0 of the centroid to y axis, i. e. :

$$Ax_0 = \int_A^B xg(x)\mathrm{d}x$$

$$\Delta = \Sigma\frac{1}{EI}\left[a\int_A^B xg(x)\mathrm{d}x + b\int_A^B g(x)\mathrm{d}x\right] = \Sigma\frac{1}{EI}(aAx_0 + bA) = \Sigma\frac{A}{EI}(ax_0 + b)$$

又从图 4.19 可知：

It can also be known from Figure 4.19 that:

$$y_c = (ax_0 + b)$$

其中 y_c 为面积 A 形心位置对应于 $f(x)$ 的竖标，所以：

Where y_c is the vertical coordinate corresponding to $f(x)$ of the centroid position of the area A, so:

$$\Delta = \Sigma\frac{Ay_c}{EI} \tag{4.14}$$

这种图形相乘计算位移的方法就叫图乘法，式（4.14）就是图乘法的最终表达式，图乘法的使用需要注意以下几点：

This method of calculating displacement by multiplying graphs is called the graph multiplication method, and Equation (4.14) is the final expression for the graph multiplication method, which requires attention for application in the following points:

（1）必须满足推导的 3 个假定（等截面、直杆、至少一个直线弯矩图形）。

(1) 3 assumptions of the derivation must be satisfied (equal sections, straight rods, and at least one straight-line moment figure).

（2）y_c 和 A 必须取自不同图形，且 y_c 必须取自直线图形。

(2) y_c and A must be taken from different graphs, and y_c must be taken from a straight-line graph.

（3）图乘有正负：当两弯矩图形在基线同侧取正，异侧取负。

(3) The graph multiplication is positive or negative: It is positive when the two moment graphs are on the same side of the baseline and negative on the opposite sides.

（4）如果直线图形由折线构成，则需要分段计算（即 y_c 不能取自折线部分）。

(4) If the straight-line graph is a broken line, it needs to be calculated in seg-

ments (y_c cannot be taken from the broken line part).

4.4.2 常见图形的面积及其形心位置

4.4.2 Areas and the Locations of Centroids for Common Graphs

图乘法使用的两个基本要素就是面积和形心位置，表4.4给出了常见的标准图形和其形心位置。

The two primary elements in the graph multiplication method are the area and the location of the centroid, and typical standard graphs and their centroid locations are given in Table 4.4.

常见标准图形的面积和形心位置 表4.4

Areas and locations of centroids of common standard graphs Table 4.4

图形长 l，宽 h Graph length l, width h	面积 (lh) Area (lh)	与 h 距离 (l) Distance from h (l)	图形长 l，宽 h Graph length l, width h	面积 (lh) Area (lh)	与 h 距离 (l) Distance from h (l)
$A=lh/2$, $l/3$	$\frac{1}{2}$	$\frac{1}{3}$	$l/3$, $A=lh/2$	$\frac{1}{2}$	$\frac{1}{3}$
$A=lh/3$, $l/4$	$\frac{1}{3}$	$\frac{1}{4}$	$3l/8$, $A=2lh/3$	$\frac{2}{3}$	$\frac{3}{8}$

从表4.4也可看出系数符合积分规律，因此，可总结成表4.5方便记忆。

The coefficients are also consistent with the law of integration as can be seen from Table 4.4, and therefore can be summarized in Table 4.5 for ease of memory.

标准图形的面积和形心位置 表4.5

Areas and locations of centroids of standard graphs Table 4.5

$F(x)$ 次数：n Times of $F(x)$: n	面积 $A_2(lh)$：Area $A_2(lh)$： $\left(1-\frac{1}{n+1}\right)\times lh$	形心距 h，长度 x_2 (l)：Centroid distance h, length x_2 (l)： $\left(1-\frac{1}{n+2}\right)\times\frac{l}{2}$
1	$\frac{1}{2}$	$\frac{1}{3}$
2	$\frac{2}{3}$	$\frac{3}{8}$

续表
Continued

$F(x)$ 次数：n Times of $F(x)$：n	面积 $A_2(lh)$： Area $A_2(lh)$： $\left(1-\frac{1}{n+1}\right)\times lh$	形心距 h，长度 x_2（l）： Centroid distance h，length x_2（l）： $\left(1-\frac{1}{n+2}\right)\times \frac{l}{2}$
3	$\frac{3}{4}$	$\frac{2}{5}$

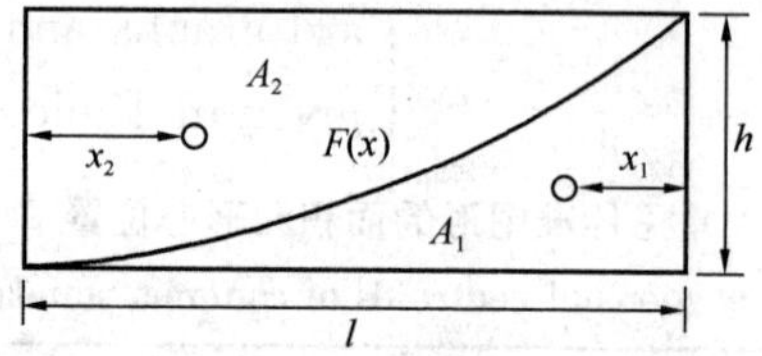

$F(x)$ 次数：n Times of $F(x)$：n	面积 $A_1(lh)$： Area $A_1(lh)$： $\frac{1}{n+1}\times lh$	形心距 h，长度 x_1（l）： Centroid distance h，length x_1（l）： $\frac{1}{n+2}\times l$
1	$\frac{1}{2}$	$\frac{1}{3}$
2	$\frac{1}{3}$	$\frac{1}{4}$
3	$\frac{1}{4}$	$\frac{1}{5}$

几点注意事项：

A few points to be noticed:

（1）标准图形可以组合，如图 4.20 所示。

(1) Standard graphs can be combined, as shown in Figure 4.20.

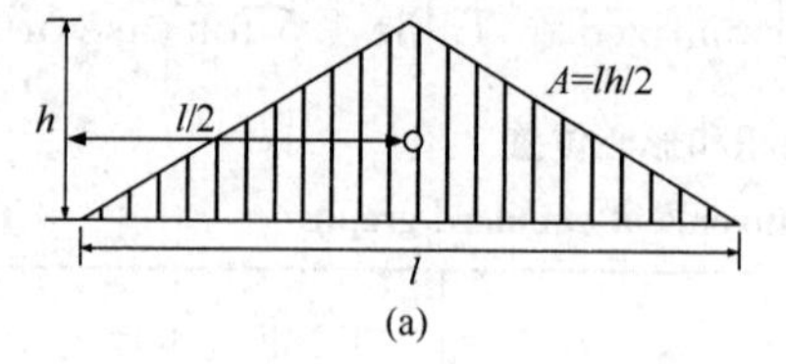

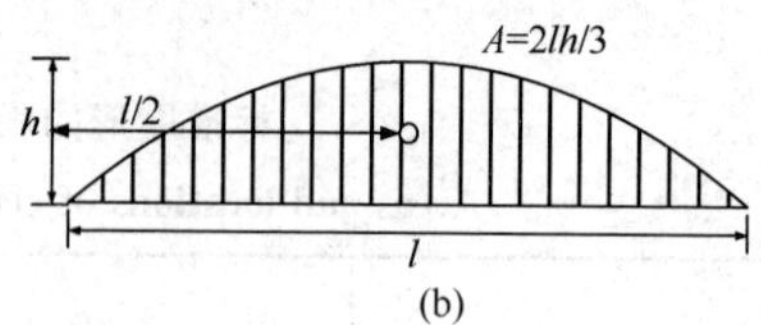

图 4.20 图形的组合
（a）三角形；（b）二次抛物线
Figure 4.20 Combination of graphs
(a) Triangle; (b) Quadratic parabola

（2）标准图形中，抛物线的顶点必须斜率为零，即此处截面剪力 $F_Q=0$，如

(2) In the standard graph, the vertex of the parabola's slope must be zero, i.e.,

图 4.21所示。

the shear force at this point is $F_Q=0$, as shown in Figure 4.21.

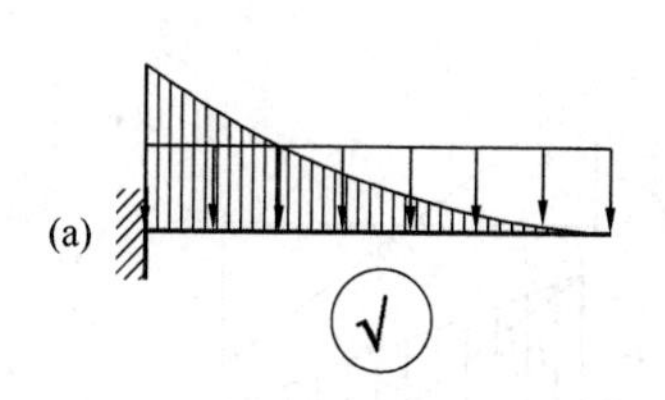

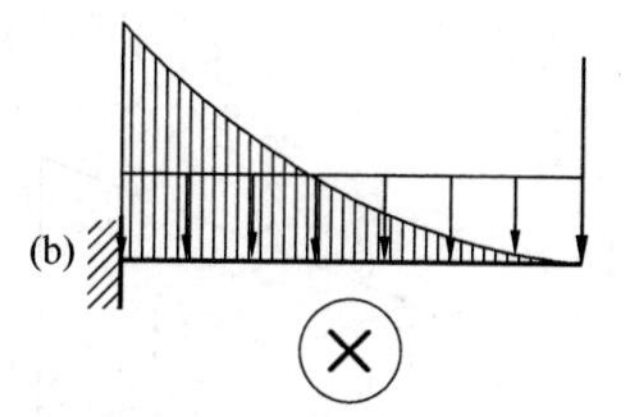

图 4.21 标准图形的判断

Figure 4.21 Judgment of a standard graph

4.4.3 图乘的分解

4.4.3 Decomposition of the Graph Multiplication

依据图乘法推导的三个假设，对于一些不直接满足假设的情况可以进行分段处理，如图 4.22 所示。

Based on the three assumptions derived from the graph multiplication method, segmentation can be performed for some cases that do not directly satisfy the assumptions, as shown in Figure 4.22.

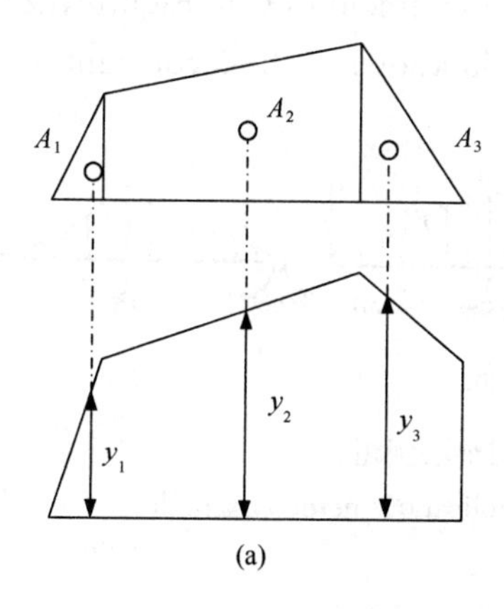

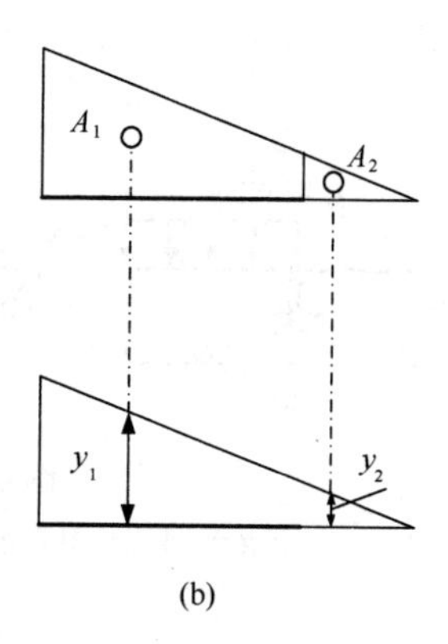

图 4.22 满足假设的分段

(a) 折线分段；(b) 变截面分段

Figure 4.22 Segmentation to satisfy the assumptions

(a) Folded segment; (b) Variable section segment

值得注意的是，图 4.22 中都分出了梯形弯矩图，这在标准图形中并未列出，虽然也可按照梯形推算面积和形心位置，但不如继续分解图形简便，如图 4.23 所示。

It is worth noting that the trapezoidal moment graph is subdivided in Figure 4.22, which is not listed in the standard figures, and although it is possible to derive the area and the location of the centroid according to the trapezoid, it is not as easy as continuing to decompose the figures, as shown in Figure 4.23.

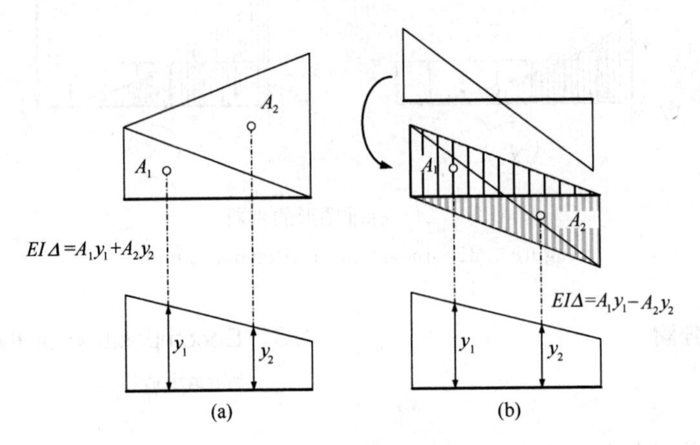

图 4.23 图乘的分解

Figure 4.23 Decomposition of the graph multiplication

【例 4.7】 用图乘法求图 4.24（a）中 C 点的竖向位移，已知 EI 为常数。

[Example 4.7] Use the graph multiplication method to find out the vertical displacement of C in Figure 4.24(a), where EI is known to be a constant.

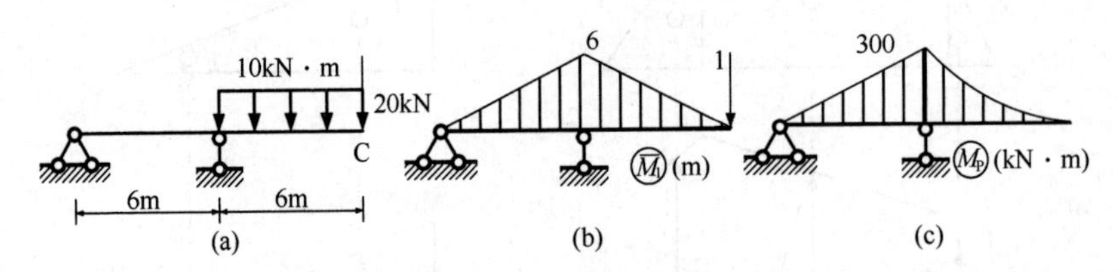

图 4.24 图乘法例题

Figure 4.24 Graph multiplication method example

【解】 首先，虚设力状态，然后分别做出荷载作用下的弯矩图（图 4.24c）和虚设单位力作用下的弯矩图（图 4.24b）。下面进行图乘计算。

[Solution] First, suppose the virtual force state and then draw the bending moment diagram under actual loads (Figure 4.24c) and the bending moment diagram under virtual unit force (Figure 4.24b), respectively. The graph multiplication method shows as follows.

第一种图乘方法如图 4.25 所示。

The first graph multiplication method is shown in Figure 4.25.

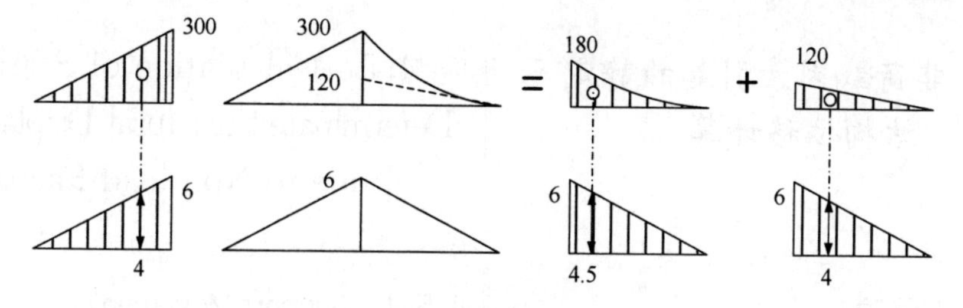

图 4.25 图乘法举例 1

Figure 4.25 Graph multiplication method example 1

拆分图乘三次，叠加结果可得：

Splitting of graph multiplication three times, and the superposition results can be obtained as follows:

$$\Delta = 6660/EI(\downarrow)$$

第二种图乘方法如图 4.26 所示。

The second graph multiplication method is shown in Figure 4.26.

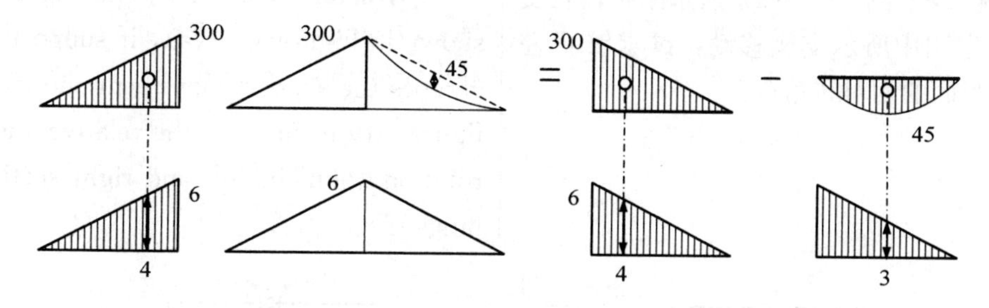

图 4.26 图乘法举例 2

Figure 4.26 Graph multiplication method example 2

叠加结果可得：

The superposition results can be obtained as follows:

$$\Delta = 6660/EI(\downarrow)$$

可见，图乘的拆分方法并不唯一，实际操作中宜选择更简便的方法。

It can be seen that there is no unique method of splitting of graph multiplication, and it is appropriate to choose a simpler method in practice.

4.5 非荷载因素引起的静定结构位移计算

4.5 Calculation of Statically Determinate Structural Displacements due to Non-load Factors

4.5.1 支座移动

4.5.1 Support Movement

静定结构由于支座移动并不产生内力也无变形，只发生刚体位移，所以式（4.7）可简化为：

The statically determinate structures do not generate internal forces nor deformations due to the movement of the support with only rigid body displacements, so Equation (4.7) can be simplified as:

$$1\times\Delta+\Sigma\overline{F}_{Ri}c_i=0$$

$$\Delta=-\Sigma\overline{F}_{Ri}c_i \tag{4.15}$$

【例 4.8】图 4.27（a）所示刚架，若支座 B 发生图中所示支座移动，试求铰 C 左右两侧截面的相对转角 θ_C。

[**Example 4.8**] For the rigid frame shown in Figure 4.27(a), if support B undergoes the support movement shown in the figure, try to find out the relative angles of rotation θ_C on the left and right sections of hinge C.

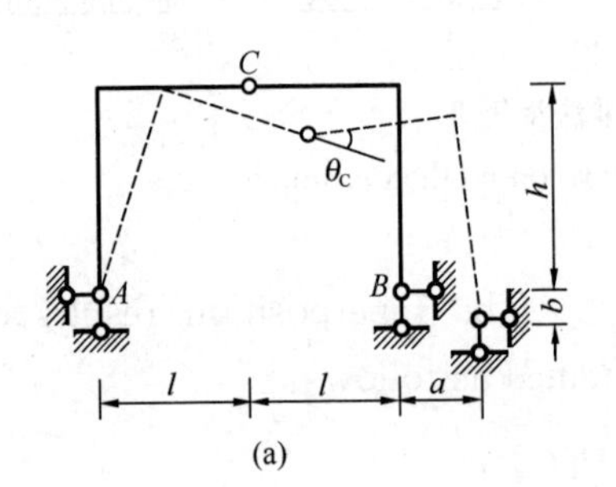

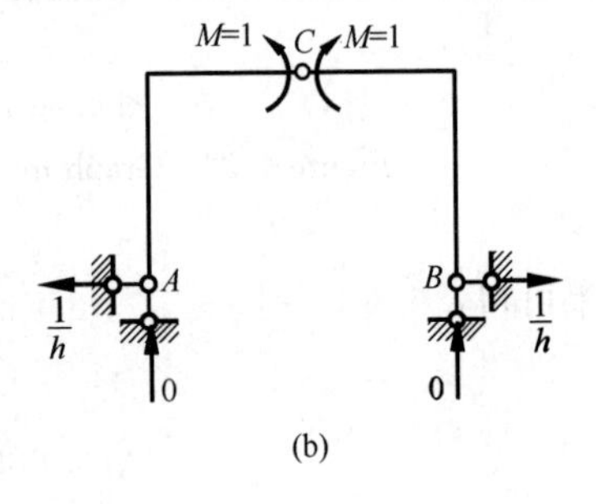

图 4.27 例 4.8

Figure 4.27 Example 4.8

【解】（1）虚设力状态：在铰 C 左右两侧截面加一对方向相反的单位力偶，如图 4.27（b）所示。

[**Solution**] (1) Virtual force state: add a pair of unit force couples in opposite directions to the left and right sections of hinge C, as shown in Figure 4.27(b).

（2）由平衡条件求出虚设单位荷载引起的支座反力，如图 4.27（b）所示。

(2) Find out the support reaction forces due to the virtual unit load by the equilibrium condition, as shown in Figure 4.27 (b).

（3）计算 θ_C。

(3) Calculate θ_C.

$$\theta_C=-\Sigma\bar{F}_{Ri}c_i=-\left(\frac{1}{h}\times a\right)=-\frac{a}{h}(\curvearrowleft\curvearrowright)$$

结果为负值，表明铰 C 左右两侧截面相对转角的实际方向与所设虚单位力偶方向相反。

The result is negative, indicating that the actual direction of the relative rotation angles of the left and right sections of the hinge C is opposite to the direction of the set virtual unit force couple.

4.5.2 温度变化

4.5.2 Temperature Variation

对于静定结构，当温度改变时并不产生内力，但由于材料的膨胀和收缩，却会使结构产生变形和位移。下面分析温度变化引起的变形计算，假设任一微段 ds，上侧升温 t_1，下侧升温 t_2。材料线膨胀系数为 α，如图 4.28 所示。

For a statically determinate structure, no internal forces are generated when the temperature changes, but deformation and displacement of the structure occur due to the material's expansion and contraction. In the following analysis, the deformation due to temperature change is calculated, assuming any micro-section ds with upper side warming t_1 and the lower side warming t_2. The coefficient of linear expansion of the material is α, as shown in Figure 4.28.

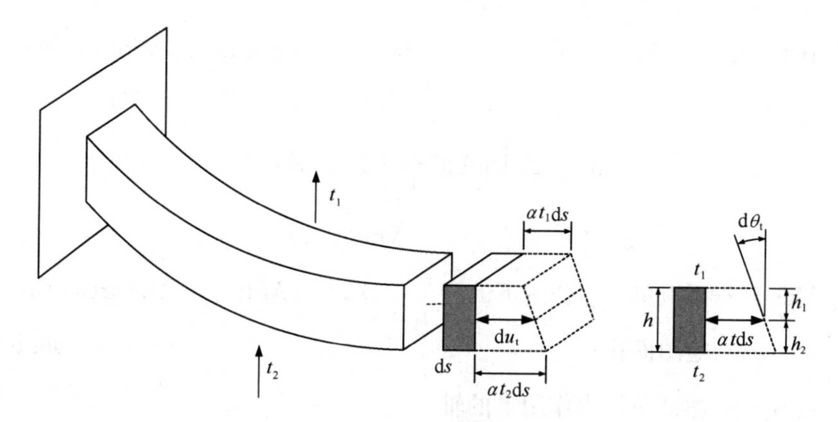

图 4.28 温度变化引起的轴向变形

Figure 4.28 Axial deformation due to temperature change

此处假设温度沿截面高度线性变化，杆轴线温度变化为：

Here it is assumed that the temperature varies linearly along the height of the section and the temperature of the rod axis varies as:

$$t = \frac{h_2}{h} t_1 + \frac{h_1}{h} t_2$$

如果杆件截面对称于形心轴，即 $h_1 = h_2 = \frac{h}{2}$，则 $t = \frac{t_1 + t_2}{2}$，ds 在轴线处的伸长量可表示为：

If the rod section is symmetrical to the centroid axis, $h_1 = h_2 = \frac{h}{2}$, then $t = \frac{t_1 + t_2}{2}$, the elongation of ds at the axis can be expressed as:

$$\mathrm{d}u_{\mathrm{t}} = \alpha t \mathrm{d}s \tag{4.16}$$

ds 在两端截面的相对转角可表示为：

Relative rotation angle of ds at both ends can be expressed as:

$$\mathrm{d}\theta_{\mathrm{t}} = \frac{\alpha t_2 \mathrm{d}s - \alpha t_1 \mathrm{d}s}{h} = \frac{\alpha \Delta_{\mathrm{t}} \mathrm{d}s}{h} \tag{4.17}$$

此外，由于假定温度沿杆长均匀分布，不可能出现剪切变形，即 $\mathrm{d}\eta = 0$，由此可得温度变化引起的位移计算公式，如式（4.18）所示：

In addition, since the temperature is assumed to be uniformly distributed along the length of the rod, no shear deformation is occurred, i. e., $\mathrm{d}\eta = 0$, which leads to the equation for the displacement due to temperature change, as shown in Equation (4.18):

$$1 \times \Delta_{\mathrm{t}} = \Sigma \int \bar{M} \mathrm{d}\theta_{\mathrm{t}} + \Sigma \int \bar{F}_{\mathrm{N}} \mathrm{d}u_{\mathrm{t}} \tag{4.18}$$

代入式（4.16），式（4.17）为：

Substituting into Equation(4.16), Equation (4.17) can be shown as:

$$\Delta_{\mathrm{t}} = \Sigma \pm \alpha A_{\bar{\mathrm{M}}} \frac{\Delta_{\mathrm{t}}}{h} + \Sigma \pm \alpha A_{\bar{\mathrm{F}}_{\mathrm{N}}} t \tag{4.19}$$

式中

$A_{\bar{\mathrm{M}}} = \int \bar{M} \mathrm{d}s$——虚拟单位力作用下的弯矩图面积；

$A_{\bar{\mathrm{F}}_{\mathrm{N}}} = \int \bar{F}_{\mathrm{N}} \mathrm{d}s$——虚拟单位力作用下的轴力图面积；

$\Delta_{\mathrm{t}} = | t_1 - t_2 |$——两侧温度变化值之差；

t——杆轴线温度变化值，计算中只取绝对值；

h——截面高度；

α——材料线膨胀系数。

Where

$A_{\bar{\mathrm{M}}} = \int \bar{M} \mathrm{d}s$——The area of bending moment diagram under virtual unit force;

$A_{\bar{\mathrm{F}}_{\mathrm{N}}} = \int \bar{F}_{\mathrm{N}} \mathrm{d}s$——The area of axial force diagram under the virtual unit force;

$\Delta_{\mathrm{t}} = | t_1 - t_2 |$——The difference of temperature change between two sides;

t——The value of the temperature change of the rod axis, and only the absolute value is taken in the calculation;

h——The height of the section;

α——The coefficient of linear expansion of the material.

注：式（4.19）中"±"的含义是：如果虚拟力状态中变形趋势与温度改变引起的变形趋势一致取正号，相反取负号。

Note: The meaning of "±" in Equation (4.19) is that a positive sign is taken if the deformation trend in the virtual force state is the same as the deformation trend caused by the temperature change, and a negative sign is taken under opposite conditions.

从式（4.19）也可以看出，温度变化引起的位移计算中，不能忽略轴向变形。

It can also be seen from Equation (4.19) that axial deformation cannot be neglected in calculating displacement due to temperature change.

【例 4.9】图 4.29（a）所示刚架施工时温度为 20℃，试求冬季当外侧温度为 −10℃，内侧温度为 0℃时 C 点的竖向位移 Δ_{CV}。已知：$l=4\text{m}$，$\alpha=1\times10^{-5}$，各杆截面均为矩形，截面高度 $h=40\text{cm}$。

［**Example 4.9**］For the rigid frame shown in Figure 4.29(a), the temperature is 20℃ during construction, try to find out the vertical displacement Δ_{CV} at point C when the outside temperature is −10℃ and the inside temperature is 0℃ in winter. It is known that $l=4\text{m}$, $\alpha=1\times10^{-5}$, and the sections of each rod are rectangular with the section height $h=40\text{cm}$.

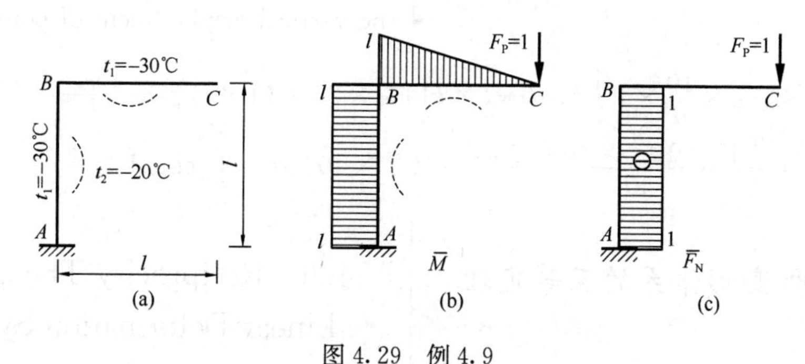

图 4.29 例 4.9

Figure 4.29 Example 4.9

【解】(1) 虚设力状态：在 C 点加一竖向单位力 $F_P=1$，如图 4.29 (b) 所示。

[Solution] (1) Virtual force state: add a vertical unit force $F_P=1$ at point C, as shown in Figure 4.29(b).

(2) 分别作刚架在虚单位荷载作用下的 $\bar{M}$ 图（图 4.29b）及 $\bar{F}_N$ 图（图 4.29c）。

(2) Draw respectively the graphs $\bar{M}$ (Figure 4.29b) and $\bar{F}_N$ (Figure 4.29c) of the rigid frame under virtual unit load.

(3) 计算杆件形心轴处温度改变 t_0 及两侧温度改变之差 Δ_t。

(3) Calculate temperature change t_0 at the centroid axis of the rod and the difference of temperature change Δ_t at two sides.

外侧温度改变为 $t_1=-10-20=-30℃$，内侧温度改变为 $t_2=0-20=-20℃$。故有：

The outside temperature change is $t_1=-10-20=-30℃$, and the inside temperature change is $t_2=0-20=-20℃$. Therefore:

$$t_0=\left|\frac{t_1+t_2}{2}\right|=\left|\frac{-30-20}{2}\right|=25℃$$

$$\Delta_t=|t_2-t_1|=|-20-(-30)|=10℃$$

(4) 计算 Δ_{CV}。Δ_t 及 $\bar{M}$ 所引起的杆件弯曲方向分别如图 4.29 (a) 和 (b) 所示，二者方向相反，故式 (4.19) 中第一项取负号；至于轴向变形的影响一项，$\bar{F}_N$ 为压力，使杆件缩短，而 t_0 为负值，也使杆件缩短，二者变形方向一致，故此项应取正号。因此，C 点的竖向位移为：

(4) Calculate Δ_{CV}. Rods' bending directions caused by Δ_t and $\bar{M}$ are shown in Figures 4.29(a) and (b), respectively, which are in the opposite directions, so the first item in Equation (4.19) is negative; as for the influence of axial deformation, $\bar{F}_N$ is pressure, which shortens the rod, and t_0 is negative, which also makes the rod shortened, the deformation directions are consistent, so it should take a positive sign. Therefore, the vertical displacement of point C is:

$$\Delta_{CV}=-\alpha\times\frac{10}{h}\times\left(\frac{1}{2}l\times l+l\times l\right)+\alpha\times 25\times(1\times l)=-\frac{15\alpha l^2}{h}+25\alpha l$$

$$=-\frac{15\times 10^{-5}\times 400^2}{40}+25\times 10^{-5}\times 400=-0.5\text{cm}(\uparrow)$$

4.6 线性变形体系的互等定理

4.6 Reciprocity Theorem for Linear Deformation Systems

线性变形体系有四个互等定理，即：

The linear deformation system has four

reciprocity theorems:

1. 功的互等定理；
2. 位移互等定理；
3. 反力互等定理；
4. 反力与位移互等定理。

1. Work reciprocity theorem;
2. Displacement reciprocity theorem;
3. Reaction force reciprocity theorem;
4. Reaction force and displacement reciprocity theorem.

其中最基本的是功的互等定理，其他三个互等定理都可由此定理推导出来。

Among them, the most basic one is work reciprocity theorem, from which the other three reciprocity theorems can be derived.

4.6.1 功的互等定理

4.6.1 Work Reciprocity Theorem

图 4.30 显示了同一结构的两种状态：图 4.30（a）为状态Ⅰ，其中内力记为 M_1，F_{N1}，F_{Q1}，图 4.30（b）为状态Ⅱ，其中内力记为 M_2，F_{N2}，F_{Q2}。

Figure 4.30 shows two states of the same structure: State Ⅰ in Figure 4.30(a), the internal forces are noted as M_1, F_{N1}, F_{Q1}, and state Ⅱ in Figure 4.30(b), the internal forces are noted as M_2, F_{N2}, F_{Q2}.

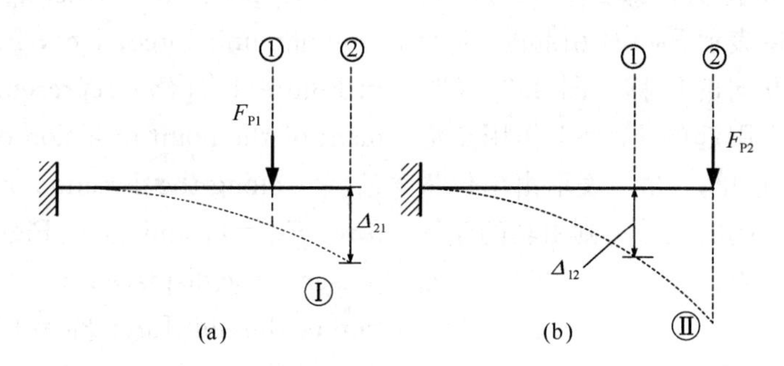

图 4.30 同一结构两种状态

Figure 4.30 Two states of the same structure

如果将状态Ⅰ视为力状态，状态Ⅱ视为位移状态，依据虚功原理，有：

If state Ⅰ is considered as force state and state Ⅱ as displacement state, based on the principle of virtual work:

$$F_{P1}\Delta_{12}=\Sigma\int\frac{M_1M_2}{EI}\mathrm{d}s+\Sigma\int\mu\frac{F_{Q1}F_{Q2}}{GA}\mathrm{d}s+\Sigma\int\frac{F_{N1}F_{N2}}{EA}\mathrm{d}s \tag{4.20}$$

如果将状态Ⅱ视为力状态，状态Ⅰ视为位移状态，依据虚功原理，有：

If state Ⅱ is considered as force state and state Ⅰ as displacement state, based on the principle of virtual work:

$$F_{P2}\Delta_{21}=\Sigma\int\frac{M_2M_1}{EI}ds+\Sigma\int\mu\frac{F_{Q2}F_{Q1}}{GA}ds+\Sigma\int\frac{F_{N2}F_{N1}}{EA}ds \qquad (4.21)$$

联立式（4.20）和式（4.21）可得：

Combining Equations (4.20) and (4.21):

$$F_{P1}\Delta_{12}=F_{P2}\Delta_{21} \qquad (4.22)$$

这表明第一状态的外力在第二状态的位移上所做的虚功，等于第二状态的外力在第一状态的位移上所做的虚功。这就是功的互等定理，也记作：$W_{12}=W_{21}$。

It shows that the virtual work done by the external force in the first state on the displacement of the second state is equal to the virtual work done by the external force in the second state on the displacement of the first state. This is the work reciprocity theorem, also noted as $W_{12}=W_{21}$.

4.6.2 位移互等定理

4.6.2 Displacement Reciprocity Theorem

位移互等定理是功的互等定理的一种特殊情况。

The displacement reciprocity theorem is a special condition of the work reciprocity theorem.

如图 4.31 所示，设在两种状态中，结构都只承受一个单位力，即 $F_{P1}=F_{P2}=1$。图 4.31（a）中 δ_{21} 表示 $F_{P1}=1$ 引起的 $F_{P2}=1$ 作用点沿 F_{P2} 方向的位移，图 4.31（b）中 δ_{12} 表示 $F_{P2}=1$ 引起的 $F_{P1}=1$ 作用点沿 F_{P1} 方向的位移，这里，“δ”表示由单位量值（如单位力）引起的位移。则由功的互等定理式（4.22）可得：

As shown in Figure 4.31, in both states, suppose the structure is subjected to only one unit force, i.e., $F_{P1}=F_{P2}=1$. δ_{21} in Figure 4.31(a) represents the displacement of the point of action of the unit force $F_{P2}=1$ along the direction of F_{P2} due to unit force $F_{P1}=1$, and δ_{12} in Figure 4.31(b) represents the displacement of the point of action of the unit force $F_{P1}=1$ along the direction of F_{P1} due to unit force $F_{P2}=1$. Here, “δ” denotes the displacement caused by a unit quantity (e.g. unit force). Then from the work reciprocity theorem of Equation (4.22), we have:

$$F_{P1}\delta_{12}=F_{P2}\delta_{21}$$

$$\because \quad F_{P1}=F_{P2}=1$$

$$\therefore \quad \delta_{12}=\delta_{21} \qquad (4.23)$$

这就是位移互等定理，它表明：在任一线性变形体系中，单位力 $F_{P2}=1$ 所引起的单位力 $F_{P1}=1$ 作用点沿 F_{P1} 方向的位移 δ_{12}，

The above is the displacement reciprocity theorem, which shows that in any linear deformation system, the displacement δ_{12} of

在数值上等于单位力 $F_{P1}=1$ 所引起的单位力 $F_{P2}=1$ 作用点沿 F_{P2} 方向的位移 δ_{21}。

the point of action of the unit force $F_{P1}=1$ along the direction of F_{P1} due to unit force $F_{P2}=1$ is numerically equal to the displacement δ_{21} of the point of action of the unit force $F_{P2}=1$ along the direction of F_{P2} due to unit force $F_{P1}=1$.

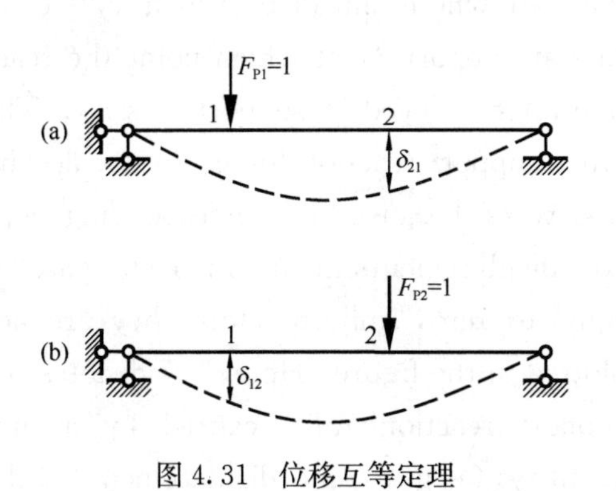

图 4.31 位移互等定理

Figure 4.31 Displacement Reciprocity Theorem

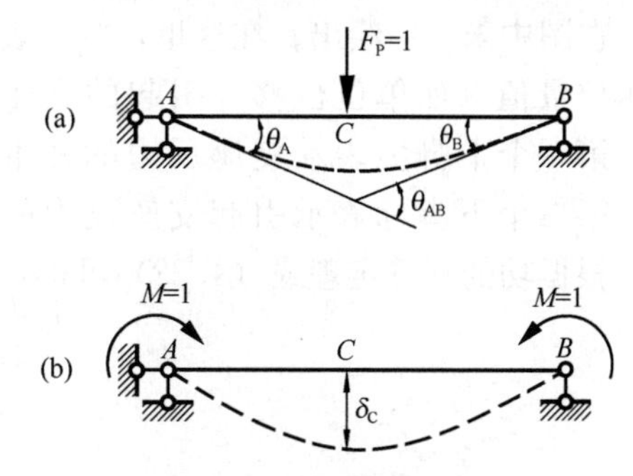

图 4.32 广义力和广义位移

Figure 4.32 Generalized forces ans displacements

显然，单位力 F_{P1} 及 F_{P2} 也可以是广义力，而 δ_{12} 和 δ_{21} 则是相应的广义位移。例如在图 4.32 所示的两种状态中，由位移互等定理可知，图 4.32（a）中 A、B 两截面的相对转角 θ_{AB} 与图 4.32（b）中 C 截面的线位移 δ_C 虽然量纲不同，但二者在数值上是互等的。

Clearly, the unit forces F_{P1} and F_{P2} can also be generalized forces, while δ_{12} and δ_{21} are the corresponding generalized displacements. For example, in the two states shown in Figure 4.32, it follows from the displacement reciprocity theorem that the relative rotation angle θ_{AB} in Figure 4.32(a) and the linear displacement δ_C at section C in Figure 4.32(b) are numerically equivalent to each other, although the magnitudes are different.

4.6.3 反力互等定理

4.6.3 Reaction Force Reciprocity Theorem

反力互等定理也是功的互等定理的一种特殊情况。它用来说明超静定结构在两个支座分别发生单位位移时，这两种状态中相应反力的互等关系。

The reaction force reciprocity theorem is also a special condition of the work reciprocity theorem. It is used to state the reciprocal equivalence of the corresponding reaction forces in two states with unit displacements of the two supports in statically indeterminate structures.

图 4.33 所示为同一线性变形体系的两

Two states of the same linear deforma-

种状态。其中图4.33（a）表示支座1发生单位位移$c_1=1$的状态，此时在支座2引起的反力为r_{21}；图4.33（b）表示支座2发生单位位移$c_2=1$的状态，此时在支座1引起的反力为r_{12}。其他支座反力因所对应的另一状态的支座位移都等于零而不做虚功，因此，在图中未一一绘出。在这里，“r”表示由单位量值（如单位位移）引起的支座反力，第一个下脚标表示支座反力的发生位置，第二个下脚标表示引起支座反力的原因。根据功的互等定理式（4.22）可得：

tion system are shown in Figure 4.33. Figure 4.33(a) represents the state where unit displacement $c_1=1$ occurs at support 1, at which point the reaction force induced in support 2 is r_{21}; Figure 4.33(b) represents the state where unit displacement $c_2=1$ occurs at support 2, at which point the reaction force induced in support 1 is r_{12}. The other support reaction forces do not do virtual work because the corresponding support displacements in the other state are all equal to zero, and therefore they are not plotted in the figure. Here, "r" denotes the support reaction force caused by a unit quantity (e.g., unit displacement), the first subscript indicates the location of the occurrence of the support reaction force, and the second subscript indicates the cause of the support reaction force. According to the Equation (4.22) of the work reciprocity theorem, we can obtain:

$$r_{12}c_1 = r_{21}c_2$$

$$\because \quad c_1 = c_2 = 1$$

$$\therefore \quad r_{12} = r_{21} \tag{4.24}$$

这就是反力互等定理，它表明：在任一线性变形体系中，支座2发生单位位移$c_2=1$所引起的支座1的反力r_{12}，在数值上等

This is the reaction force reciprocity theorem, which shows that in any linear deformation system, the reaction force r_{12} of

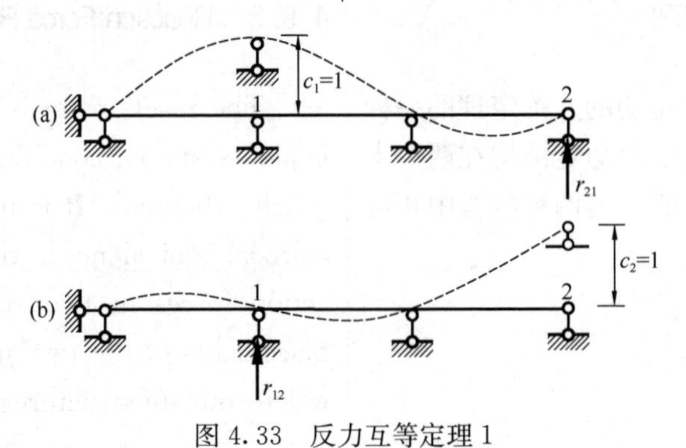

图4.33 反力互等定理1
Figure 4.33 Reaction force reciprocity theorem 1

于支座 1 发生单位位移 $c_1=1$ 所引起的支座 2 的反力 r_{21}。

bearing 1 caused by the occurrence of a unit displacement $c_2=1$ in bearing 2 is numerically equal to the reaction force r_{21} caused by the occurrence of a unit displacement $c_1=1$ in bearing 1.

图 4.34 所示为反力互等定理的另一例子，在图示的两种状态中，由反力互等定理可知，反力 r_{12} 与反力矩 r_{21} 虽然量纲不同，但在数值上具有互等的关系。

Another example of the reaction force reciprocity theorem is shown in Figure 4.34. Under the two states, from the reaction force reciprocity theorem, the reaction forces r_{12} and r_{21} are numerically equivalent to each other, although the magnitudes are different.

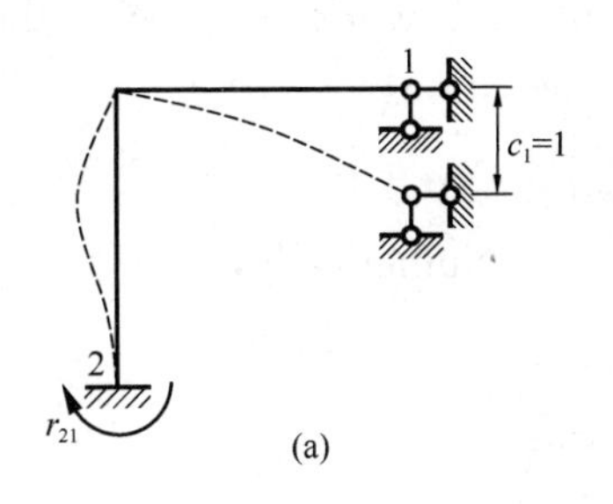

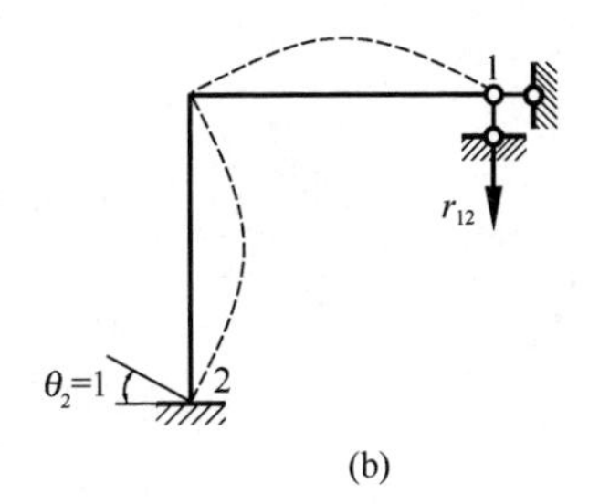

图 4.34 反力互等定理 2

Figure 4.34 Reaction force reciprocity theorem 2

需要指出，由于支座移动在静定结构中不引起反力，故反力互等定理只有应用于超静定结构才有意义。此外，这里所说的支座也可以换成别的约束，因此，单位支座位移可以换成与该约束相应的单位广义位移，而支座反力则可以换成与该约束相应的广义力。

It should be noted that since the support movement does not cause reaction forces in a stationary structure, the reaction force reciprocity theorem is only relevant when applied to a super-stationary structure. Moreover, the supports referred to here can be replaced by other constraints so that the unit support displacement can be replaced by the unit generalized displacement corresponding to that constraint, while the support reaction force can be replaced by the generalized force corresponding to that constraint.

4.6.4 反力与位移互等定理

4.6.4 Reaction Force and Displacement Reciprocity Theorem

这一定理是功的互等定理的又一特殊情

This theorem is another special condi-

况，它说明了线性变形体系一种状态的反力与另一种状态的位移具有互等关系。

tion of the work reciprocity theorem, which states that the reaction force of one state of a linear deformation system is reciprocally related to the displacement of another state.

以图4.35所示两种状态为例，图4.35（a）表示当单位力 $F_{P2}=1$ 作用于2点时，支座1的反力矩为 r_{12}；图4.35（b）表示当支座1沿 r_{12} 方向发生一单位转角 $\theta_1=1$ 时，点2沿 F_{P2} 方向的位移为 δ_{21}。对此两种状态应用功的互等定理，有：

Take the two states shown in Figure 4.35 as an example, Figure 4.35(a) shows that when a unit force $F_{P2}=1$ acts on point 2, the reaction moment of support 1 is r_{12}; Figure 4.35(b) shows that when a unit angle of rotation $\theta_1=1$ occurs along the direction of r_{12} for support 1, the displacement of point 2 along the direction of F_{P2} is δ_{21}. Apply the work reciprocity theorem to these two states:

$$r_{12}\theta_1 + F_{P2}\delta_{21} = 0$$

因在数值上，

Numerically,

$$\theta_1 = F_{P2} = 1$$

故得：

So:

$$r_{12} = -\delta_{21} \tag{4.25}$$

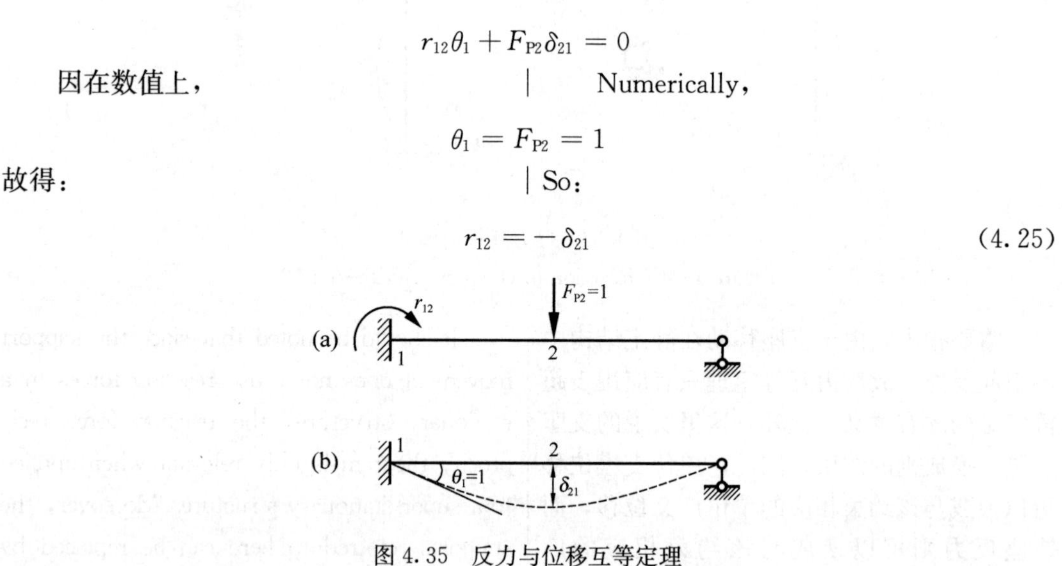

图4.35 反力与位移互等定理

Figure 4.35 The reaction force and displacement reciprocity theorem

此即为反力与位移互等定理，它表明：在任一线性变形体系中，作用于点2处的单位力 $F_{P2}=1$ 所引起的支座1的反力 r_{12}，在数值上等于支座1沿 r_{12} 方向发生单位位移 $c_1=1$ 所引起的点2处沿 F_{P2} 方向的位移 δ_{21}，但符号相反。

This is the reaction force and displacement reciprocity theorem, which shows that in any linear deformation system, the reaction force r_{12} at support 1 caused by the unit force $F_{P2}=1$ acting at point 2, is numerically equal to the displacement δ_{21} along the direction of F_{P2} at point 2 caused by the unit displacement $c_1=1$ occurring at support 1 along the direction of r_{12}, but of opposite sign.

思 考 题

4.1 结构位移有哪两大类？

4.2 产生位移的因素主要有哪些？位移都是由变形引起的吗？

4.3 结构位移计算的目的是什么？

4.4 什么是线性变形体系？它有哪些应用条件？

4.5 虚功与实功有何不同？

4.6 如何理解虚功中广义力与广义位移的相应关系？

4.7 刚体体系虚功原理在静定结构中有哪两种形式的应用？它们有何区别？

4.8 单位位移法和单位荷载法的特点分别是什么？

4.9 刚体体系虚功原理与变形体系虚功原理有何异同？

4.10 请写出结构位移计算的一般公式，并说明其中每一项的物理意义。

4.11 如何理解结构位移计算的一般公式（式4.7）应用的普遍性和广泛性？

Questions

4.1 What are the two major categories of structural displacement?

4.2 What are the main factors of displacement? Are displacements all caused by deformations?

4.3 What is the purpose of the structure displacement calculation?

4.4 What is the linear deformation system? What are its application conditions?

4.5 What is the difference between virtual work and real work?

4.6 How to understand the corresponding relationship between generalized force and generalized displacement in virtual work?

4.7 What are the two kinds of application for virtual work principle of rigid body system in statically determinate structure? What is the difference between them?

4.8 What are the characteristics of unit displacement method and unit load method, respectively?

4.9 What are the similarities and differences between the virtual work principle of rigid body system and that of deformation system?

4.10 Please write down the general formula for structural displacement calculation and explain the physical significance of each item.

4.11 How to understand the universality and extensiveness of the application of the general formula (Equation 4.7) for structural displacement calculation?

4.12　请根据图4.36所示虚设力状态，求出位移。

4.12　Please find out the intended displacements according to the supposed force states shown in Figure 4.36.

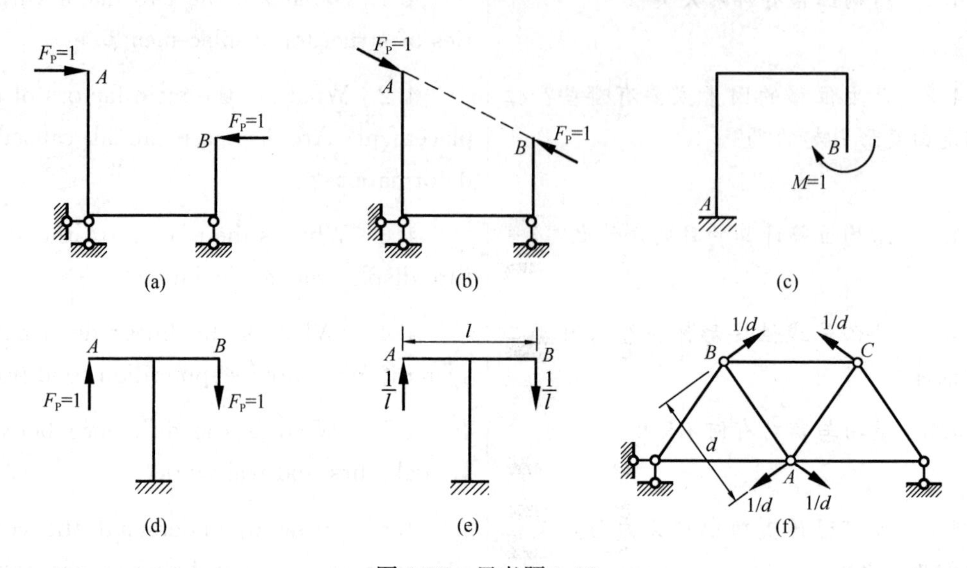

图4.36　思考题4.12

Figure 4.36　Question 4.12

4.13　如何判定所求位移Δ的实际方向？

4.13　How to determine the actual direction of the displacement Δ?

4.14　在式（4.10）中，$\overline{F}_N$、$\overline{F}_Q$、$\overline{M}$与F_{NP}、F_{QP}、M_P的物理意义分别是什么？该式能否适用于非弹性体系？

4.14　In Equation (4.10), what are the physical meanings of $\overline{F}_N$, $\overline{F}_Q$, $\overline{M}$ and F_{NP}, F_{QP}, M_P respectively? Can this be applied to inelastic systems?

4.15　图乘法的应用条件是什么？求变截面梁、曲梁和拱的位移时，能否采用图乘法？求等截面梁、刚架位移时，在$\overline{F}_N$和F_{NP}以及$\overline{F}_Q$和F_{QP}之间能否应用图乘法？

4.15　What are the application conditions of graph multiplication method? Can graph multiplication method be used to calculate the displacement of variable cross-section beam, curved beam and arch? Can graph multiplication method be applied between $\overline{F}_N$, F_{NP} and $\overline{F}_Q$, F_{QP} when calculating the displacement of beam with equal section and rigid frame?

4.16　在应用图乘法时，Ay_0的正负号如何确定？

4.16　How to determine the sign of Ay_0 when applying graph multiplication method?

4.17 图4.37所示图乘计算是否正确？为什么？

4.17 Are the graph multiplication calculations shown in Figure 4.37 correct? Why?

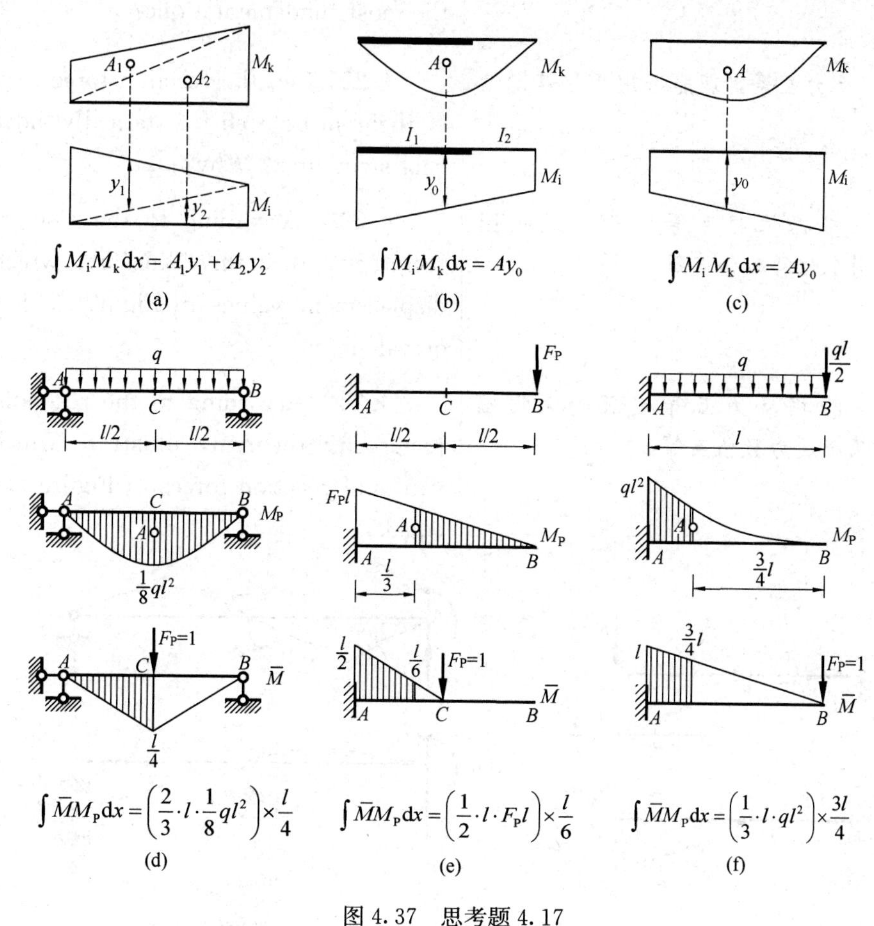

图4.37 思考题4.17

Figure 4.37 Question 4.17

4.18 式（4.17）中的$\overline{F}_{Ri}$与c_i的物理意义各是什么？如何确定二者乘积的正负号？

4.18 What are the physical meanings of $\overline{F}_{Ri}$ and c_i in Equation (4.17)? How to determine the sign of the multiplication between them?

4.19 式（4.18）、式（4.19）中的正负号如何确定？

4.19 How to determine the signs in Equation (4.18) and Equation (4.19)?

4.20 位移计算公式（4.10）是否适用于超静定结构？式（4.17）、式（4.18）、式（4.19）呢？

4.20 Is Equation (4.10) for displacement calculation applicable to statically indeterminate structure? What about Equation (4.17), Equation (4.18), and Equation (4.19)?

4.21 线性变形体系的互等定理有哪几个？其中最基本的是哪个互等定理？

4.21 Which are the reciprocity theorems of linear deformation system? What is the most fundamental one?

4.22 反力互等定理能否用于超静定结构？为什么？

4.22 Can the reaction force reciprocity theorem be used for statically indeterminate structures? Why?

4.23 根据位移互等定理，说明图4.38中哪两个位移数值互等。

4.23 According to the displacement reciprocity theorem, illustrate which two displacement values in Figure 4.38 are equivalent.

4.24 根据反力互等定理，说明图4.39中哪两个反力数值互等。

4.24 According to the reaction force reciprocity theorem, illustrate which two values of reaction forces in Figure 4.39 are equal.

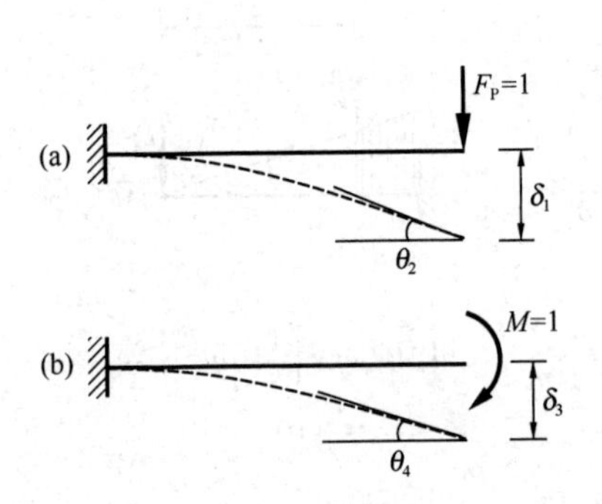

图4.38 思考题4.23

Figure 4.38 Question 4.23

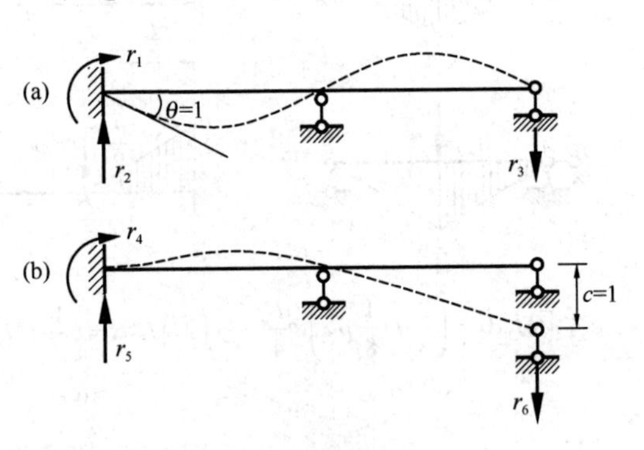

图4.39 思考题4.24

Figure 4.39 Question 4.24

习 题

Exercises

4.1 利用虚位移原理求图4.40所示静定结构指定的支座反力和截面内力：（a）F_{RD}、F_{QA}、M_A、M_C；（b）F_{RB}、F_{QB}^L、F_{QB}^R、M_B；（c）支座B的水平推力F_H和杆AC的轴力F_{NAC}。

4.1 Use the principle of virtual displacement to calculate the specified reaction forces and section internal forces of statically determinate structures in Figure 4.40: (a) F_{RD}、F_{QA}、M_A、M_C; (b) F_{RB}、F_{QB}^L、F_{QB}^R、M_B; (c) The horizontal thrust F_H of support B and the axial force F_{NAC} of rod AC.

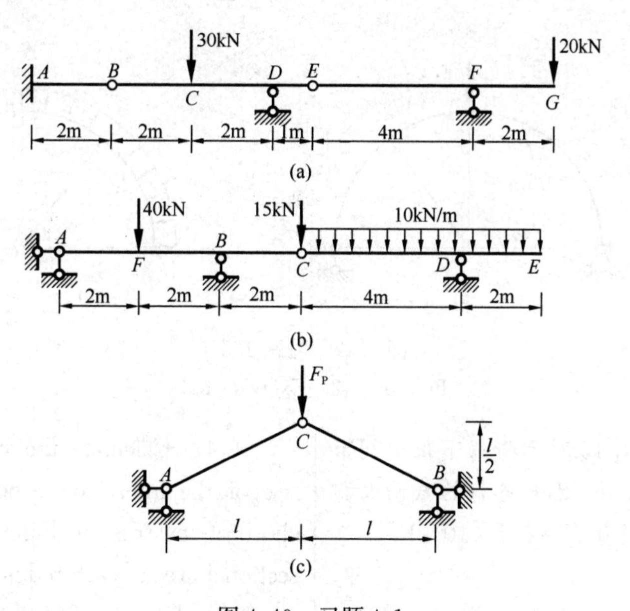

图 4.40 习题 4.1

Figure 4.40 Exercise 4.1

4.2 用积分法计算图 4.41 所示单跨静定梁的位移（忽略剪切变形的影响）：(a) Δ_{CV}；(b) θ_A；(c) Δ_{BV} and θ_B；(d) Δ_{BV}。

4.2 Calculate the displacements of single-span statically determinate beams in Figure 4.41 by integral method (neglecting the effect of shear deformation)：(a) Δ_{CV}；(b) θ_A；(c)Δ_{BV} and θ_B；(d)Δ_{BV}.

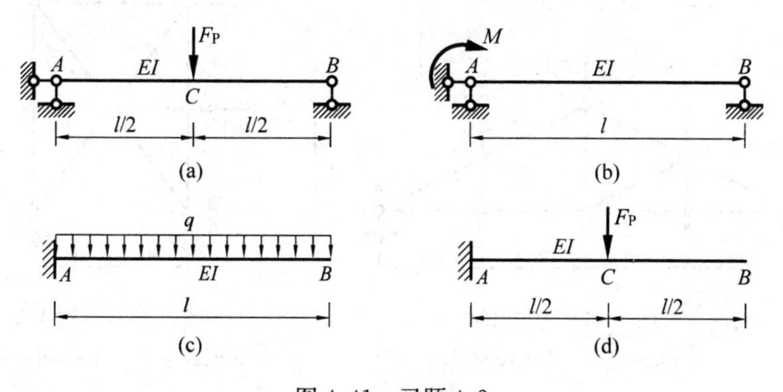

图 4.41 习题 4.2

Figure 4.41 Exercise 4.2

4.3 用积分法计算图 4.42 所示圆弧曲杆的位移（不考虑曲率的影响和 F_Q、F_N 的影响）：(a) Δ_{BH}；(b) Δ_{BV}。

4.3 Calculate the displacements of circular-arc curved rods in Figure 4.42 by integral method (without considering the influence of curvature and the influences of F_Q、F_N)：(a) Δ_{BH}；(b) Δ_{BV}.

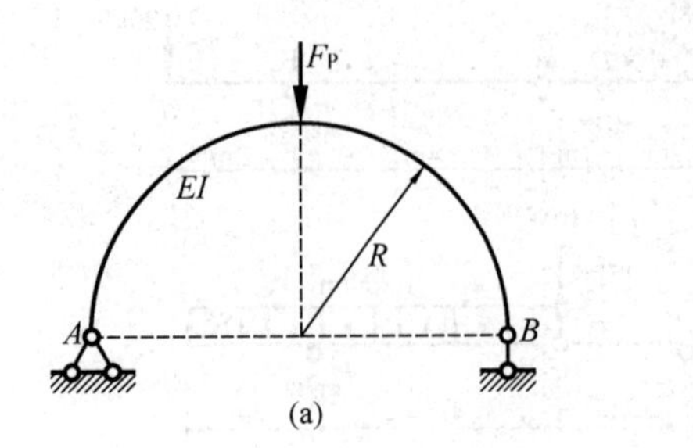

图 4.42　习题 4.3

Figure 4.42　Exercise 4.3

4.4　计算图 4.43 所示桁架下弦中间节点 D 的竖向位移 Δ_{DV}。已知各杆的截面面积 $A=12\text{cm}^2$，弹性模量 $E=2.1\times10^8\text{kPa}$。

4.4　Calculate the vertical displacement Δ_{DV} of the intermediate node D at the lower chord of the truss in Figure 4.43. The cross-sectional area of each rod is $A=12\text{cm}^2$, and the elastic modulus $E=2.1\times10^8\text{kPa}$.

4.5　计算图 4.44 所示桁架节点 G 的水平位移 Δ_{GH}。已知各杆的截面面积 $A=20\text{cm}^2$，弹性模量 $E=2.1\times10^8\text{kPa}$。

4.5　Calculate the horizontal displacement Δ_{GH} of the truss node G shown in Figure 4.44. The cross-sectional area of each rod is $A=20\text{cm}^2$, and the elastic modulus $E = 2.1\times10^8\text{kPa}$.

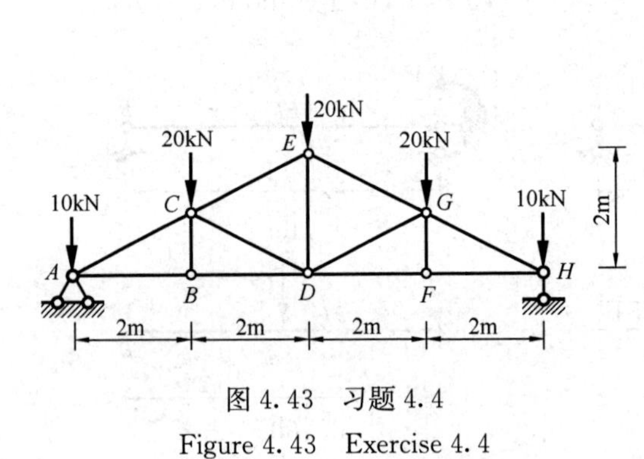

图 4.43　习题 4.4

Figure 4.43　Exercise 4.4

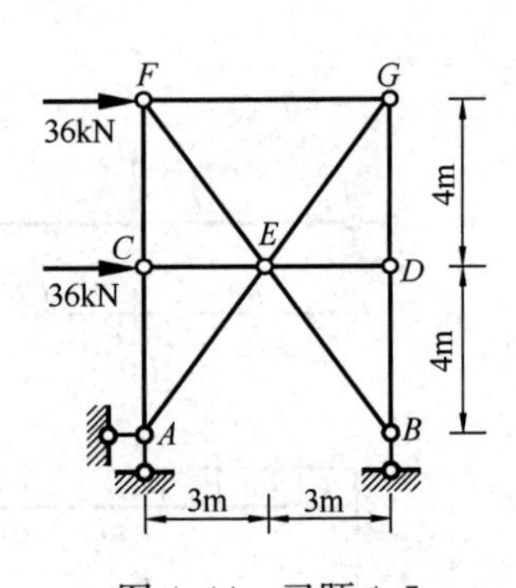

图 4.44　习题 4.5

Figure 4.44　Exercise 4.5

4.6　计算图 4.45 所示桁架 BC 杆与 BD 杆的相对转角 θ。各杆 EA=常数。

4.6　Calculate the relative rotation angle θ between the rod BC and the rod BD of the truss shown in Figure 4.45. EA= constant in each rod.

4.7　用图乘法解习题 4.2。

4.7　Use graph multiplication method to solve exercise 4.2.

4.8　求图 4.46 所示阶形柱顶点 B 的

4.8　Calculate the horizontal displace-

水平位移Δ_{BH}和转角θ_B。

ment Δ_{BH} and the rotation angle θ_B at the vertex B of the stepped column in Figure 4.46.

4.9 用图乘法求图 4.47 所示梁 C 截面的竖向位移Δ_{CV}。已知 $E=2.1\times10^8$kPa，$I=10000$cm^4。

4.9 Try to use graph multiplication method to calculate the vertical displacement Δ_{CV} of section C in Figure 4.47. $E=2.1\times10^8$kPa，$I=10000$cm^4.

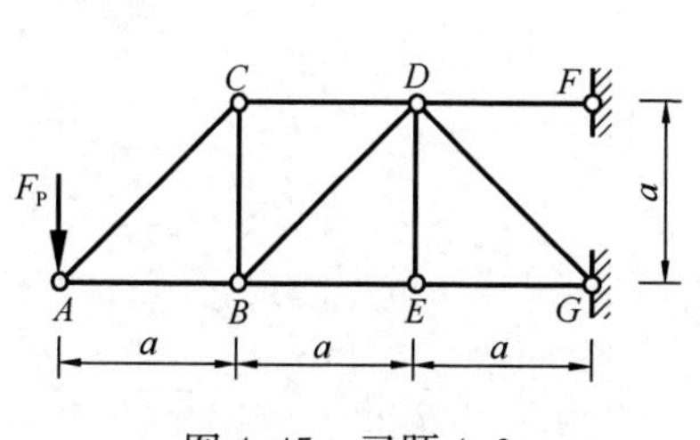

图 4.45 习题 4.6

Figure 4.45 Exercise 4.6

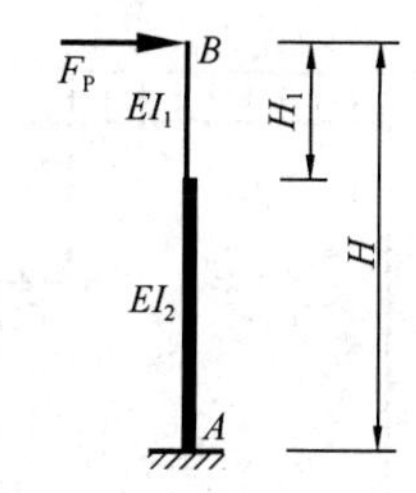

图 4.46 习题 4.8

Figure 4.46 Exercise 4.8

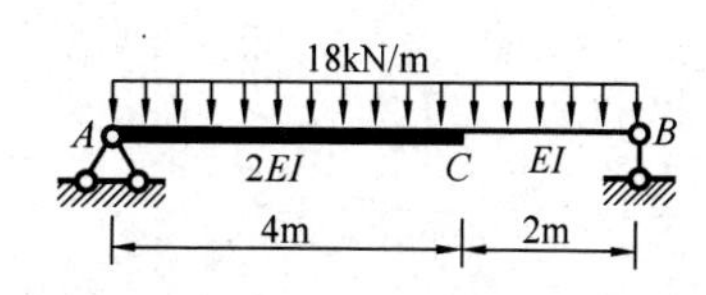

图 4.47 习题 4.9

Figure 4.47 Exercise 4.9

4.10 求图 4.48 所示伸臂梁 C 点的竖向位移Δ_{CV}。已知 $EI=2.0\times10^8$kN·cm^2。

4.10 Calculate the vertical displacements Δ_{CV} of point C in the cantilever beams in Figure 4.48. $EI=2.0\times10^8$kN·cm^2.

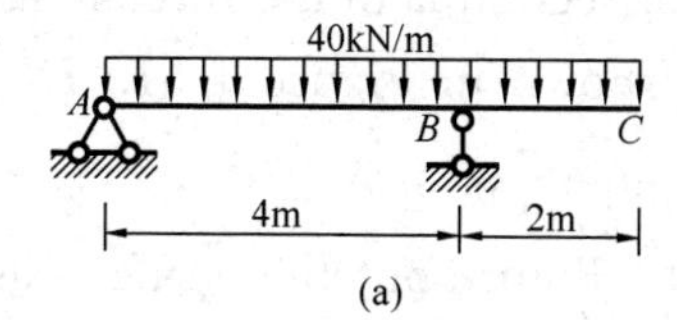

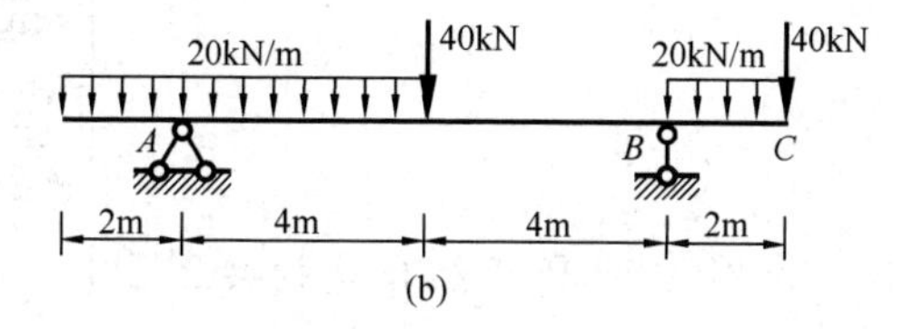

图 4.48 习题 4.10

Figure 4.48 Exercise 4.10

4.11 求图 4.49 所示简支刚架 B 截面的水平位移Δ_{BH}和转角θ_B。

4.11 Calculate the horizontal displacement Δ_{BH} and the rotation angle θ_B of the section B of the simply-supported rigid frame shown in Figure 4.49.

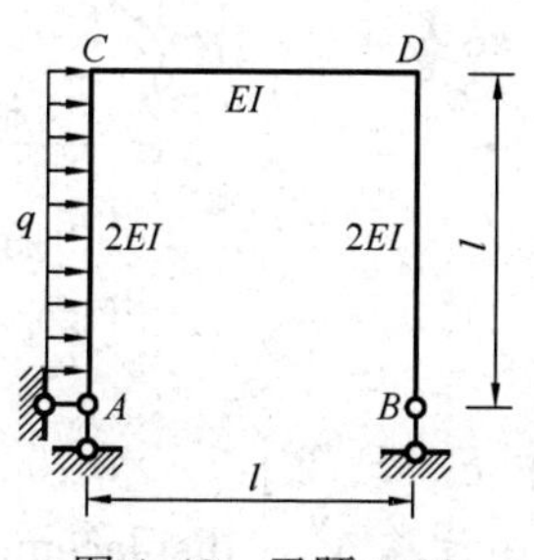

图 4.49 习题 4.11

Figure 4.49 Exercise 4.11

4.12 求图4.50所示悬臂刚架 B、C 两点的相对水平位移 Δ_{BC}^{H}。已知各杆惯性矩 $I=28600\text{cm}^4$，弹性模量 $E=2.06\times10^8\text{kPa}$。

4.12 Calculate the relative horizontal displacement Δ_{BC}^{H} of points B and C of the cantilever rigid frame shown in Figure 4.50. The moment of inertia of each rod $I=28600\text{cm}^4$, and the modulus of elasticity $E=2.06\times10^8\text{kPa}$.

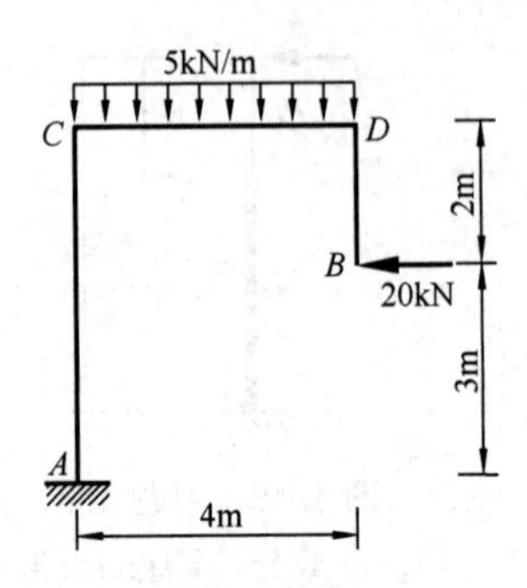

图4.50 习题4.12

Figure 4.50 Exercise 4.12

4.13 求图4.51所示三铰刚架点 E 的水平位移 Δ_{EH} 和截面 B 的转角 θ_B。$EI=$常数。

4.13 Calculate the horizontal displacement Δ_{EH} of point E and the rotation angle θ_B of section B in the three-hinged rigid frame shown in Figure 4.51. $EI=$ constant.

4.14 求图4.52所示组合结构梁上 D 点的竖向位移 Δ_{DV}。已知 $E=2.1\times10^8\text{kPa}$，$I=8600\text{cm}^4$，$A=25\text{cm}^2$。

4.14 Figure out the vertical displacement Δ_{DV} of point D on the composite structure beam in Figure 4.52. $E=2.1\times10^8$ kPa, $I=8600\text{cm}^4$, $A=25\text{cm}^2$.

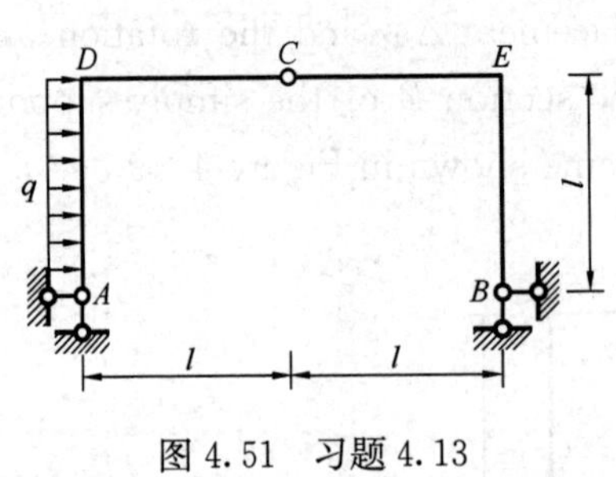

图4.51 习题4.13

Figure 4.51 Exercise 4.13

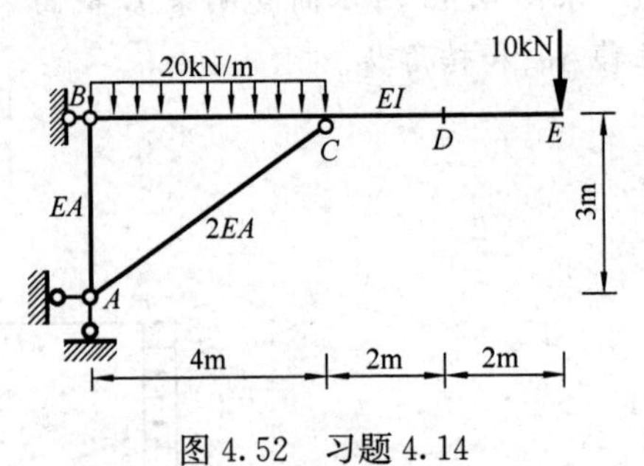

图4.52 习题4.14

Figure 4.52 Exercise 4.14

4.15 求图4.53所示刚架中 E、F 两截面的相对水平位移 Δ_{EF}^{H}、相对竖向位移 Δ_{EF}^{V} 及相对转角 θ_{EF}。E、F 是切口两侧的截面，

4.15 Calculate the relative horizontal displacement Δ_{EF}^{H}, relative vertical displacement Δ_{EF}^{V} and relative rotation angle θ_{EF} of

EI＝常数。

sections E and F of the rigid frame shown in Figure 4.53. E and F are the sections on both sides of the incision, EI=constant.

4.16 图 4.54 所示简支刚架支座 B 下沉 b，试求 D 点的竖向位移 Δ_{DV} 和水平位移 Δ_{DH}。

4.16 Figure 4.54 shows the settlement b of the support B in the simply-supported rigid frame, try to find the vertical displacement Δ_{DV} and horizontal displacement Δ_{DH} of point D.

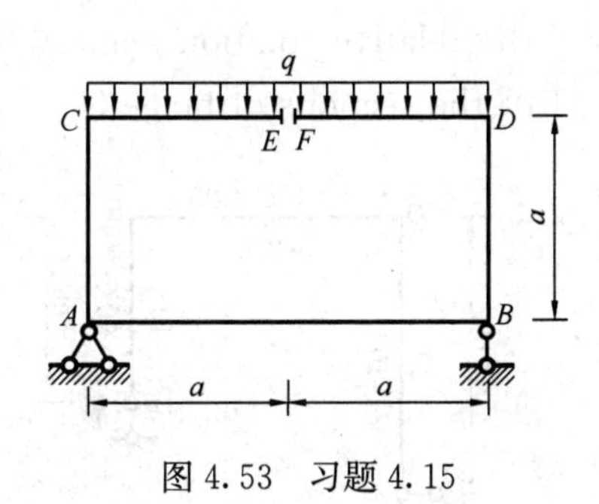

图 4.53 习题 4.15

Figure 4.53 Exercise 4.15

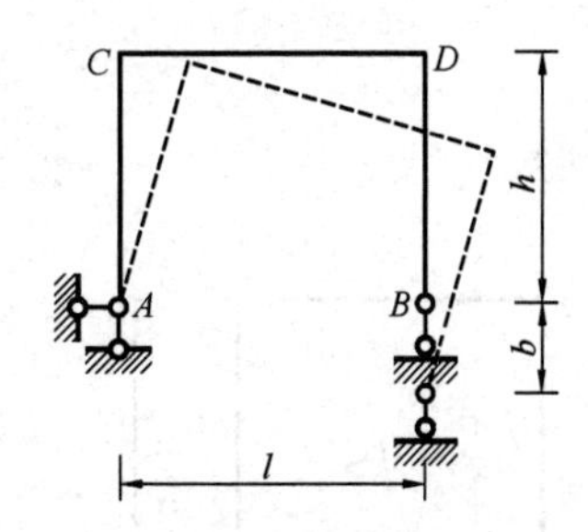

图 4.54 习题 4.16

Figure 4.54 Exercise 4.16

4.17 在图 4.55 所示刚架中，l=4m，支座 A 发生给定的位移 a=2cm、b=4cm、φ=0.01rad，试求 B 点的水平位移 Δ_{BH} 和铰 C 左右两侧截面的相对转角 θ_C。

4.17 In the rigid frame in Figure 4.55, l=4m, support A has displacements of a=2cm, b=4cm, φ=0.01rad, try to calculate horizontal displacement Δ_{BH} of point B and the relative rotation angle θ_C on both sides of the section C.

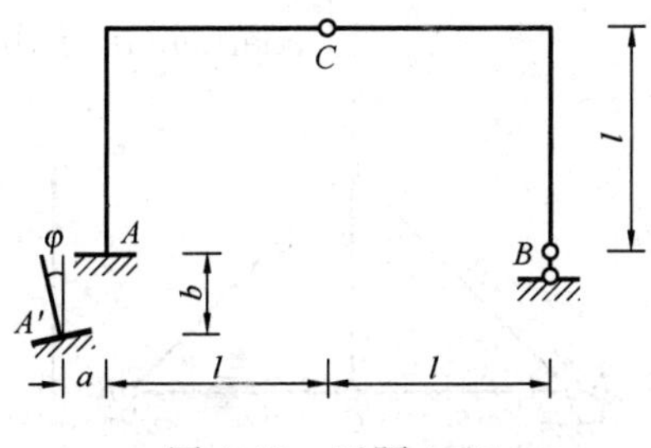

图 4.55 习题 4.17

Figure 4.55 Exercise 4.17

4.18 图 4.56 所示悬臂刚架内部温度升高 t℃，求 D 点的竖向位移 Δ_{DV}、水平位移 Δ_{DH} 和转角 θ_D。材料的线膨胀系数为 α，各杆截面均为矩形，且高度 h 相同。

4.18 Figure 4.56 shows that the internal temperature of the cantilever rigid frame increases by t℃, calculate the vertical displacement Δ_{DV}, horizontal displacement Δ_{DH} and rotation angle θ_D at point D. The linear expansion coefficient of the material is α, and the sections of each rod are rectangular with the same height h.

4.19　三铰刚架温度改变如图 4.57 所示，各杆均为矩形截面，截面高度相同，$h=40\text{cm}$，材料线膨胀系数 $\alpha=0.00001$。求 E 点的水平位移 Δ_{EH} 和铰 C 左右两侧截面的相对转角 θ_C。

4.19　Figure 4.57 shows the temperature changes in three-hinged frame, each rod has a rectangular section with the same height, $h=40\text{cm}$, the material linear expansion coefficient is $\alpha=0.00001$. Find out the horizontal displacement Δ_{EH} of point E and the relative rotation angle θ_C on both sides of the sections of hinge C.

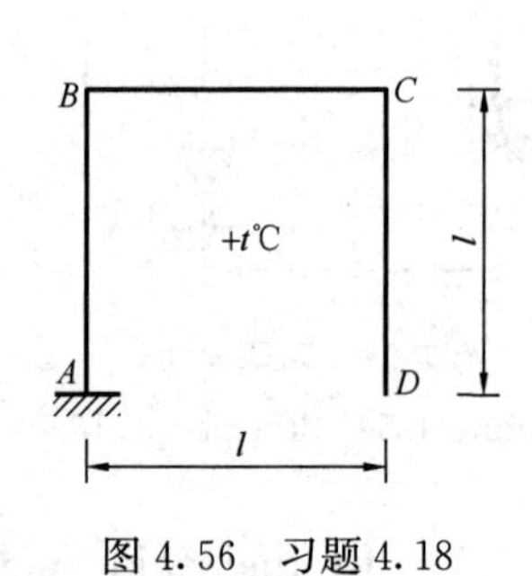

图 4.56　习题 4.18

Figure 4.56　Exercise 4.18

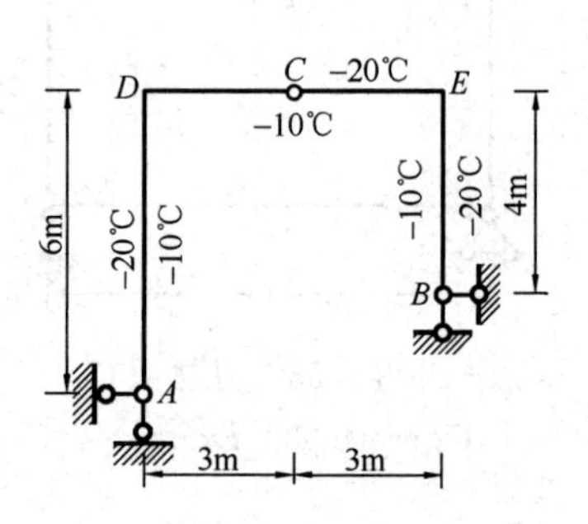

图 4.57　习题 4.19

Figure 4.57　Exercise 4.19

4.20　在图 4.58 所示桁架中，若只有 AD 杆的温度上升 t℃，试求节点 C 的竖向位移 Δ_{CV}。已知材料的线膨胀系数为 α。

4.20　In the truss shown in Figure 4.58, only the temperature of rod AD rises by t℃, try to find out the vertical displacement Δ_{CV} of node C. The linear expansion coefficient of the material is α.

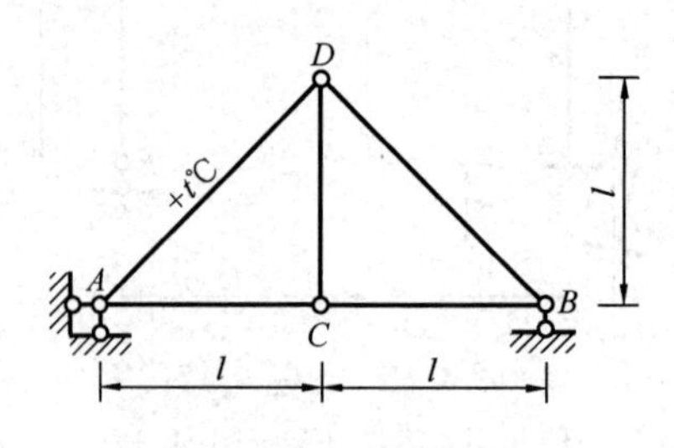

图 4.58　习题 4.20

Figure 4.58　Exercise 4.20

4.21　已知图 4.59（a）所示结构在支座 B 下沉 $c=1$ 时，D 点的竖向位移 $\Delta_{DV}=\frac{11}{16}$，试作图 4.59（b）所示结构的弯矩图。

4.21　In Figure 4.59(a), when the settlement of support B is $c=1$, the vertical displacement of point D is $\Delta_{DV}=\frac{11}{16}$, try to draw the bending moment diagram in Figure 4.59(b).

4.22 在图 4.60 所示桁架中，因制造误差，杆 AB 比原来设计长度短 4cm，试求因此引起的节点 C 的竖向位移 Δ_{CV}。

4.22 In the truss in Figure 4.60, because of manufacturing error, rod AB is shorter than the original design length of 4cm, try to calculate the vertical displacement Δ_{CV} of the node C.

（提示：应用公式 4.7，只有 AB 杆有轴向变形且已知。）

(Note: By applying Equation 4.7, only rod AB has axial deformation and it is known.)

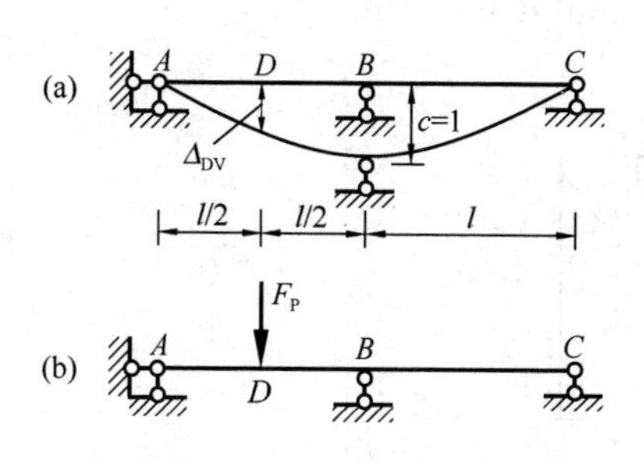

图 4.59 习题 4.21

Figure 4.59 Exercise 4.21

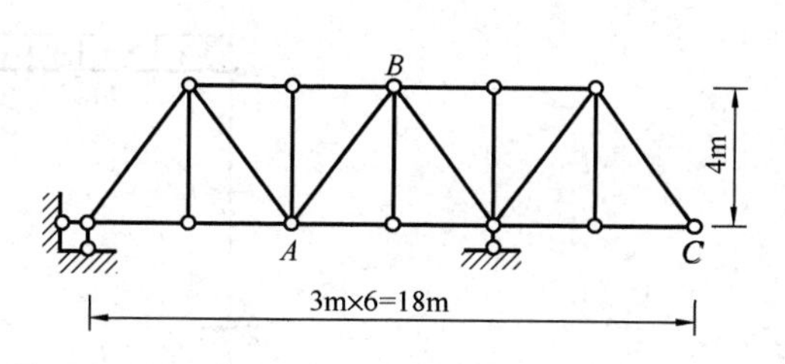

图 4.60 习题 4.22

Figure 4.60 Exercise 4.22

4.23 求图 4.61 所示弹性支承梁 D 点的竖向位移 Δ_{DV} 和截面 C 的转角 θ_C。设 EI = 常数。已知弹性支杆的刚度系数 $k=\dfrac{3EI}{l^3}$，弹性转动支座的刚度系数 $k_\theta=\dfrac{48EI}{l}$。

4.23 Find out the vertical displacement Δ_{DV} of point D and the rotation angle θ_C of the section C in the elastic support beam in Figure 4.61. EI = constant. The stiffness coefficient $k=\dfrac{3EI}{l^3}$ of the elastic strut and the stiffness coefficient $k_\theta=\dfrac{48EI}{l}$ of the elastic rotating support are known.

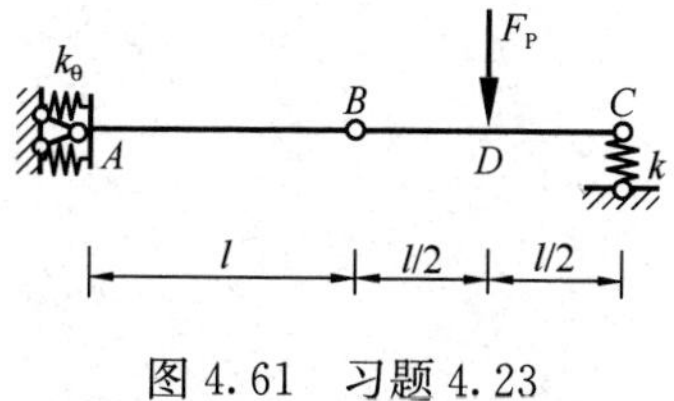

图 4.61 习题 4.23

Figure 4.61 Exercise 4.23

（提示：位移计算公式中增加弹性约束

(It is suggested that the virtual work term of elastic restraint force is added in the

力的虚功项：$\Sigma\frac{\bar{F}_R F_{RP}}{k}$。）

displacement calculation formula: $\Sigma\frac{\bar{F}_R F_{RP}}{k}$.）

4.24 求图4.62所示刚架C点的全位移Δ_C。已知$I_2:I_1=5:4$，E为常数。

（提示：可先分别求出Δ_{CH}与Δ_{CV}，再合成；或者沿任意方向α加一单位力，求Δ_C在此方向上的投影，其极大值即为全位移。）

4.24 Find out the total displacement Δ_C of point C in the rigid frame in Figure 4.62. $I_2:I_1=5:4$, and E is a constant.

(It is indicated that Δ_{CH} and Δ_{CV} could be calculated separately first and then synthesized; Or add a unit force along any direction α, and calculate the projection of Δ_C in this direction, and the maximum value is the total displacement.)

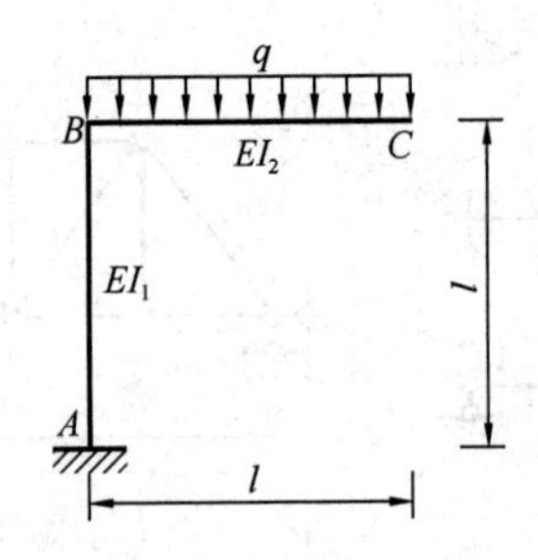

图4.62 习题4.24

Figure 4.62 Exercise 4.24

第5章 Chapter 5

力法 Force Method

要点

➢ 超静定结构的组成和超静定次数
➢ 力法的基本概念
➢ 超静定刚架和排架
➢ 超静定桁架和组合结构
➢ 利用对称性求解对称超静定结构
➢ 非荷载作用下的超静定结构计算
➢ 超静定结构的位移计算
➢ 超静定结构计算的校核
➢ 超静定结构的特性

Keys

➢ Composition of statically indeterminate structure and degree of indeterminacy
➢ The basic concept of force method
➢ Statically indeterminate rigid frame and bent frame
➢ Statically indeterminate truss and composite structure
➢ Using symmetry to solve symmetric statically indeterminate structure
➢ Calculation of statically indeterminate structure under non-load actions
➢ Displacement calculation of statically indeterminate structure
➢ Checking of statically indeterminate structure calculation
➢ The properties of statically indeterminate structure

在结构的静力问题中，内力和位移必须满足平衡条件和变形连续条件（物理条件，几何条件）。对于静定结构来说，平衡条件是求解结构反力和内力的唯一条件，而超静定结构必须同时满足以上两个条件，问题才能获得全部解答。两者的根本区别在于超静定结构有多余约束的存在。正因如此，超静定结构抵抗突然破坏的防护能力，内力状态以及结构的刚度，稳定性等都要比静定结构有所改善和提高，所以超静定结构在实际的

In the static problem of a structure, the internal force and displacement must satisfy the equilibrium condition and the deformation continuity condition(physical condition, geometric condition). For a statically determinate structure, the equilibrium condition is the only condition for solving the structural reaction and internal forces, while a statically indeterminate structure must satisfy both of the two conditions for

土木工程中有着更加广泛的应用。

the problem to be fully solved. The fundamental difference between them is the existence of redundant constraints in statically indeterminate structure. Because of that, the protection ability of statically indeterminate structure against sudden damage, the internal force state, and the stiffness and stability is improved and enhanced compared with the statically determinate structure, so the statically indeterminate structure has a wider application in actual civil engineering.

力法的思路就是以多余约束为突破口，将其用未知力代替，然后借助变形连续条件进行求解，本章内容将基于此展开。

The idea of the force method is to take the redundant constraint as a breakthrough, replace it with an unknown force, and then solve it with the help of the deformation continuity condition, on which the contents of this chapter will be based.

5.1　超静定结构的组成和超静定次数

5.1 Composition of Statically Indeterminate Structure and Degree of Indeterminacy

5.1.1　超静定结构的基本概念

5.1.1 Basic Concepts of Statically Indeterminate Structure

如果一个结构的支座反力和各截面内力都可以由静力平衡条件唯一确定，该结构称为**静定结构**。而单靠静力平衡条件不能确定全部反力和内力的结构，便称为**超静定结构**。例如图 5.1（a）所示的连续梁，它的水平反力虽可由静力平衡条件求出，但其竖向反力只凭静力平衡条件就无法确定，因此也就不能进一步求出其全部内力。又如图 5.1（b）所示的加劲梁，虽然它的反力可由静力平衡条件求得，但却不能确定杆件的内力。因此，这两个结构都是超静定结构。

If the support reaction force and the internal force of each section of a structure can be uniquely determined by the static equilibrium conditions, the structure is called a **statically determinate structure.** The structure that the static equilibrium conditions alone cannot determine all the reaction forces and internal forces is called a **statically indeterminate structure.** For example, the continuous beam shown in Figure 5.1(a), its horizontal reaction force can be found by the static equilibrium conditions, but its vertical reaction force cannot be determined by the static equilibrium conditions alone, so not all internal forces can be further determined. Another example is the stiffened beam shown in Figure 5.1(b), although the static equilibrium condition can find its reaction force, it cannot determine the internal force of the rod. Therefore, both of these structures are statically indeterminate structures.

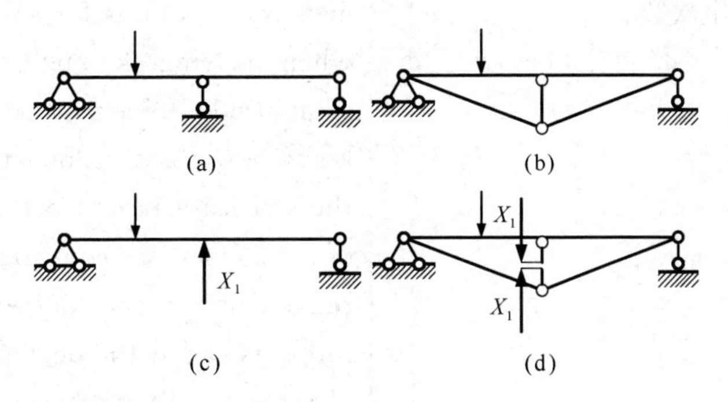

图 5.1 超静定结构

Figure 5.1 Statically indeterminate structure

分析以上两个结构的几何组成，可知它们都具有多余约束。多余约束中产生的力称为**多余未知力。**图 5.1（c）所示的连续梁中，其多余未知力 X_1就是该支座的支座反力 X_1。图 5.1（d）所示的加劲梁，可认为其多余未知力为该杆的轴力 X_1。超静定结构在去掉多余约束后，就变成静定结构。

Analyzing the geometric composition of the above two structures, it is known that they both have redundant constraints. The force generated in the redundant restraint is called the **redundant unknown force.** In the continuous beam shown in Figure 5.1(c), the redundant unknown force X_1 is the support reaction force X_1 of the support. In the stiffened beam shown in Figure 5.1(d), the redundant unknown force can be considered as the axial force X_1 of the rod. Statically indeterminate structure becomes a statically determinate structure after the redundant restraints are removed.

5.1.2 超静定次数的确定

在超静定结构中，由于具有多余未知力，平衡方程的数目少于未知数的数目，所以仅靠平衡方程无法求解全部未知力，必须靠位移条件建立补充方程来求解。因此用力法求解超静定结构时，必须先了解该结构有多少个多余未知力，以此来确定补充方程的数目。多余联系或多余未知力的数目，称为超静定结构的超静定次数。

确定超静定次数最直接的方法是解除多余约束，使原来的超静定结构成为静定结构，去除的约束数就是该结构的超静定次数。

从超静定结构上解除多余约束，使结构变为静定结构的方式通常有以下几种：

（1）去掉支座的一根支杆或切断一根链杆相当于去掉一个约束（图 5.2a）。

（2）去掉一个铰支座或一个简单铰相当于去掉两个约束（图 5.2b）。

（3）去掉一个固定支座或将刚性连接切

5.1.2 Determination of the Degree of Indeterminacy

In a statically indeterminate structure, the number of equilibrium equations is less than the number of unknowns because of the redundant unknown forces, so the equilibrium equations alone cannot solve for all the unknown forces, and the displacement conditions must be used to establish supplementary equations for solution. Therefore, when solving a statically indeterminate structure by force method, it is necessary to know how many redundant unknown forces the structure has to determine the number of supplementary equations. The number of redundant contacts or redundant unknown forces is called the degree of indeterminacy of the statically indeterminate structure.

The most direct way to determine the degree of indeterminacy is to remove the redundant contacts so that the original statically indeterminate structure becomes a statically determinate structure, the number of constraints removed is the degree of indeterminacy.

There are several ways as follows of releasing redundant links from statically indeterminate structure and turning the structure into a statically determinate one:

(1) Removing a strut from support or cutting a chain link is equivalent to removing a link (Figure 5.2a).

(2) Removing a hinge support or a simple hinge is equivalent to removing two links (Figure 5.2b).

(3) Removing a fixed support or cut-

断相当于去掉三个约束（图 5.2c）。

ting off a rigid link is equivalent to removing three links (Figure 5.2c).

（4）将固定支座改为铰支座或将刚性连接改为铰连接相当于去掉一个约束（图 5.2d）。

(4) Changing a fixed support to a hinge support or changing a rigid joint to a hinge joint is equivalent to removing one link (Figure 5.2d).

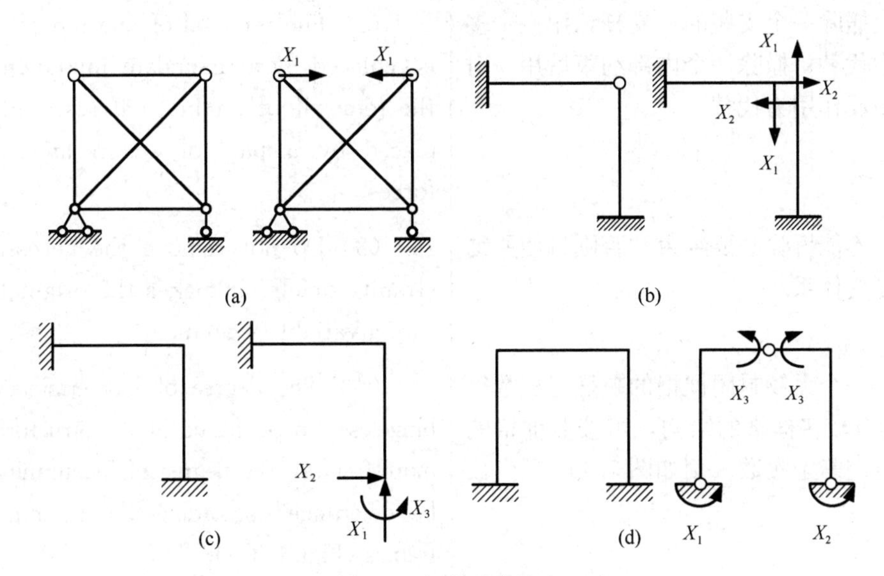

图 5.2 解除多余约束

Figure 5.2 Removal of redundant constraints

此外，也可用第 2 章计算自由度的方法来推算多余约束：因为结构自由度 $S=0$，而多余约束数 n=结构自由度 S—计算自由度 W，故 $n=-W$，如图 5.3 所示。

In addition, the method of calculating degrees of freedom in Chapter 2 can also be used to derive the redundant constraints: Because the structural degrees of freedom $S=0$ and the number of redundant constraints n=structural degrees of freedom S-computational degrees of freedom W, so $n=-W$, as shown in Figure 5.3.

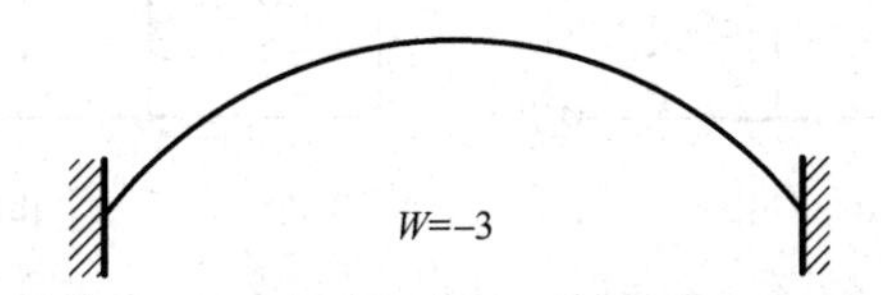

图 5.3 通过计算自由度来确定超静定次数

Figure 5.3 Determination of degree of indeterminacy by calculating degrees of freedom

在解除多余约束时，应该注意：

When removing redundant links, it should be noted that:

（1）同一个结构可用不同的方式撤除多余约束，但其超静定数目不变。

(1) One structure can be withdrawn from redundant restraints in different ways, but its degree of indeterminacy remains unchanged.

（2）撤除一个支座的一支杆，用一个多余未知力代替；撤除一个内部约束后用一对作用力与反作用力代替。

(2) The removal of one rod of support is replaced by a redundant unknown force; the removal of an internal restraint is replaced by a pair of action and reaction forces.

（3）不能拆除必要约束，否则将使原结构变为可变体系。

(3) Do not remove the necessary restraint, or else it makes the original structure a variable system.

（4）一个无铰封闭框格的超静定次数等于3。具有较多框格的结构，可按照框格的数目确定超静定次数（例如图5.4）。

(4) The degree of indeterminacy for a hingeless closed frame is 3. Structures with more frames, the degree of indeterminacy can be determined according to the number of frames (Figure 5.4).

（5）地基应视作一个开口刚片。

(5) The foundation should be treated as a rigid open body.

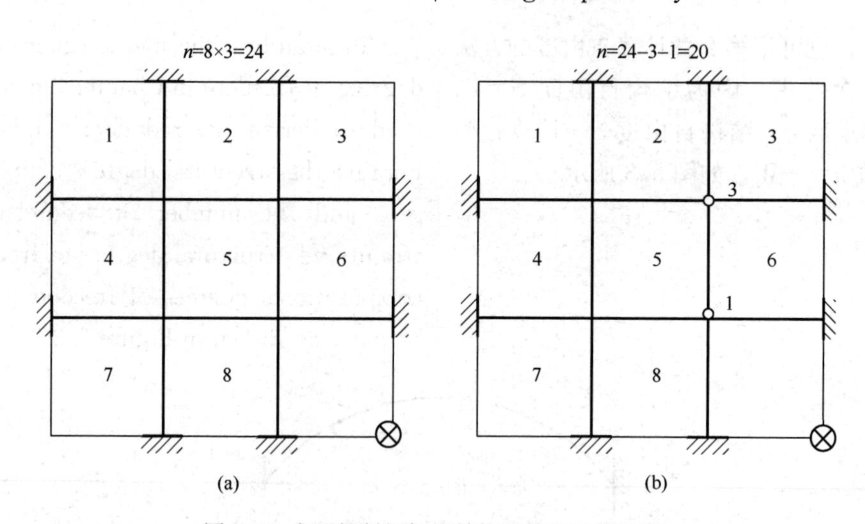

图5.4　含无铰封闭框格结构超静定次数计算

Figure 5.4　Calculation of degree of indeterminacy for structures containing hingeless closed frames

5.2 力法的基本概念

5.2.1 力法的基本思路

本节采用一个简单算例来说明力法的基本思路，图 5.5（b）是含有一个多余约束的超静定结构，若以多余未知力代替多余约束（图 5.5c），则可得到缺少多余约束的简单结构，称之为**基本结构**，如图 5.5（d）所示（注意基本结构必须几何不变），基本结构在原荷载和多余未知力作用下的体系称为**基本体系**，如图 5.5（c）所示。

5.2 The Basic Concepts of Force Method

5.2.1 The Conceptual Framework of Force Method

This section uses a simple example to illustrate the basic idea of the force method. Figure 5.5(b) is a statically indeterminate structure containing one redundant constraint, and if the redundant unknown force is used instead of the redundant constraint (Figure 5.5c), a simple structure out of the redundant constraint can be obtained, which is called the **basic structure**, as shown in Figure 5.5(d) (Note that the basic structure must be geometrically invariant), and the system of the basic structure under the original load and the redundant unknown force is called the **basic system**, as shown in Figure 5.5(c).

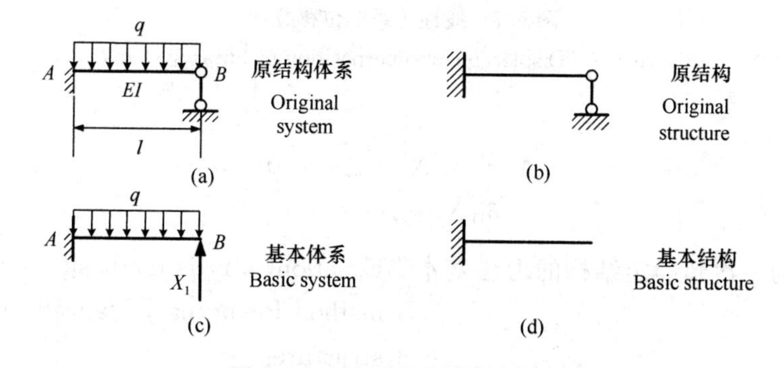

图 5.5 力法原理中的基本概念

Figure 5.5 The basic concepts of force method

由于未知力数量（4 个）大于平衡方程总数（3 个），仅凭静力平衡条件无法求解 X_1，因此引入位移条件：B 支座竖向位移为零，如图 5.6 所示（所有符号第一个下标为位置，第二个下标为原因，如 Δ_{1P} 表示荷载 P 引起的第 1 个方向的位移）。

Since the number of unknown forces (4) is greater than the number of equilibrium equations (3), X_1 cannot be solved only by static equilibrium conditions, so the displacement condition is introduced: The vertical displacement of support B is zero, as shown in Figure 5.6 (The first subscript of all symbols is the position, and the second subscript is the cause. For example, Δ_{1P} indicates the displacement in the first direction caused by load P).

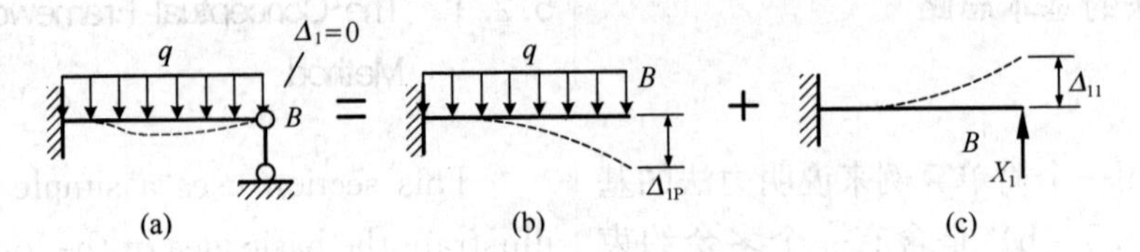

图 5.6 变形条件（或位移条件）

Figure 5.6 Deformation condition (or displacement condition)

$$\Delta_1 = 0$$
$$\Delta_1 = \Delta_{11} + \Delta_{1P} = 0$$

又因为线性体系的位移可以叠加，如图 5.7所示，有：

Because the displacement of the linear system can be superimposed, as shown in Figure 5.7, then:

图 5.7 线性体系的位移叠加

Figure 5.7 Displacement superposition of linear systems

$$\Delta_1 = \delta_{11} X_1 + \Delta_{1P} = 0$$
$$\delta_{11} X_1 + \Delta_{1P} = 0 \tag{5.1}$$

式（5.1）即为一次超静定结构的力法基本方程。

Equation(5.1) is the basic equation of force method for primary statically indeterminate structures.

$$\delta_{11} = \frac{l^3}{3EI}$$
$$\Delta_{1P} = -\frac{q l^4}{8EI}$$
$$\therefore \quad X_1 = \frac{3ql}{8}$$

由于求解结果为正，所以方向和图 5.8一致，即向上，至此未知力解出，最终弯矩

Since the solution result is positive, the direction is consistent with that in Figure

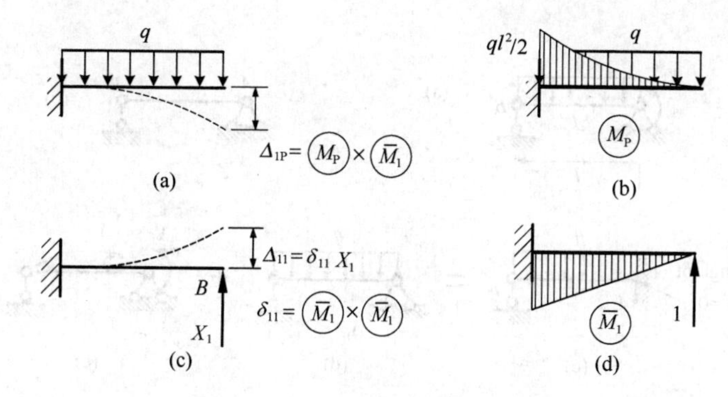

图 5.8 力法方程中系数的求解

Figure 5.8 Solution of the coefficients in the force method equation

图可以叠加而得，如图 5.9 所示。剪力图可快速由弯矩图得到，如图 5.10 所示。

5.8, i.e. upward. So far, the unknown force is solved, and the final bending moment diagram can be obtained by superposition, as shown in Figure 5.9. The shear force diagram can be obtained by the bending moment diagram quickly, as shown in Figure 5.10.

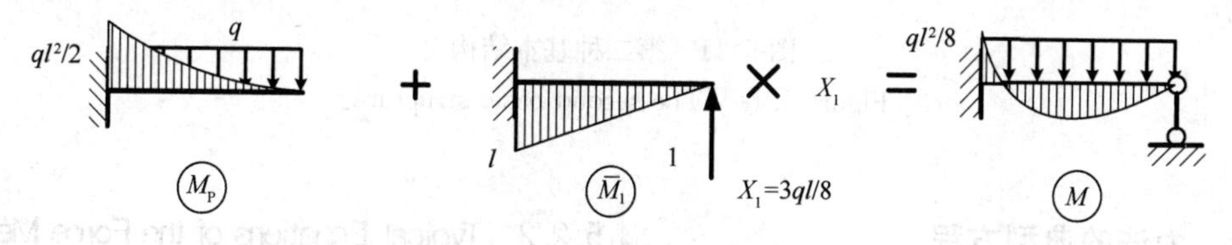

图 5.9 总弯矩图的叠加

Figure 5.9 Superposition of bending moment diagrams

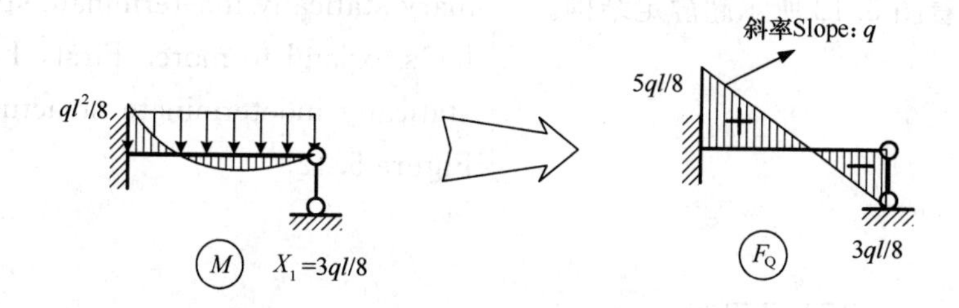

图 5.10 由弯矩图可快速得到剪力图

Figure 5.10 The shear force diagram can be obtained by the bending moment diagram quickly

值得注意的是，基本结构的选取可以为无穷多种，但计算结果均一致。下面进行举例说明，仍以图 5.5（a）为对象，如图 5.11所示。

It is worth noting that the selection of basic structures can be infinite, but the calculation results are consistent. The following is an example that takes Figure 5.5(a) as the object, which is shown in Figure 5.11.

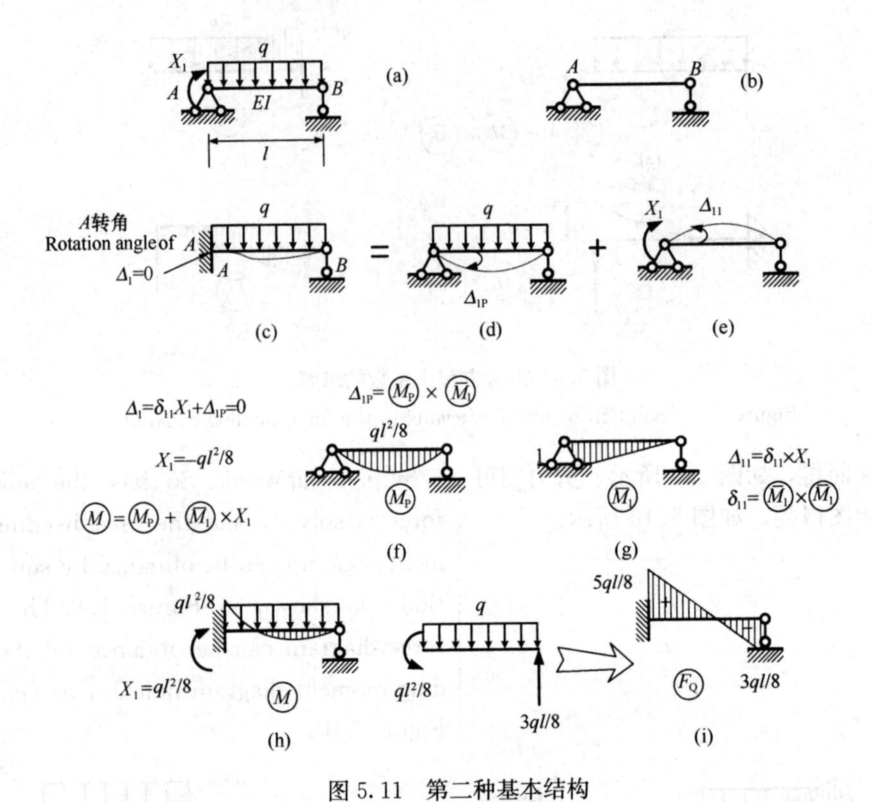

图 5.11　第二种基本结构

Figure 5.11　The second basic structure

5.2.2　力法的典型方程

以上讨论的是一次超静定结构，下面进行扩展，首先看图 5.12 所示超静定结构。

5.2.2　Typical Equations of the Force Method

The above discussion is about the primary statically indeterminate structure, now let's expand to more. First, Let's see the statically indeterminate structure shown in Figure 5.12.

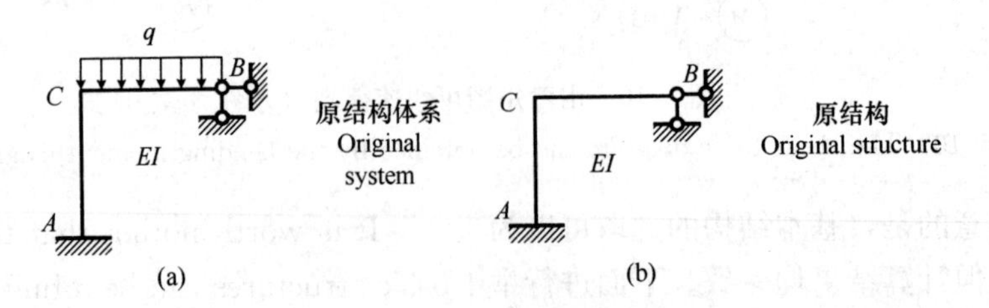

图 5.12　力法求解二次超静定结构(一)

Figure 5.12　Force method for solving quadratic statically indeterminate structure(One)

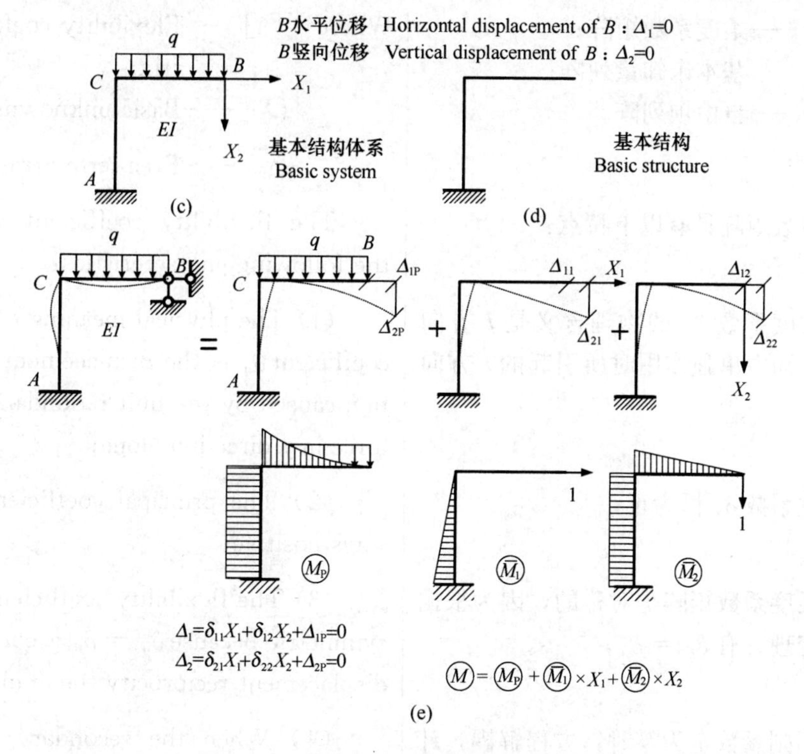

图 5.12 力法求解二次超静定结构(二)

Figure 5.12 Force method for solving quadratic statically indeterminate structure(Two)

继续拓展，可得 n 次超静定结构的力法基本方程：

The basic equations of force method for n-order statically indeterminate structures can be obtained by further expansion:

式 (5.2) 即为力法基本方程，可写成矩阵形式：$[\delta]\{X\}+\{\Delta_P\}=\{\Delta\}$，如式 (5.3)所示：

Equation(5.2) is the basic equation of force method, which can be written in matrix form: $[\delta]\{X\}+\{\Delta_P\}=\{\Delta\}$, as shown in Equation (5.3):

$$\left.\begin{array}{c}\delta_{11}X_1+\delta_{12}X_2+\cdots+\delta_{1n}X_n+\Delta_{1P}=\Delta_1\\ \delta_{21}X_1+\delta_{22}X_2+\cdots+\delta_{2n}X_n+\Delta_{2P}=\Delta_2\\ \cdots\cdots\\ \delta_{n1}X_1+\delta_{n2}X_2+\cdots+\delta_{nn}X_n+\Delta_{nP}=\Delta_n\end{array}\right\} \tag{5.2}$$

$$\begin{bmatrix}\delta_{11} & \delta_{12} & \cdots & \delta_{1i} & \delta_{1j} & \cdots & \delta_{1n}\\ \delta_{21} & \delta_{22} & \cdots & \delta_{2i} & \delta_{2j} & \cdots & \delta_{2n}\\ \vdots & & \ddots & & \vdots & & \vdots\\ \delta_{i1} & \delta_{i2} & \cdots & \delta_{ii} & \delta_{ij} & \cdots & \delta_{in}\\ \delta_{j1} & \delta_{j2} & \cdots & \delta_{ji} & \delta_{jj} & \cdots & \delta_{jn}\\ \vdots & & \ddots & & \vdots & & \vdots\\ \delta_{n1} & \delta_{n2} & \cdots & \delta_{ni} & \delta_{nj} & \cdots & \delta_{nn}\end{bmatrix}\begin{bmatrix}X_1\\X_2\\\vdots\\X_i\\X_j\\\vdots\\X_n\end{bmatrix}+\begin{bmatrix}\Delta_{1P}\\\Delta_{2P}\\\vdots\\\Delta_{iP}\\\Delta_{jP}\\\vdots\\\Delta_{nP}\end{bmatrix}=\begin{bmatrix}\Delta_1\\\Delta_2\\\vdots\\\Delta_i\\\Delta_j\\\vdots\\\Delta_n\end{bmatrix} \tag{5.3}$$

式中 $[\delta]$——柔度系数矩阵；
$\{X\}$——基本未知量列阵；
$\{\Delta_P\}$——自由项列阵。

Where $[\delta]$——Flexibility coefficient matrix;
$\{X\}$——Basic unknowns array;
$\{\Delta_P\}$——Free term array.

柔度系数矩阵具有以下特点：

The flexibility coefficient matrix has the following characteristics:

(1) 柔度系数 δ_{ij} 的物理含义是 j 方向单位多余未知力单独作用时所引起的 i 方向上的位移；

(1) The physical meaning of flexibility coefficient δ_{ij} is the displacement in i direction caused by the unit redundant unknown force in j direction alone;

(2) 主系数 δ_{ii} 恒为正；

(2) The principal coefficient δ_{ii} is always positive;

(3) 柔度系数矩阵是对称的，因为根据位移互等定理，有 $\delta_{ij}=\delta_{ji}$；

(3) The flexibility coefficient matrix is symmetric because $\delta_{ij}=\delta_{ji}$ according to the displacement reciprocity theorem;

(4) 当副系数全为零时，方程解耦，计算量小。

(4) When the secondary coefficients are all zero, the equation is decoupled and the amount of calculation is small.

力法典型方程是表示位移条件，故又称为**柔度方程**，力法又被称为**柔度法**。力法解题步骤可总结如下：

The typical equation of the force method is to represent the displacement condition, which is also known as the **flexibility equation**, and the force method is also known as the **flexibility method**. The steps of the force method solution can be summarized as follows:

(1) 确定超静定次数 n，选取合适的基本体系；

(1) Determine the number of superstatic times n and select a suitable basic system;

(2) 根据变形协调条件建立力法基本方程；

(2) Establish the basic equations of the force method according to the deformation coordination conditions;

(3) 计算柔度系数和自由项；

(3) Calculate flexibility coefficients and free terms;

(4) 解方程，得到多余未知力；

(4) Solve the equations to obtain the redundant unknown forces;

（5）利用叠加法绘制内力图。

(5) Plot the internal forces diagram by superposition method.

5.3 超静定梁、刚架

5.3 Statically Indeterminate Beam and Rigid Frame

5.3.1 超静定梁

5.3.1 Statically Indeterminate Beam

【例 5.1】用力法求解图 5.13（a）所示超静定梁弯矩图，已知 EI 为常数，$l=4$m。

[**Example 5.1**] Use force method to get the bending moment diagram of statically indeterminate beam shown in Figure 5.13 (a), EI is known to be constant and $l=4$m.

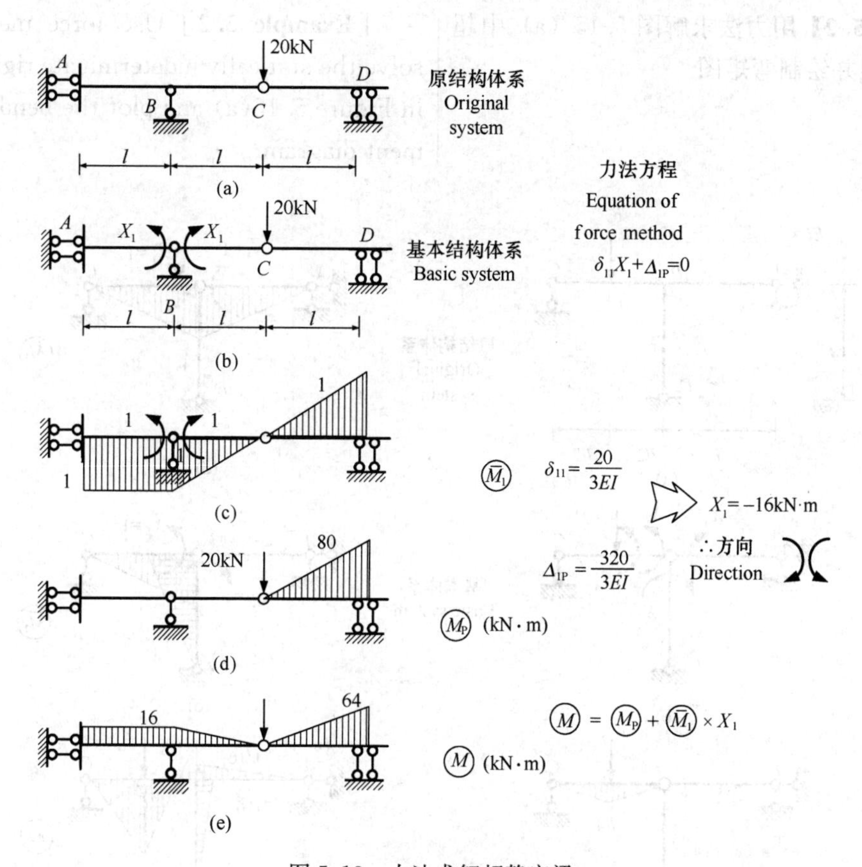

图 5.13 力法求解超静定梁

Figure 5.13 Force method for solving statically indeterminate beam

需要注意的是，虽然初学者更容易凭直观选择图 5.14 作为基本体系，也同样能求解正确答案，但仍然建议选择支座铰化的方

It should be noted that although it is easier for beginners to intuitively choose Figure 5.14 as the basic system and get the

式来构造基本体系，这样会简化多跨超静定梁的求解过程。

correct answer as well, it is still recommended to choose the hinging of the support to establish the basic system, which will simplify the process of solving the statically indeterminate multi-span beam.

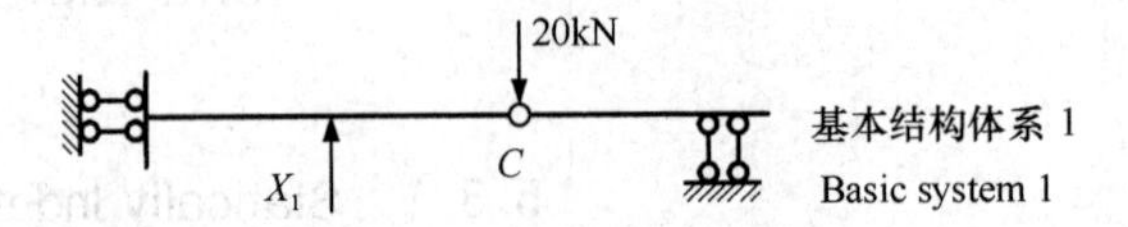

图 5.14 另外一种基本体系

Figure 5.14 Another basic system

5.3.2 超静定刚架

5.3.2 Statically Indeterminate Rigid Frame

【例 5.2】 用力法求解图 5.15（a）中超静定刚架并绘制弯矩图。

[**Example 5.2**] Use force method to solve the statically indeterminate rigid frame in Figure 5.15(a) and plot the bending moment diagram.

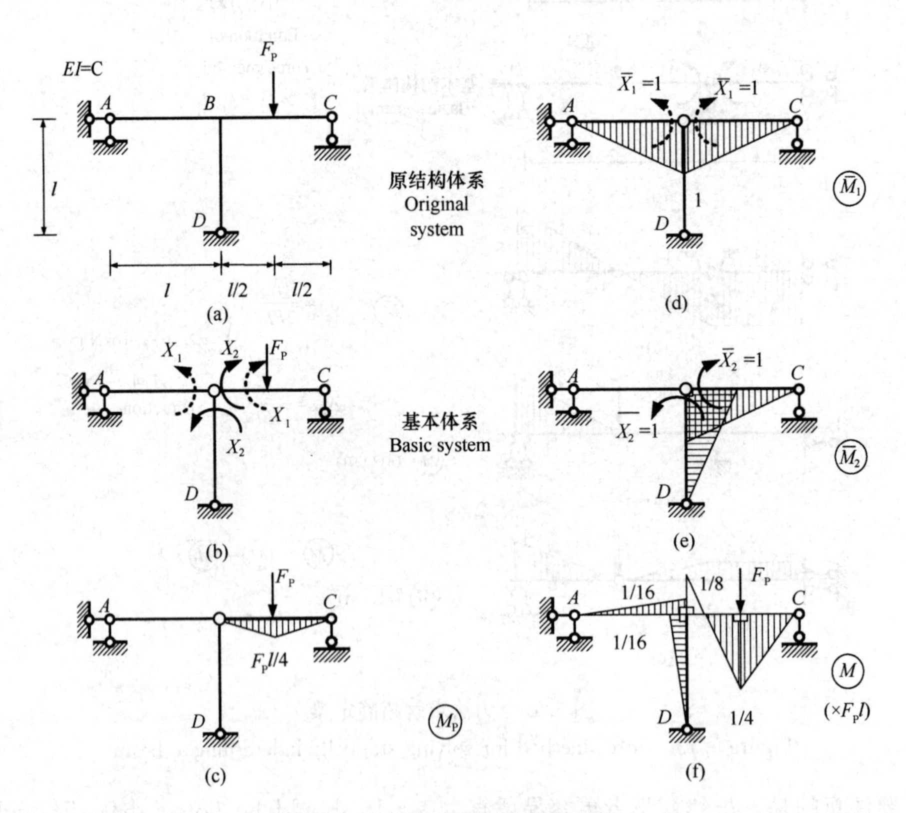

图 5.15 用力法求解超静定刚架

Figure 5.15 Solving statically indeterminate rigid frame by force method

【解】(1) 取基本体系如图 5.15 (b) 所示，此处应注意 B 节点是复刚节点，变成复铰节点需要提供 $n-1$ 对多余未知弯矩 (n 为复刚节点所连杆件数量)。

[**Solution**] (1) Take the basic system as shown in Figure 5.15(b). It should be noted that node B is a compound rigid node, to be a compound hinge node, it needs to provide $n-1$ pairs of excess unknown bending moments (n is the number of rods attached to the compound rigid node).

(2) 建立力法基本方程：$[\delta]\{X\}+\{\Delta_P\}=\{0\}$，原来 B 为刚节点，无相对转动，故等式右边为零。

(2) Establish the basic equations of the force method: $[\delta]\{X\}+\{\Delta_P\}=\{0\}$, the original node B is a rigid node with no relative rotation, so the right-hand side of the equation is zero.

(3) 计算系数和自由项：用图乘可得：

(3) Calculate coefficients and free terms: Use graph multiplication:

$$\delta_{11}=\frac{2l}{3EI},\delta_{22}=\frac{2l}{3EI},\delta_{12}=\delta_{21}=\frac{l}{3EI},\Delta_{1P}=\frac{Fl^2}{16EI},\Delta_{2P}=\frac{Fl^2}{16EI}$$

(4) 解方程可得：

(4) Solve the equation:

$$X_1=X_2=-\frac{Fl}{16EI}$$

(5) 叠加法得到最终弯矩图，如图 5.15 (f) 所示。

(5) The final bending moment diagram is obtained by the superposition method, as shown in Figure 5.15(f).

5.4 超静定桁架、超静定排架和超静定组合结构

5.4 Statically Indeterminate Truss, Statically Indeterminate Bent Frame and Statically Indeterminate Composite Structure

5.4.1 超静定桁架

5.4.1 Statically Indeterminate Truss

5.4.1.1 拆除多余链杆

5.4.1.1 Removal of Redundant Links

超静定桁架的计算通常是“拆”或者“切”多余链杆，两者物理含义不同，但最终的力法方程回归一致，计算结果也相同，下面举例说明。

The statically indeterminate truss is usually calculated by “dismantling” or “cutting” the redundant links. The physical meanings of them are different, but the final force method equation regresses consistent, and the calculation results are the same as well. The following examples are given.

对于图5.16所示超静定桁架，观察可得为1次超静定，不妨以CD为多余链杆。如果拆去，以多余未知力X_1代替，则变形协调条件为CD相对靠近的值，此处要注意$\Delta_1=-X_1/(EA/l)$，负号表示X_1作用于CD杆，是拉伸效果，与Δ_1方向相反。因此，力法方程为：

For the statically indeterminate truss in Figure 5.16, it can be observed that it is statically indeterminate for one time. Link CD may be used as the redundant link. If it is removed and replaced by the redundant unknown force X_1, the deformation coordination condition is the value that C and D get close. Here, pay attention to $\Delta_1=-X_1/(EA/l)$. A negative sign indicates that X_1 acts on the rod CD, which is the stretching effect with the opposite direction of Δ_1. Therefore, the force method equation is:

$$\delta_{11}X_1+\Delta_{1P}=-\frac{X_1}{EA}l$$

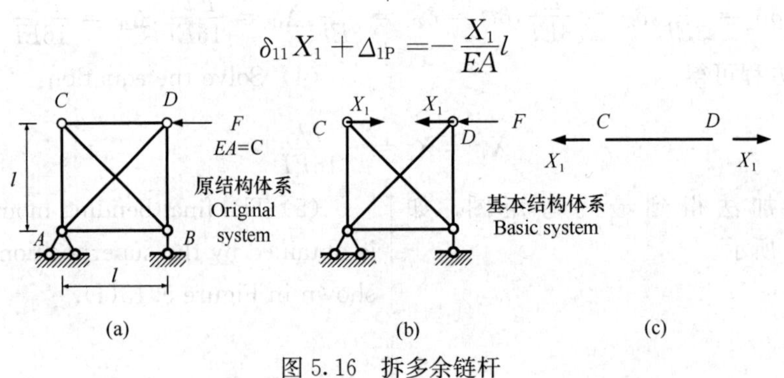

图5.16　拆多余链杆

Figure 5.16　Dismantling redundant links

依次求出基本体系在F、X_1作用下所有杆件的轴力，如图5.17所示。

Calculate the axial forces successively of all members of the basic system under the actions of F and X_1, as shown in Figure 5.17.

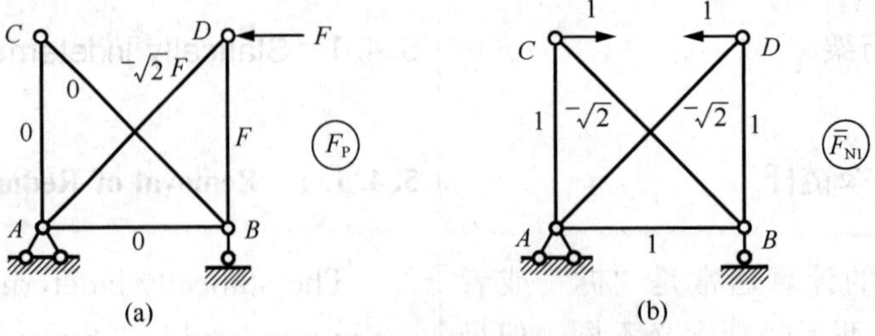

图5.17　拆多余链杆后的力法求解

(a) F_P; (b) $\overline{F}_{N1}$

Figure 5.17　The force method solution after dismantling redundant links

(a) F_P; (b) $\overline{F}_{N1}$

通过表 5.1 计算系数和自由项。

Calculate coefficients and free terms by Table 5.1.

超静定桁架柔度系数、自由项计算表 **表 5.1**

Calculation of flexibility coefficients and free terms of statically indeterminate truss

Table 5.1

杆件 Rod	l	EA	$\overline{F}_{N1}$	F_P	$\frac{\overline{F}_{N1}^2}{EA}l$	$\frac{\overline{F}_{N1}F_Pl}{EA}$	$F_N=\overline{F}_{N1}X_1+F_P$
AB	1	1	1	0	1	0	$-0.396F$
AC	1	1	1	0	1	0	$-0.396F$
AD	$\sqrt{2}$	1	$-\sqrt{2}$	$-\sqrt{2}F$	$2\sqrt{2}$	$2\sqrt{2}F$	$0.854F$
BC	$\sqrt{2}$	1	$-\sqrt{2}$	0	$2\sqrt{2}$	0	$0.56F$
BD	1	1	1	F	1	F	$0.604F$
CD	1	1					$-0.396F$
Σ					$3+4\sqrt{2}$	$(2\sqrt{2}+1)F$	

$$\delta_{11}=(3+4\sqrt{2})\frac{l}{EA},\ \Delta_{1P}=(2\sqrt{2}+1)F\frac{l}{EA},$$

$$X_1=-0.396F$$

5.4.1.2 切开多余链杆

5.4.1.2 Cutting Redundant Links

仍以上节为例，如果切开 CD 杆，也可以作为基本体系，如图 5.18 所示。

Still taking the above section as an example, if the link CD is cut, it can also be used as the basic system, as shown in Figure 5.18.

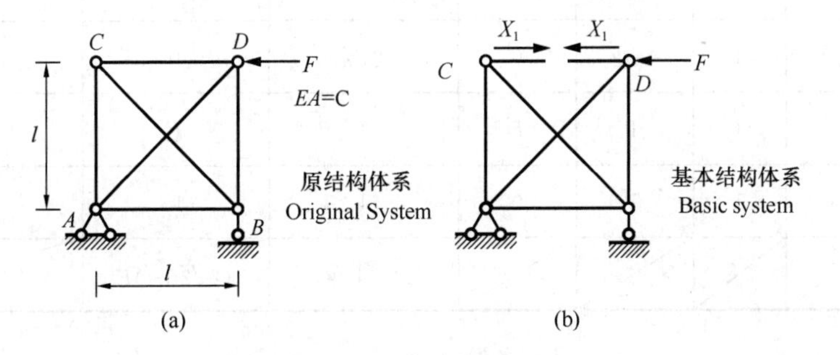

图 5.18 切开多余链杆

Figure 5.18 Cutting redundant links

此时，变形协调条件变为切口两侧的相对靠近值，显然这一值为零（图 5.19），故力法方程变为：

At this time, the deformation coordination condition becomes the relatively close value on both sides of the notch. Obviously, this value is zero (Figure 5.19), so the force method equation becomes:

$$\delta_{11} X_1 + \Delta_{1P} = 0$$

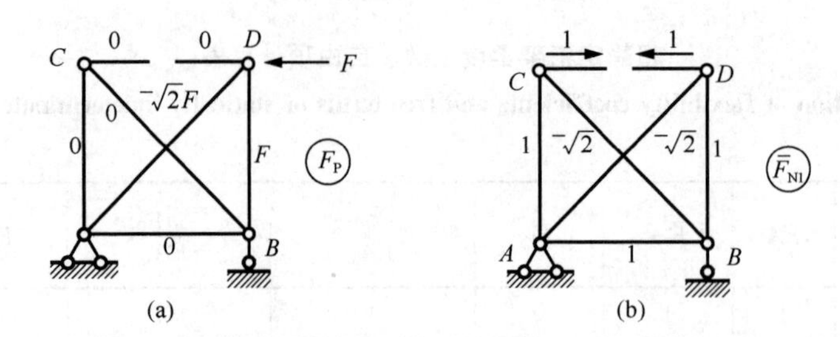

图 5.19　切开多余链杆后的力法方程求解

(a) F_P；(b) $\overline{F}_{N1}$

Figure 5.19　The force method solution after cutting redundant links

(a) F_P; (b) $\overline{F}_{N1}$

通过表 5.2 计算系数和自由项。

Calculate coefficients and free terms by Table 5.2.

超静定桁架柔度系数、自由项计算表　　表 5.2

Calculation of flexibility coefficients and free terms of statically indeterminate truss

Table 5.2

杆件 Rod	l	EA	$\overline{F}_{N1}$	F_P	$\frac{\overline{F}_{N1}^2}{EA}l$	$\frac{\overline{F}_{N1}F_P l}{EA}$	$F_N=\overline{F}_{N1}X_1+F_P$
AB	1	1	1	0	1	0	$-0.396F$
AC	1	1	1	0	1	0	$-0.396F$
AD	$\sqrt{2}$	1	$-\sqrt{2}$	$-\sqrt{2}F$	$2\sqrt{2}$	$2\sqrt{2}F$	$0.854F$
BC	$\sqrt{2}$	1	$-\sqrt{2}$	0	$2\sqrt{2}$	0	$0.56F$
BD	1	1	1	F	1	F	$0.604F$
CD	1	1	1	0	1	0	$-0.396F$
Σ					$4+4\sqrt{2}$	$(2\sqrt{2}+1)F$	

$$\delta_{11} = (4+4\sqrt{2})\frac{l}{EA},\ \Delta_{1P} = (2\sqrt{2}+1)F\frac{l}{EA},$$

$$X_1 = -0.396F$$

可以看出，计算结果是一致的。

It can be seen that the calculation results are consistent.

两种方法的比较见表 5.3。

Table 5.3 shows the comparison of the two methods.

拆和切的比较 **表 5.3**

Comparison of dismantling and cutting **Table 5.3**

	变形协调条件 Deformation coordination condition	Δ_1	力法方程 Force method equation	δ_{11}
拆 Dismantling	*CD* 相对位移靠近量 The value for *CD* getting closer	$\neq 0$	$\delta_{11}X_1+\Delta_{1P}=-\dfrac{X_1}{EA/l}$	$(3+4\sqrt{2})\dfrac{l}{EA}$
切 Cutting	切开截面两侧的相对位移 Relative displacement on both sides of notch	$=0$	$\delta_{11}X_1+\Delta_{1P}=0$	$(4+4\sqrt{2})\dfrac{l}{EA}$

从方程和系数来看，两个方程只是进行了移项，本质是一样的。

From the point of view of equations and coefficients, the two equations have only shifted terms and are essentially the same.

5.4.2 超静定排架

5.4.2 Statically Indeterminate Bent Frame

排架结构多见于厂房，包括屋架、柱和基础，如图 5.20 所示。

Bent frame structures are mostly found in plants, including roof frames, columns and foundations, as shown in Figure 5.20.

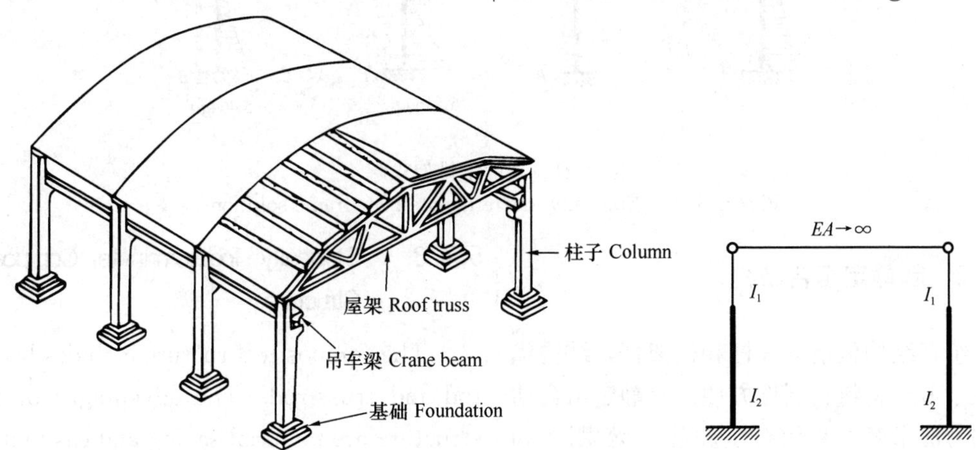

图 5.20 排架及其计算简图

Figure 5.20 Bent frame and its calculation sketch

计算超静定排架时，一般把横梁作为多余约束而切断，下面举例说明。

The crossbeam is generally cut off as a redundant constraint when calculating the statically indeterminate bent frame, as illustrated by the following example.

【例 5.3】 绘制图 5.21（a）所示排架在风荷载作用下的弯矩图。

［Example 5.3］ Draw the bending moment diagram for the bent frame shown in Figure 5.21 (a) under wind loads.

【解】 求解过程如图 5.21（b）～（e）所示。

［Solution］ The solving process is shown in Figures 5.21 (b) - (e).

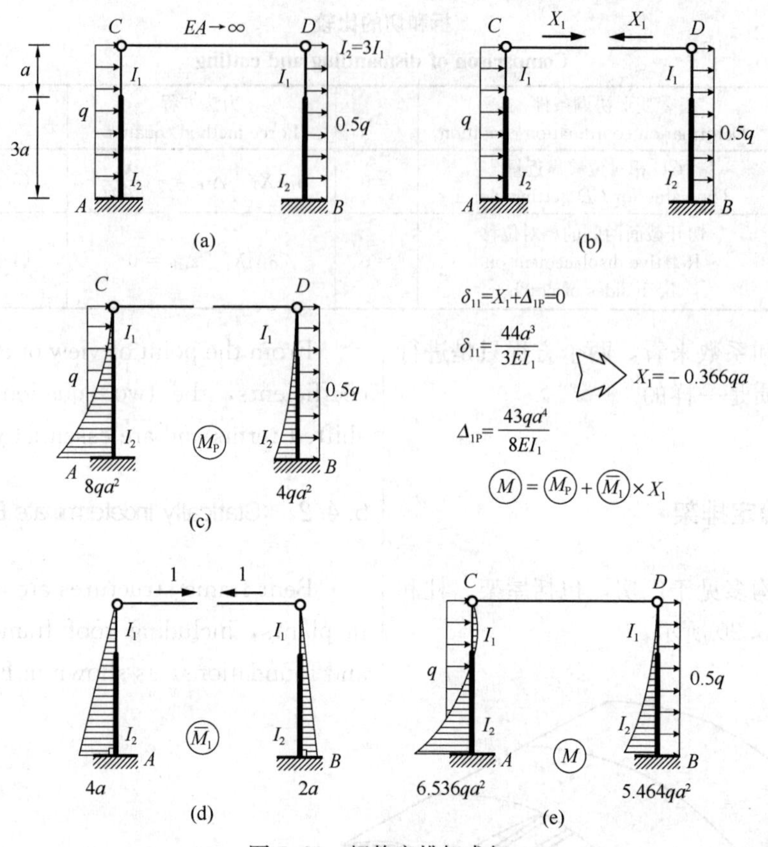

图 5.21　超静定排架求解

Figure 5.21　Statically indeterminate bent frame solution

5.4.3　超静定组合结构

组合结构包括梁式杆和桁架杆，结构优点在于节约材料、制作方便。超静定组合结构的力法求解过程综合了超静定梁/刚架和超静定桁架，主要体现在柔度系数和自由项的计算上：

5.4.3　Statically Indeterminate Composite Structure

The composite structure includes beam-rod and truss-rod. The advantages of the structure are material saving and easy fabrication. The force method solution process of the statically indeterminate composite structure integrates statically indeterminate beam/rigid frame and statically indeterminate truss，which is mainly calculating flexibility coefficients and free terms：

$$\delta_{ii} = \Sigma\int \frac{\overline{M}_i^2}{EI}\mathrm{d}x + \Sigma\frac{\overline{F}_{\mathrm{N}i}^2}{EA}l$$

$$\delta_{ij} = \Sigma\int \frac{\overline{M}_i\,\overline{M}_j}{EI}\mathrm{d}x + \Sigma\frac{\overline{F}_{\mathrm{N}i}\,\overline{F}_{\mathrm{N}j}}{EA}l$$

$$\Delta_{i\mathrm{P}} = \Sigma\int \frac{\overline{M}_i\,M_{\mathrm{P}}}{EI}\mathrm{d}x + \Sigma\frac{\overline{F}_{\mathrm{N}i}\,F_{\mathrm{NP}}}{EA}l$$

【例 5.4】用力法求解超静定组合结构 C 截面弯矩，如图 5.22 所示。$A=I/(1\mathrm{bm}^2)$

[**Example 5.4**] Solve the bending moment of section C in statically indeterminate composite structure by the force method, as shown in Figure 5.22. $A=I/(1\mathrm{bm}^2)$.

【解】求解过程如图 5.22 所示。

[**Solution**] The solving process is shown in Figure 5.22.

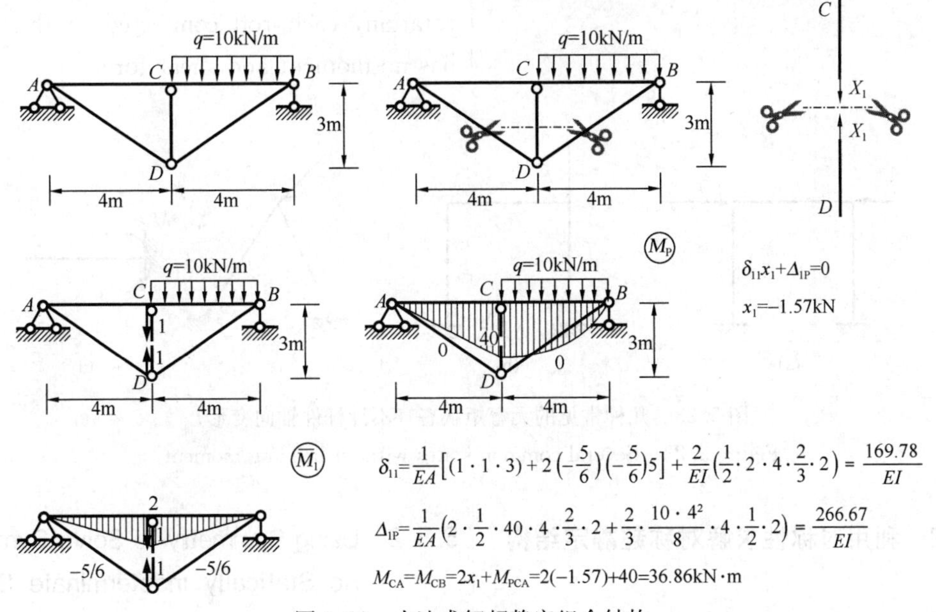

图 5.22 力法求解超静定组合结构

Figure 5.22 Force method for solving statically indeterminate composite structure

5.5 力法求解过程中的一些技巧

5.5 Some Tips for Using the Force Method

5.5.1 无弯矩状态的判定

5.5.1 Determination of the State Without Bending Moment

不考虑轴向变形的前提下：

Under the condition without considering axial deformation:

(1) 一对等大反向集中力沿某直杆作用时，只有该杆有轴力作用（图 5.23a）。

(1) When a pair of concentrated force with equal values but opposite directions acting along a certain straight rod, only that rod has axial forces (Figure 5.23a).

(2) 集中力沿某柱轴线作用时，只有该柱承受压力，其余杆件无内力（图 5.23b）。

(2) When the concentrated force acting along axis of a certain rod, only that rod

is under pressure with no internal forces on the other rods (Figure 5.23b).

(3) 集中力作用在没有线位移的节点上或集中力偶作用在不转动的节点上时，汇交于节点的各杆无弯矩和剪力。

(3) When concentrated force acting on a node without linear displacement or concentrated couple acting on a node with no rotation, each rod connected to this node has no moment and shear force.

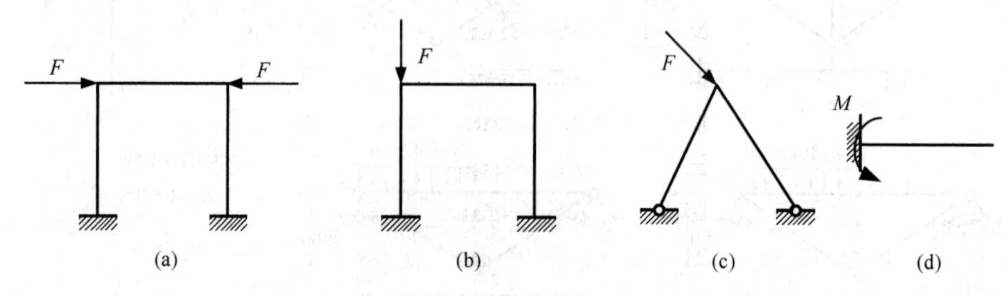

图 5.23 几种常见的无弯矩状态（不计杆件轴向变形）

Figure 5.23 Several common states without bending moment

5.5.2 利用对称性求解对称超静定结构

5.5.2 Using Symmetry to Solve Symmetric Statically Indeterminate Structure

由于超静定结构的求解过程较为复杂，计算量大，因此，在对称超静定结构的求解过程中可以利用对称性达到简化计算的目的。

Since the solution process of the statically indeterminate structure is more complicated and computationally intensive, symmetry can be used in the solution process of the symmetric statically indeterminate structure to simplify the calculation.

5.5.2.1 利用对称性选取基本结构

5.5.2.1 Selection of Basic Structures by Symmetry

如图 5.24 (a) 所示对称结构，为 3 次超静定，采用图 5.24 (b) 所示基本体系，计算可以看出，有 4 个柔度系数等于 0，X_3 直接解耦，从而减小了计算量。

As shown in Figure 5.24 (a), the symmetrical structure is cubic statically indeterminate, basic system is shown in Figure 5.24 (b), it can be seen that 4 flexibility coefficients are equal to 0, X_3 is directly decoupled, thus reducing the amount of calculation.

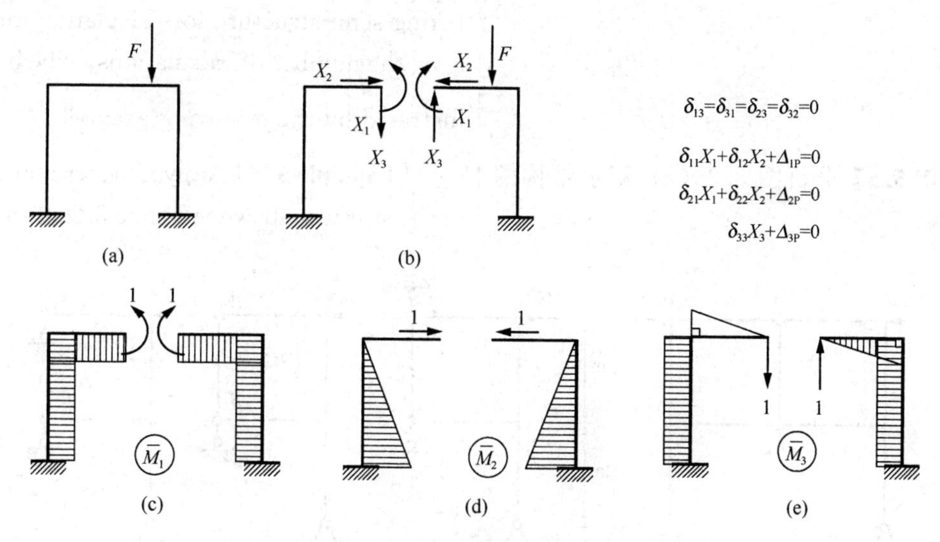

图 5.24 对称结构的基本体系

Figure 5.24 The basic system of symmetric structure

5.5.2.2 多余未知力分组

如图 5.25（a）所示，将单个未知力进行线性组合，可以使得一些单位内力图互相正交（如图 5.25c、图 5.25d），从而使负系数为零，简化计算。

5.5.2.2 Grouping of Redundant Unknown Forces

As shown in Figure 5.25 (a), the linear combination of a single unknown force can make some unit internal force diagrams orthogonal (as shown in Figures 5.25c, d), so that the negative coefficient is zero and the calculation is simplified.

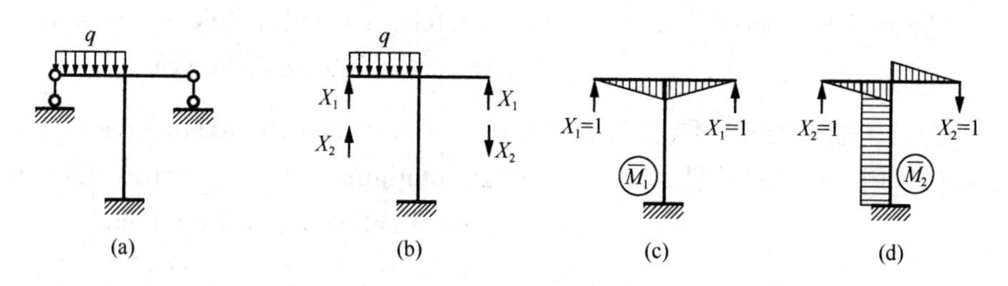

图 5.25 多余未知力分组

Figure 5.25 Grouping of redundant unknown forces

5.5.2.3 取半结构

根据对称结构在对称/反对称荷载下的受力和变形特点，取半结构进行计算，可以降低计算次数，下面举例说明。

5.5.2.3 Taking Semi-structure

According to the force and deformation characteristics of the symmetrical structure under symmetrical/anti-symmetrical loads, taking semi-structure for calculation can reduce the number of calculations, which is illustrated by the following example.

【例 5.5】 分析图 5.26(a) 所示结构的半结构。

[Example 5.5] Analyze the semi-structure of the structure shown in Figure 5.26 (a).

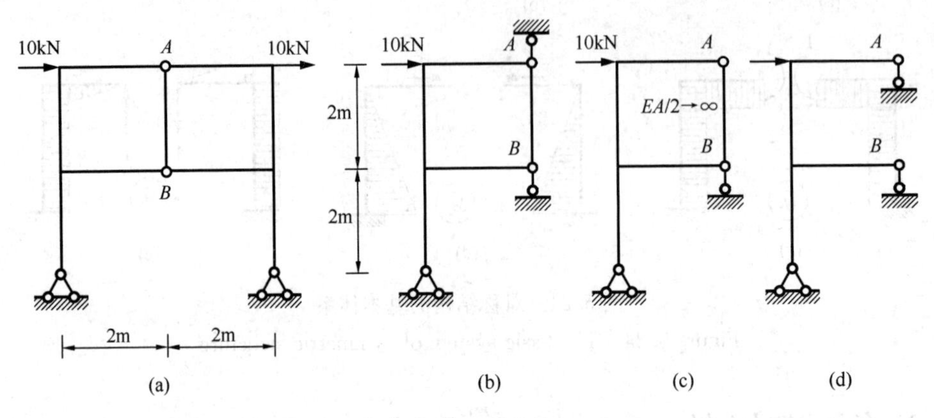

图 5.26 半结构

Figure 5.26 Semi-structure

【解】(1) 该对称结构在反对称荷载作用下，AB 没有竖向位移，故可以在 AB 位置增加竖向链杆，如图 5.26 (b) 所示。

[Solution] (1) Under the anti-symmetrical loads, rod AB of the symmetrical structure has no vertical displacement, so the vertical link can be added at AB, as shown in Figure 5.26 (b).

(2) 进一步，如果 AB 杆件无伸缩，可以去掉一个竖向链杆，如图 5.26 (c) 所示。

(2) Further, if the rod AB does not stretch, a vertical link can be removed, as shown in Figure 5.26 (c).

(3) 结合 AB 轴力为零（图 5.27），可以去掉 AB 杆，用一个链杆代替。

(3) When the axial force of rod AB is zero (Figure 5.27), the rod AB can be removed and replaced by a link.

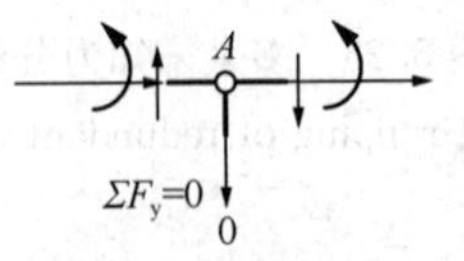

图 5.27 节点局部受力图

Figure 5.27 Local force diagram of node

(4) 可得半结构如图 5.26 (d) 所示。

(4) The available semi-structure can be obtained in Figure 5.26 (d).

5.5.3 合理选择基本结构

5.5.3 Reasonable Choice of Basic Structure

5.5.3.1 减少弯矩图覆盖

5.5.3.1 Reducing Overlay of the Bending Moment Diagram

前文有提到多跨超静定梁选择支座铰化的基本结构可减少计算量。实际上，将单位弯矩图限制在局部范围内是一种解题思路，弯矩图覆盖越少，计算越简单。

The previous sections have mentioned that the selection of the basic structure of support hinging for multi-span statically indeterminate beams can reduce the calculation. In fact, limiting the unit moment diagram to a local extent is a solution idea, and the less the moment diagram covers, the more straightforward the calculation.

可选用的手段有铰化、将单位未知力分组（图 5.28）。

Optional means are hinging, grouping unknown unit forces (Figure 5.28).

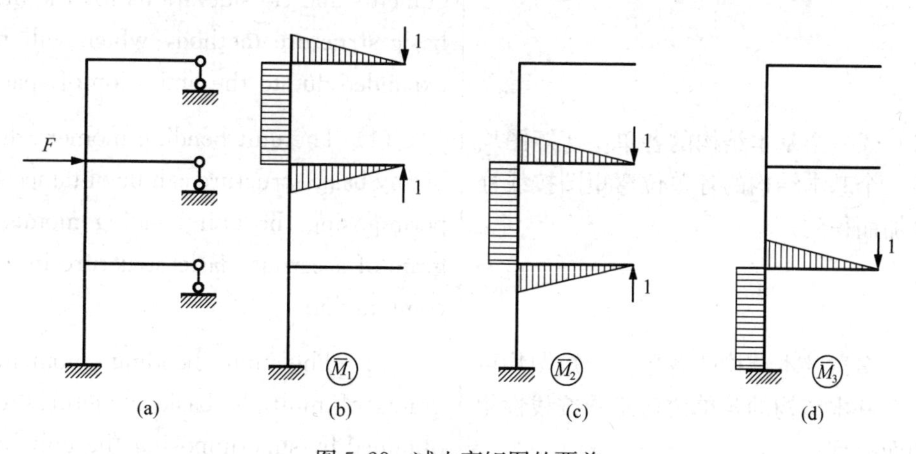

图 5.28 减少弯矩图的覆盖

Figure 5.28 Reducing overlay of the bending moment diagram

5.5.3.2 多重基本结构法

5.5.3.2 Multiple Basic Structure Method

【例 5.6】 分析图 5.29（a）所示超静定刚架。

[Example 5.6] Analyze the statically indeterminate rigid frame in Figure 5.29 (a).

【解】 如图 5.29 所示。

[Solution] The solution can be shown in Figure 5.29.

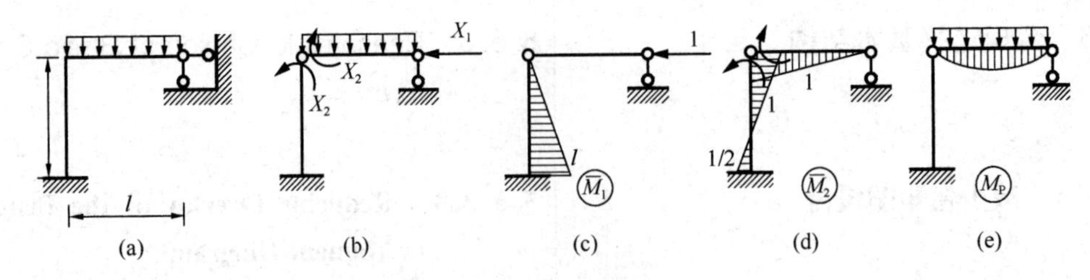

图 5.29　基本结构的选取

Figure 5.29　Selection of the basic structure

注：基本结构按图 5.29（c）和（d）选取的优势在于：柔度系数解耦，即 $\delta_{12}=\delta_{21}=0$。

Note: The advantages for selection of basic structures in Figures 5.29（c）and（d）is: Flexibility coefficients can be decoupled, i.e., $\delta_{12}=\delta_{21}=0$.

进一步，根据前人研究，选用多重基本结构，能够进一步简化计算（以下 4 点包括多重基本结构法的论证和注意事项，限于篇幅，不再展开）。

Furthermore, based on previous researches, the choice for multiple basic structures, can further simplify the calculation (The following four points are the arguments and considerations for the multiple basic structure method, which will not be expanded due to the limitation of space).

（1）任一个基本结构的各单位弯矩图均可由某一个基本结构的各单位弯矩图按线性组合叠加而得到；

（1）The unit bending moment diagram of any basic structure can be obtained by superimposing the unit bending moment diagram of a certain basic structure in a linear combination;

（2）多重基本结构的各单位弯矩图均可由某一个基本结构的各单位弯矩图按线性组合叠加而得到；

（2）The unit bending moment diagrams of multiple basic structures can be obtained by superimposing the unit bending moment diagrams of a particular basic structure in the linear combination;

（3）采用多重基本结构的单位弯矩图进行力法计算，与采用某一个基本结构的计算结果相同；

（3）The force method calculation using the unit bending moment diagram of the multiple basic structures is the same as that of a particular basic structure.

（4）采用多重基本结构法时，各单位弯矩图应是线性无关的。

（4）When using the multiple basic structure method, each unit bending moment diagram should be linearly independent.

【例 5.7】依然以图 5.29（a）结构进行举例，说明多重基本结构法的使用。

【解】求解过程如图 5.30 所示。

[**Example 5.7**] Still take the structure in Figure 5.29 (a) as an example to illustrate the use of the multiple basic structure method.

[**Solution**] The solution process is shown in Figure 5.30.

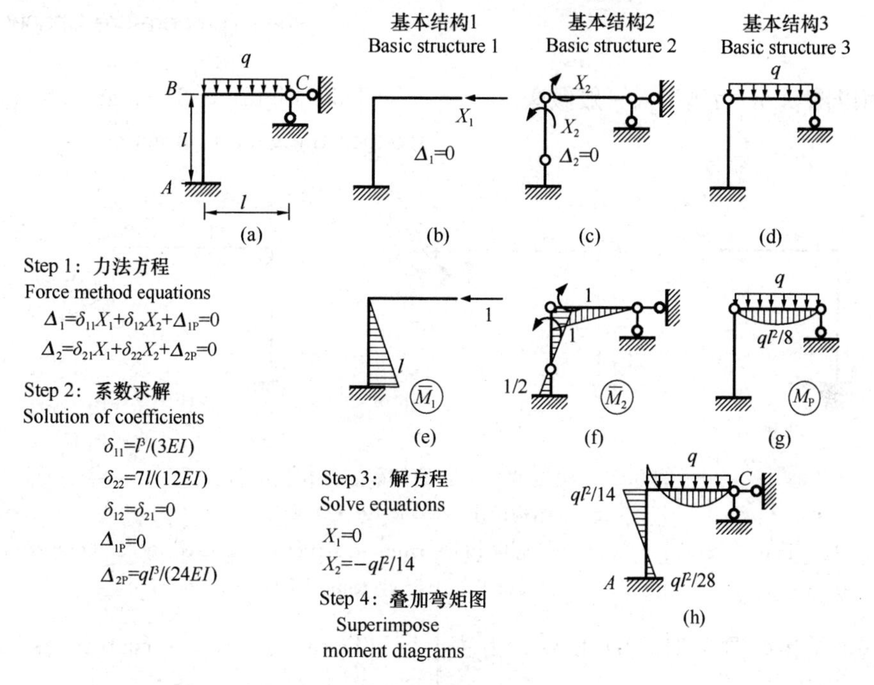

图 5.30　多重基本结构法

Figure 5.30　Multiple basic structure method

5.6　非荷载因素下超静定结构自内力计算

5.6　Calculation of Self-internal Forces of Statically Indeterminate Structure under Non-load Factors

超静定结构由于具有多余约束，因此，即使没有荷载作用，但只要有发生位移的因素，如支座移动、温度变化、材料收缩、制造误差等，都可以产生内力，即非荷载因素在超静定结构中也会产生内力，这种内力称为**自内力**。

The statically indeterminate structure has redundant restraints, so even if there is no load action, but as long as there are factors that occur displacement, such as support movement, temperature change, material shrinkage, manufacturing error, etc., they can generate internal forces, that is, non-load factors in the statically indeterminate structure will also generate internal forces, such internal forces are called **self-internal force.**

5.6.1 温度改变时超静定结构的内力计算

5.6.1 Calculation of Internal Forces of Statically Indeterminate Structure when Temperature Changes

下面以图 5.31 为例引入一般做法。

Take Figure 5.31 as an example to introduce the general practice.

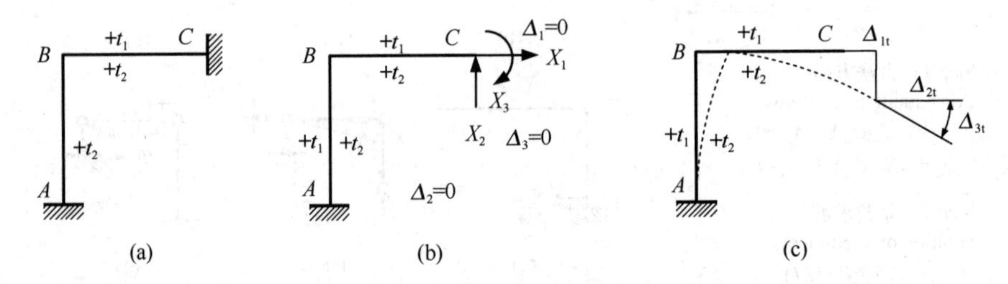

图 5.31 温度变化引起的超静定结构内力计算

(a) 原结构；(b) 基本体系；(c) 变形

Figure 5.31 The internal forces of statically indeterminate structures caused by temperature changes

(a) Original structure; (b) Basic system; (c) Deformation

取基本结构如图 5.31（b）所示，力法方程为 $[\delta]\{X\}+\{\Delta_t\}=\{\Delta\}$，

Take the basic structure as shown in Figure 5.31 (b), and the force method equation is $[\delta]\{X\}+\{\Delta_t\}=\{\Delta\}$,

$$\Delta_{it}=\Sigma(\pm)\int \overline{M}_i \frac{\alpha \Delta t}{h} ds + \Sigma(\pm)\int \overline{F}_{Ni}\alpha t_0 ds$$

由于基本结构是静定的，所以其位移计算与前面章节一致。将未知力 $\{X\}$ 求出后，内力图叠加即可。

Since the basic structure is statically determinate, the displacement calculation is consistent with the previous section. After the unknown force $\{X\}$ is calculated, the internal force diagram can be superimposed.

【例 5.8】图 5.32（a）所示刚架，浇筑混凝土时温度 15℃，冬季混凝土外皮温度 −35℃，内皮温度 15℃，求温度变化引起的自内力。已知各杆 EI 为常数，截面尺寸为矩形，400mm×600mm，混凝土弹性模量为 $E=2\times10^{10}$ Pa，温度膨胀系数为 $\alpha=1\times10^{-5}$。

[Example 5.8] For the rigid frame shown in Figure 5.32 (a), the temperature during concrete pouring is 15℃, the concrete skin temperature in winter is −35℃, and the inner skin temperature is 15℃. Calculate the internal force caused by temperature changes. It is known that EI of each rod is constant, and the section size is rectangular with the size of 400mm×600mm, the concrete elastic modulus is $E=2\times10^{10}$ Pa, temperature expansion coefficient is $\alpha=1\times10^{-5}$.

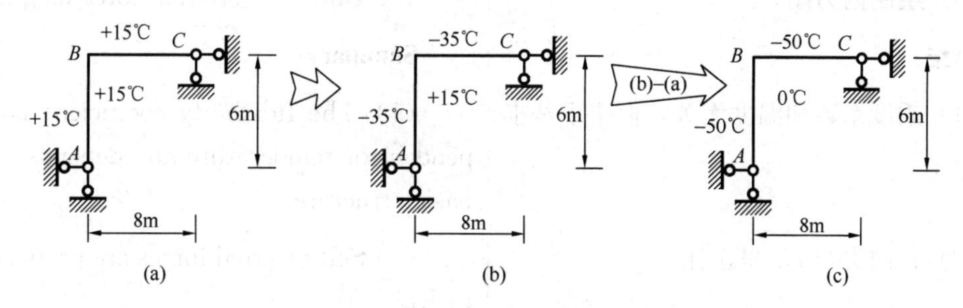

图 5.32 温度变化的计算

(a) 浇筑时温度；(b) 计算时温度；(c) 温差

Figure 5.32 Calculation of temperature changes

(a) Temperature at pouring; (b) Temperature at calculation; (c) Temperature difference

【解】采用力法进行计算。

[Solution] The following calculation is by the force method.

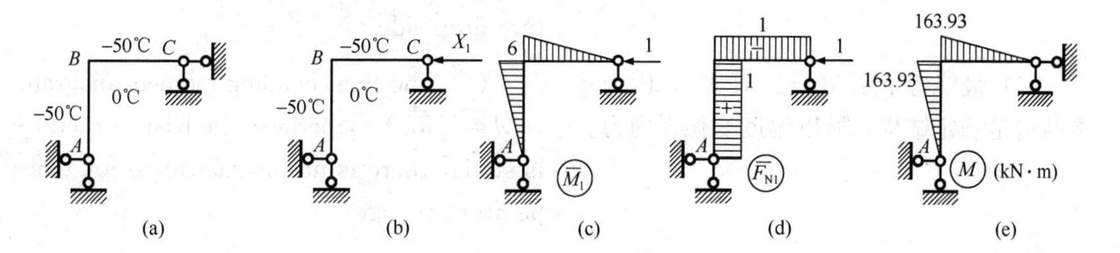

图 5.33 力法求解温度变化的超静定结构

Figure 5.33 Force method for solving statically indeterminate structure with temperature changes

(1) 取基本结构，如图 5.33（b）所示；

(1) Take the basic structure, as shown in Figure 5.33 (b);

(2) 列力法方程：$[\delta]\{X\}+\{\Delta_t\}=\{\Delta\}$，为一次超静定：

(2) List the force method equation: $[\delta]\{X\}+\{\Delta_t\}=\{\Delta\}$, it is primary statically indeterminate:

$$\delta_{11}X+\Delta_{1t}=0$$

(3) 求解系数和自由项：

(3) Solve coefficients and free terms:

$$\delta_{11}=\int\frac{\overline{M}_1^2}{EI}\mathrm{d}x=\frac{168}{EI}$$

$$\Delta_{1t}=\Sigma(\pm)\int\overline{M}_1\frac{\alpha\Delta t}{h}ds+\Sigma(\pm)\int\overline{F}_{N1}\alpha t_0 ds=-3187.5\alpha$$

（4）解方程可得：

(4) Solve the equation and the result is:

$$X_1=18.97EI\alpha$$

（5）绘制内力图。

(5) Draw the internal force diagram.

小结：

Summary:

（1）柔度系数和温度无关，取决于基本结构。

(1) The flexibility coefficient is independent of temperature and depends on the basic structure.

（2）自内力与 EI 呈正比。

(2) Self-internal forces are proportional to EI.

（3）弯矩图画在相对降温侧（可利用此性质快速定性弯矩图，如图 5.34 所示）。工程实际中，混凝土温降侧可能出现裂缝。

(3) The bending moment diagram is drawn on the relative cooling side (which can be quickly and qualitatively determine the bending moment diagram, as shown in Figure 5.34). In engineering practice, cracks may appear on the concrete temperature drop side.

（4）最终弯矩图 $M=\Sigma\overline{M}_i X_i$，因为基本结构是静定结构，所以温度变化无内力。

(4) The final bending moment diagram is $M=\Sigma\overline{M}_i X_i$, because the basic structure is static, there is no internal force for temperature change.

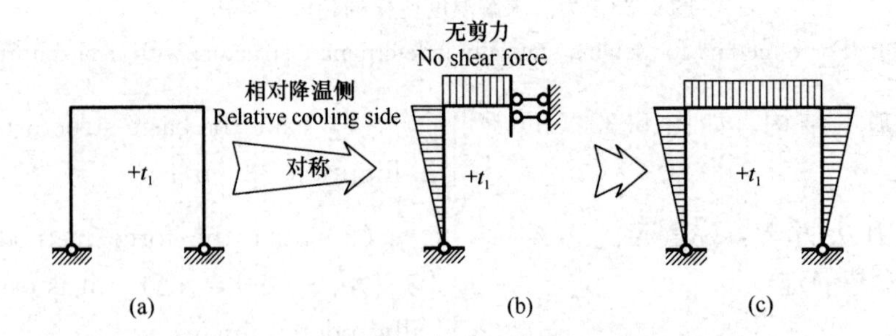

图 5.34　快速绘制超静定结构温度变化引起的弯矩图

Figure 5.34　Fast plotting bending moment diagram due to temperature changes in statically indeterminate structure

5.6.2 支座移动引起超静定结构的内力计算

5.6.2 Calculation of Internal Forces of Statically Indeterminate Structure due to Support Movement

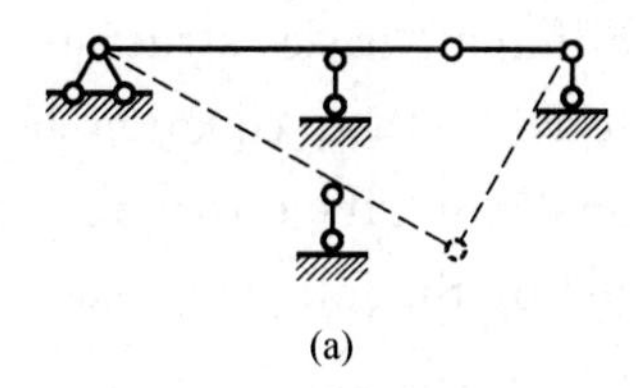

(a)

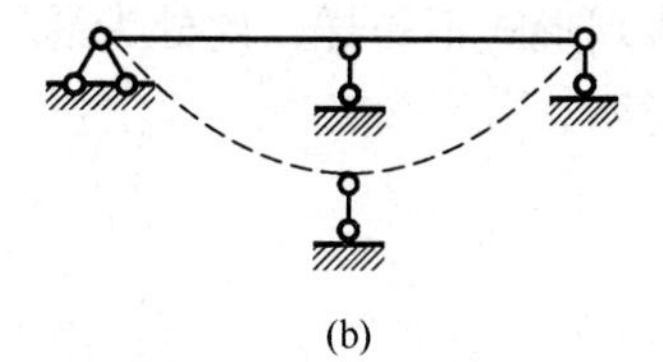

(b)

图 5.35 支座移动

(a) 静定结构；(b) 超静定结构

Figure 5.35 Support movement

(a) Statically determinate structure; (b) Statically indeterminate structure

从图 5.35 可知：

(1) 静定结构支座移动：有位移，无变形，无内力；

(2) 超静定结构支座移动：有位移，有变形，有内力。

From Figure 5.35, it can be seen that:

(1) Support movement of statically determinate structure: leads to displacement, but no deformation and internal forces;

(2) Support movement of statically indeterminate structure: leads to displacement, deformation and internal forces.

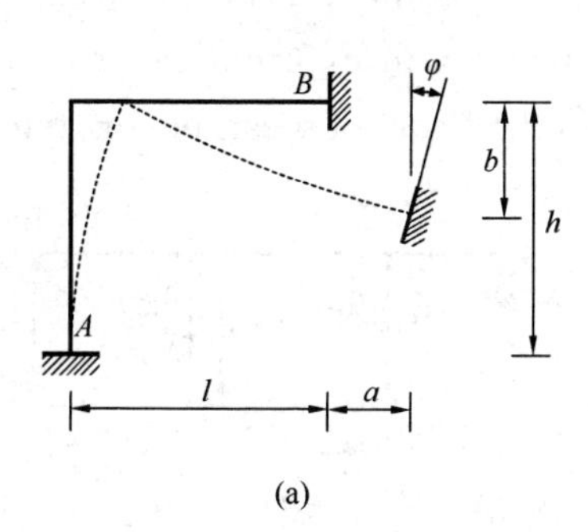

(a)

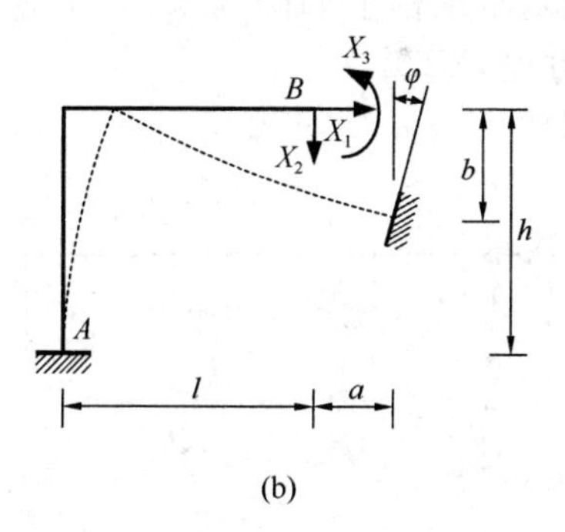

(b)

图 5.36 基本体系

Figure 5.36 Basic system

如图 5.36 所示，力法的方程可写为：

As shown in Figure 5.36, the equation of the force method can be written as:

$$[\delta]\{X\}+\{\Delta_c\}=\{\Delta\}\begin{cases}\delta_{11}X_1+\delta_{12}X_2+\delta_{13}X_3+\Delta_{1c}=a\\ \delta_{21}X_1+\delta_{22}X_2+\delta_{23}X_3+\Delta_{2c}=b\\ \delta_{31}X_1+\delta_{32}X_2+\delta_{33}X_3+\Delta_{3c}=-\varphi\end{cases}$$

然后依次求解系数与自由项、解方程、叠加弯矩图即可。

Then solve the coefficients and free terms, solve the equation and superimpose the bending moment diagram in turn.

值得说明的是，不同基本结构的选取会使得力法方程的表现形式不一样，此处为易错点，下面举例说明。

It is worth stating that the selection of different basic structures will make the force method equation behave differently, and this is the error-prone point, as illustrated by the following example.

【例 5.9】分析图 5.37（a）中支座位移引起的自内力。

［**Example 5.9**］Analyze the self-internal forces caused by support displacement in Figure 5.37 (a).

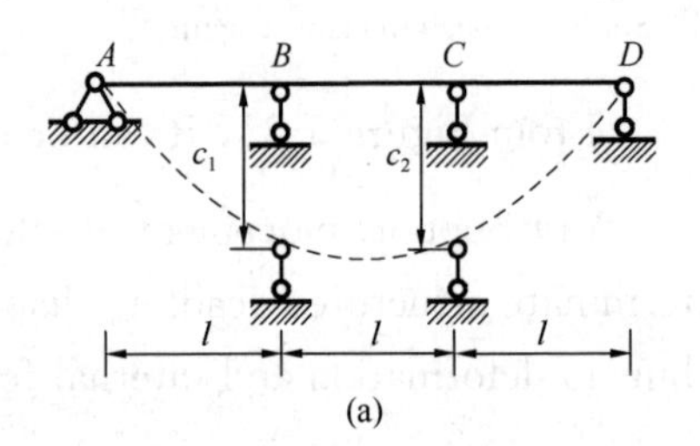

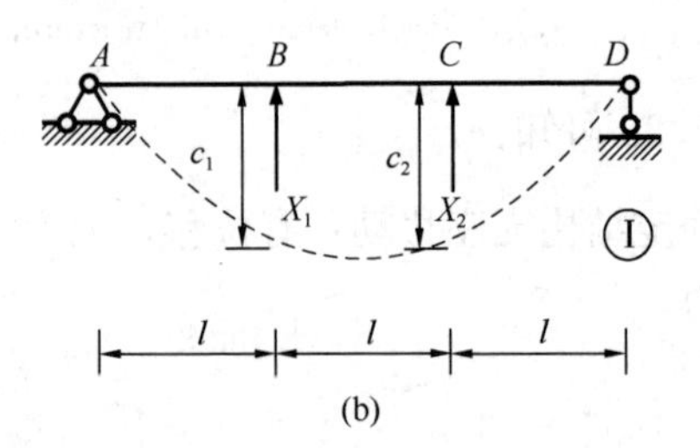

图 5.37 支座位移

Figure 5.37 Support movement

【解】取基本结构Ⅰ如图 5.37（b）所示，则力法方程可写为：

［**Solution**］Take the basic structure Ⅰ as shown in Figure 5.37 (b), and the force method equation can be written as:

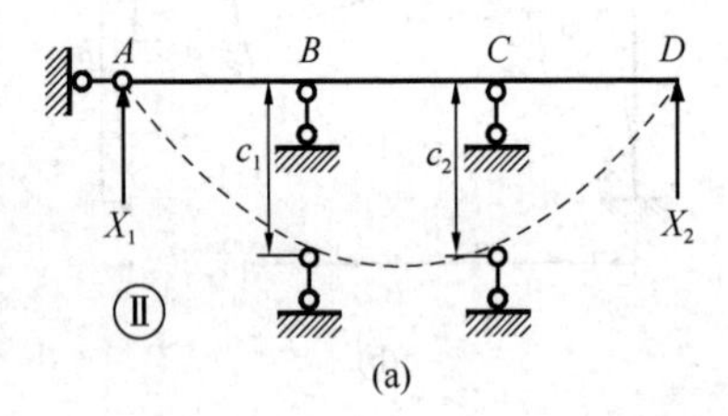

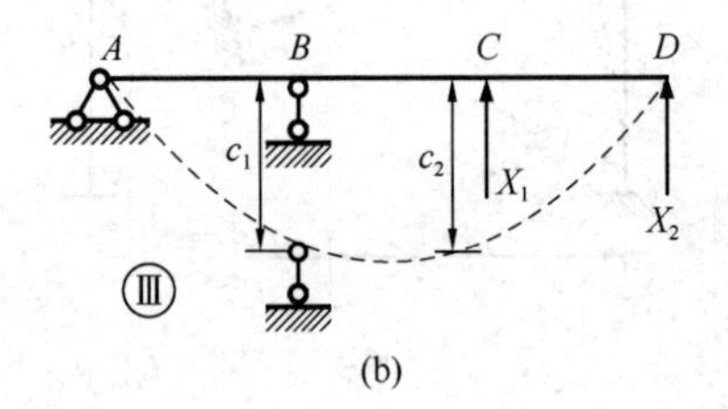

图 5.38 不同的基本结构

Figure 5.38 Different basic structures

$$\begin{cases}\delta_{11}x_1+\delta_{12}x_2=-c_1\\ \delta_{21}x_1+\delta_{22}x_2=-c_2\end{cases}$$

取基本结构Ⅱ如图 5.38（a）所示，力法方程可写为：

Take the basic structure Ⅱ as shown in Figure 5.38 (a), and the force method equation can be written as:

$$\begin{cases}\delta_{11}X_1+\delta_{12}X_2+\Delta_{1c}=0\\ \delta_{21}X_1+\delta_{22}X_2+\Delta_{2c}=0\end{cases}$$

取基本结构Ⅲ如图 5.38（b）所示，力法方程可写为：

Take the basic structure Ⅲ as shown in Figure 5.38 (b), and the force method equation can be written as:

$$\begin{cases}\delta_{11}X_1+\delta_{12}X_2+\Delta_{1c}=-c_1\\ \delta_{21}X_1+\delta_{22}X_2+\Delta_{2c}=0\end{cases}$$

下面以基本结构Ⅱ为例说明自由项 Δ_{ic} 的计算方法：

The calculation method of the free term Δ_{ic} is illustrated by taking the basic structure Ⅱ as an example.

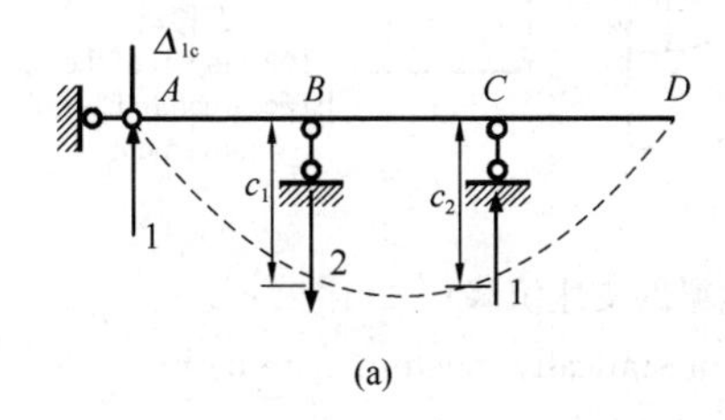

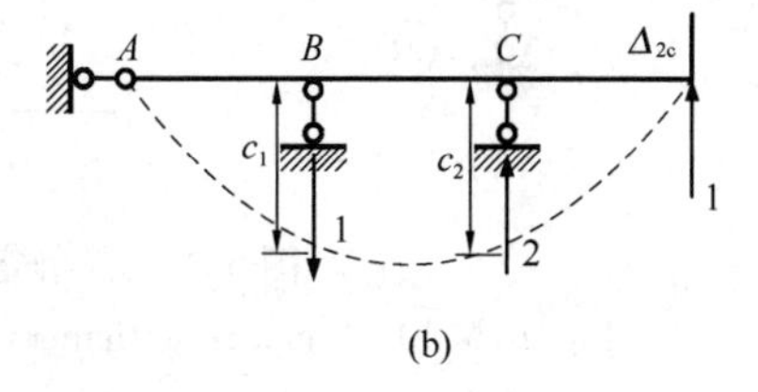

图 5.39 求解自由项

Figure 5.39 Solving the free term

基于虚功原理，由图 5.39（a）可得：

Based on virtual work principle, from Figure 5.39 (a):

$$\Delta_{1c}=-\overline{F}_R\cdot c=-(2\times c_1-1\times c_2)=c_2-2c_1$$

由图 5.39（b）可得：

From Figure 5.39 (b):

$$\Delta_{2c}=-\overline{F}_R\cdot c=-(-1\times c_1+2\times c_2)=c_1-2c_2$$

【例 5.10】用力法求解图 5.40 中超静定结构弯矩图。

[Example 5.10] Solve the bending moment diagram of statically indeterminate structure in Figure 5.40 by force method.

【解】解题过程如图 5.40 所示。

[Solution] The problem solving process is shown in Figure 5.40.

由于本题结构具有对称性，图 5.40（f）给出了半结构。请思考：

Since the structure of this example is symmetrical, Figure 5.40 (f) shows the semi-structure. Please consider:

（1）如果和对称轴重合的竖向链杆发生了沉降 Δ，那么在半结构计算中这个 Δ 是否要除以 2？

(1) If the vertical link coincident with the axis of symmetry has a settlement Δ, should this Δ be divided by 2 in the semi-structural calculation?

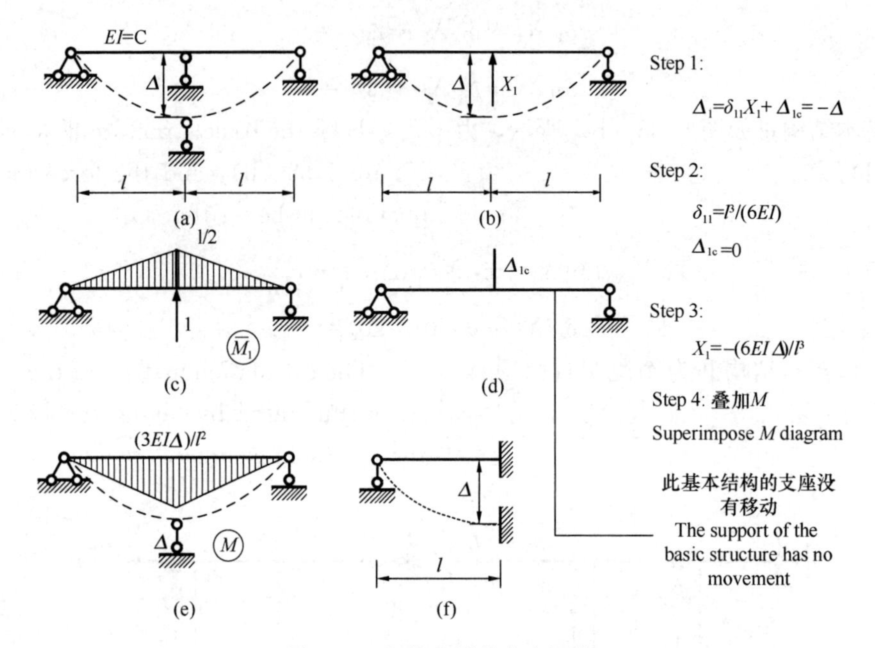

图5.40　超静定梁的支座沉降

Figure 5.40　Support settlement of statically indeterminate beam

（2）如果和对称轴重合的是一个竖向力F，那么在半结构计算中这个F是否要除以2？

(2) If a vertical force F coincides with the axis of symmetry, should this F be divided by 2 in the semi-structural calculation?

5.7　超静定结构的位移计算

5.7　Displacement Calculation of Statically Indeterminate Structure

5.7.1　超静定结构位移计算的基本思路

5.7.1　Basic Idea of Displacement Calculation for Statically Indeterminate Structure

首先，位移计算的根本出发点依然是虚功原理。

First, the fundamental starting point of displacement calculation remains the principle of virtual work.

$$1\times\Delta+\Sigma\overline{F}_{Ri}c_i=\Sigma\int\overline{M}\mathrm{d}\theta+\Sigma\int\overline{F}_{N}\mathrm{d}u+\Sigma\int\overline{F}_{Q}\mathrm{d}\eta$$

$$\Delta=\Sigma\int\frac{\overline{M}M_{P}}{EI}\mathrm{d}s+\Sigma\int\frac{\overline{F}_{N}F_{NP}}{EA}\mathrm{d}s+\Sigma\int k\frac{\overline{F}_{Q}F_{QP}}{GA}\mathrm{d}s-\Sigma\overline{F}_{Ri}c_i$$

在本章中，超静定结构的内力已经可以解出，那么根据单位荷载法就可以求得位移，但此时有一个难题，就是单位荷载如果作用于原结构，那么将再次求解超静定结构，虽然可以求得结果，但是计算量大，能否有更好的方法（图 5.41）?

In this chapter, the internal force of the statically indeterminate structure has been solved, then the displacement can be found according to the unit load method, but at this time, there is a difficult problem that if the unit load acts on the original structure, then it will solve the statically indeterminate structure again, although the result can be found, but the calculation is large, is there a better method (Figure 5.41)?

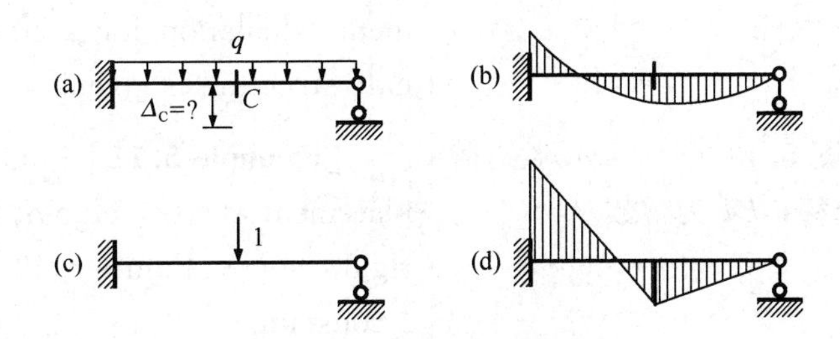

图 5.41 超静定结构位移计算

Figure 5.41 Displacement calculation for statically indeterminate structures

由于结构的变形是唯一的，既然图 5.42（a）和图 5.42（b）的变形一致，那么计算 C 截面竖向位移时，显然图 5.42（d）的计算过程要比图 5.42（c）要简单得多。

Since the deformation of the structure is unique, and the deformation of Figures 5.42 (a) and (b) are the same, when calculating the vertical displacement of section C, it is obvious that the calculation process of Figure 5.42 (d) is much simpler than Figure 5.42 (c).

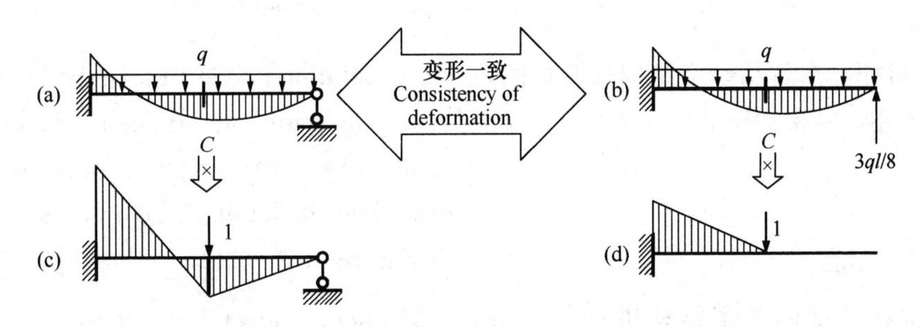

图 5.42 超静定结构的位移求解

Figure 5.42 Displacement solution for statically indeterminate structure

因此，在计算超静定结构的位移时，建议将单位荷载施加于静定的基本结构上。进

Therefore, when calculating the displacement of a statically indeterminate

一步，由于基本结构有无穷多个，所以建议选择单位荷载影响范围最小的那个基本结构进行计算。

structure, it is recommended that the unit load be applied to the statically determinate basic structure. Further, since there are infinite basic structures, selecting the one with the smallest influence range of unit load is recommended for the calculation.

5.7.2　荷载作用下超静定结构的位移计算

5.7.2　Displacement Calculation of Statically Indeterminate Structure under Loads

本节将对超静定结构的位移计算进行举例说明。

In this section, an example of displacement calculation for a statically indeterminate structure is given.

【例 5.11】求图 5.43（a）所示刚架横梁中点 D 的竖向位移，EI 为常数。

[**Example 5.11**] Find the vertical displacement at the midpoint D of the beam in rigid frame in Figure 5.43 (a), where EI is a constant.

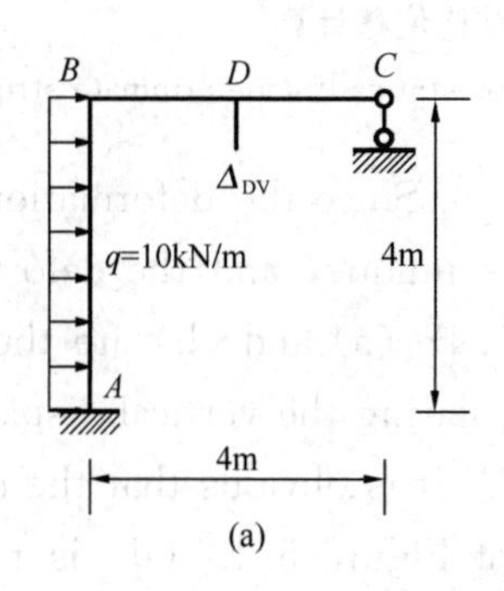

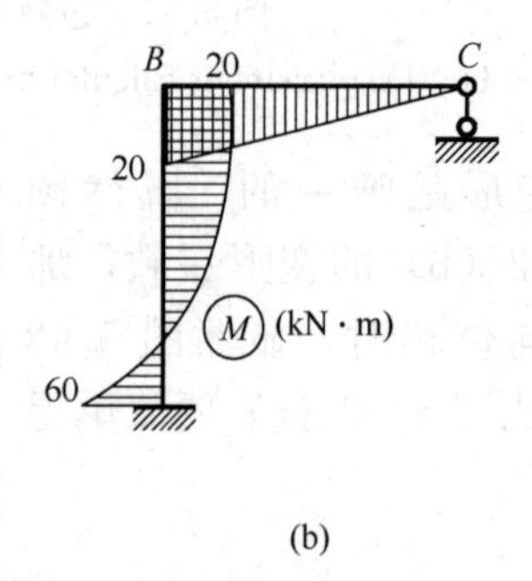

图 5.43　超静定结构的内力

Figure 5.43　Internal forces in statically indeterminate structure

【解】首先，求解超静定结构的弯矩图，过程不再详述，结果如图 5.43（b）所示。

[**Solution**] First of all, the bending moment diagram of statically indeterminate structure is solved. The process will not be described in detail. The results are shown in Figure 5.43 (b).

然后选择合适的基本结构进行位移计算。

Then, select the appropriate basic structure for displacement calculation.

显然，对比图 5.44（a），图 5.44（b）的基本结构更易于求解位移，图乘可得 $\Delta_{DV}=$

Obviously, compared with Figure 5.44(a), the basic structure in Figure 5.44(b) is easier to

20/(EI) (↓)。

solve the displacement, $\Delta_{DV}=20/(EI)$ (↓) can be obtained by graph multiplication.

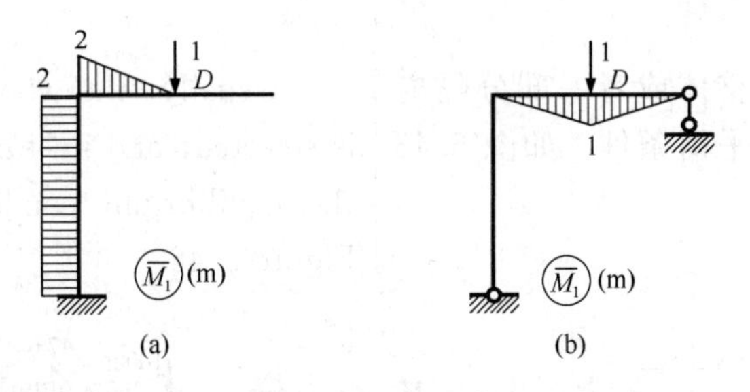

图 5.44 基本结构的选择

Figure 5.44 Selection of basic structure

图 5.45 给出了两种较简单的求解组合超静定结构的思路，一种不计算桁架杆，另一种不计算梁式杆。

Two simple ideas for solving the combined statically indeterminate structure are given in Figure 5.45, of which one does not calculate the truss rods and the other one does not calculate the beam bars.

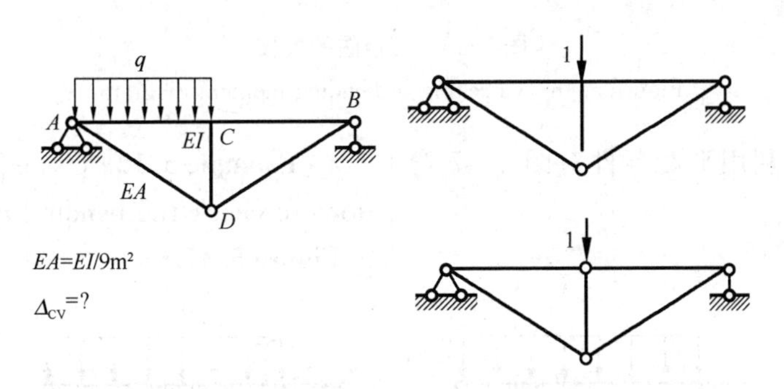

图 5.45 组合超静定结构的位移计算

Figure 5.45 Displacement calculation of combined statically indeterminate structure

5.8 超静定结构的计算校核

5.8 Checking of Statically Indeterminate Structure Calculation

由于超静定结构的计算过程较为复杂，因此，有必要进行校核。注意，校核不是重新计算，而是从平衡条件和变形条件两个方面进行验证。

Since the calculation process of statically indeterminate structure is more complicated, it is necessary to check its accuracy. Note that the checking is not a recalculation but a verification from both equilibrium conditions and deformation conditions.

5.8.1 平衡条件的校核

5.8.1 Checking of Equilibrium Conditions

通常选取结构的整体或者一部分隔离体，其受力均应满足平衡条件，如图 5.46 所示。

Usually, the whole structure or a part is selected, and its forces should all satisfy the equilibrium conditions, as shown in Figure 5.46.

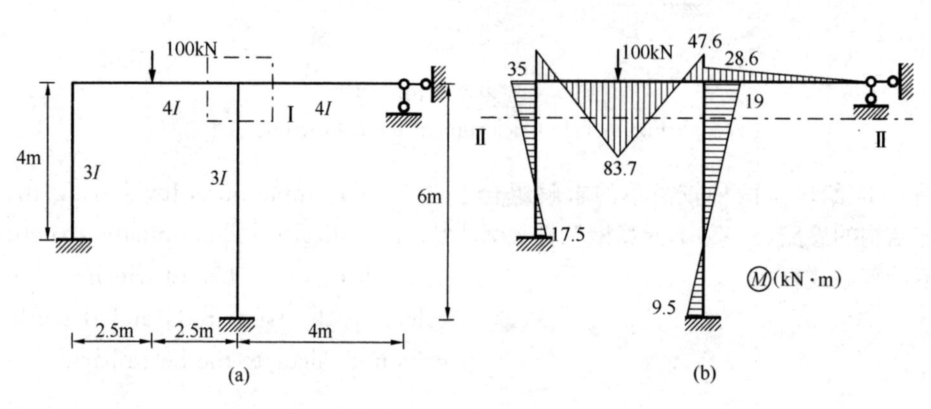

图 5.46 弯矩图的校核

Figure 5.46 Checking of bending moment diagram

【例 5.12】 利用平衡条件对图 5.47 弯矩图进行校核。

[Example 5.12] Use equilibrium conditions to check the bending moment diagram in Figure 5.47.

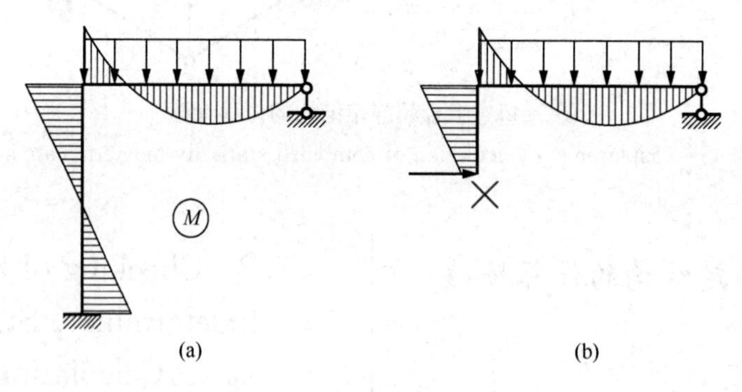

图 5.47 平衡条件的校核

Figure 5.47 Checking of equilibrium condition

【解】 截取隔离体如图 5.47（b）所示，水平方向力无法平衡，故弯矩图错误。

[Solution] The intercepted isolation is shown in Figure 5.47 (b). The forces in the horizontal direction cannot be balanced, so the bending moment diagram is wrong.

5.8.2 变形条件的校核

从图 5.48（a）和图 5.48（b）的变形条件来看，由于图乘结果一定小于零，这意味着刚节点两侧发生了相对转动，故错误。从图 5.48（c）和图 5.48（d）来看，由于图乘结果一定大于零，说明右侧有竖向位移，这与此处存在支座矛盾，故弯矩图也是错误的。

5.8.2 Checking of Deformation Conditions

From the deformation conditions of Figure 5.48 (a) and Figure 5.48 (b), the graph multiplication result must be less than zero, which means that relative rotation occurs on both sides of this rigid node, so it is wrong. From Figure 5.48 (c) and Figure 5.48 (d), the graph multiplication result must be greater than zero, which means that the right side has vertical displacement, contradicting to the existence of support, so the bending moment diagram is also wrong.

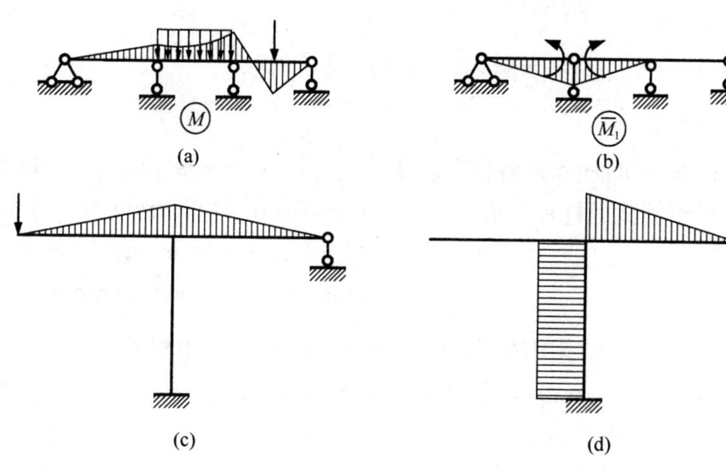

图 5.48 变形条件的校核

Figure 5.48 Checking of deformation conditions

5.8.3 无铰封闭框的校核

对于一个具有封闭框架的结构，可以利用封闭框架上任一截面相对转角等于零的条件来校核。

如图 5.49（a）所示具有封闭框架刚架的内力图，利用任一截面相对转角为零的条件校核时，$\overline{M}$ 图在封闭框架的所有截面的纵坐标都为 1，因此，有：

5.8.3 Checking of Hingeless Closed Frame

For a structure with a closed frame, the calculation checking can be operated by the condition that the relative angle of rotation of any section on the closed frame equals zero.

Figure 5.49 (a) is the internal force diagram of rigid frame with closed frame. When using the condition that the relative angle of rotation of any section on the closed frame equals zero to check the calculation, the vertical coordinates of $\overline{M}$ in all sections of the closed frame equal 1. Therefore:

$$\oint \frac{M}{EI} \mathrm{d}s = 0$$

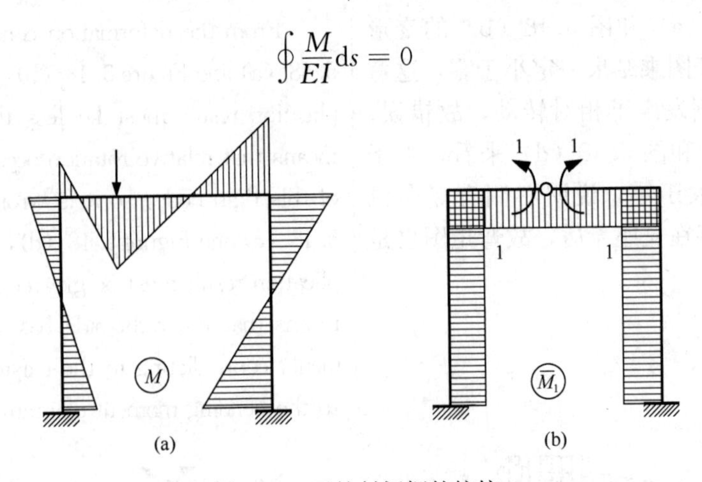

图 5.49 无铰封闭框的校核

Figure 5.49 Checking of hingeless closed frame

这意味着各杆 M 图面积除以各杆 EI 的代数和等于零，对于等截面杆件，有：

This means that the algebraic sum of the area of M diagram of each rod divided by EI of each rod is equal to zero, for a constant cross-section member:

弯矩图的框内面积＝弯矩图的框外面积

Areas inside the frame of the bending moment diagram = Areas outside the frame of the bending moment diagram

而这一结论可以进行开发利用，如图 5.50所示。

This conclusion can be developed and utilized, as shown in Figure 5.50.

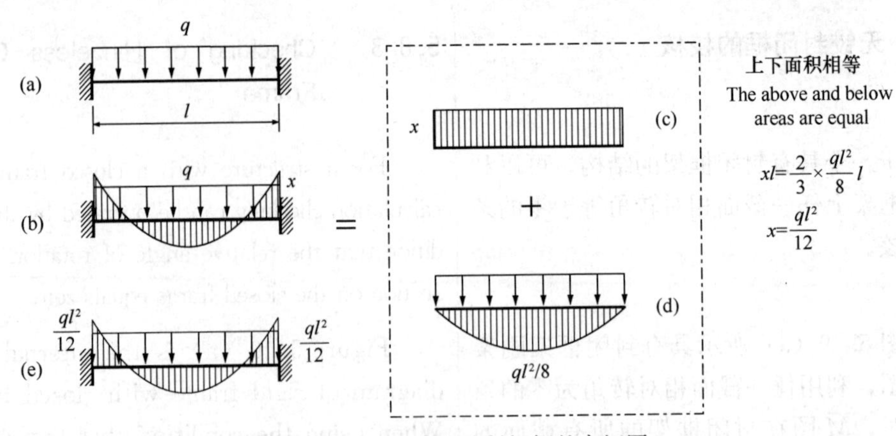

图 5.50 快速求无铰封闭框的弯矩图

Figure 5.50 Fast solving of the bending moment diagram of the hingeless closed frame

5.9 超静定结构的特点

超静定结构与静定结构对比，具有以下一些重要特性：

(1) 超静定结构具有多余约束。

从几何组成角度看，超静定结构是有多余约束的几何不变体系。由于存在多余约束，相应地就有多余未知力，因此，从受力角度来看，超静定结构的反力和内力仅凭静力平衡条件不能唯一确定，必须同时考虑变形条件后才能得到唯一解答。

从抵抗突然破坏的观点来看，静定结构有一个约束被破坏时，就成为几何可变体系，因而丧失承载能力。但是超静定结构却与其不同，当多余约束破坏时，结构仍为几何不变体系，因而还具有一定的承载能力。因此，超静定结构具有较强的防护能力。在设计工作中，选择结构形式时，应注意这一特性。

5.9 The Properties of Statically Indeterminate Structure

The statically indeterminate structure has some important characteristics compared to the statically determinate structure as follows:

(1) Statically indeterminate structure has redundant constraints.

In perspective of geometric composition, the statically indeterminate structure is a geometric stable system with redundant constraints. As there are redundant connections, there are redundant unknown forces accordingly. Therefore, from point of view of the force, the reaction and internal forces of the statically indeterminate structure cannot be uniquely determined by the static equilibrium conditions alone but must be uniquely determined by considering the deformation conditions simultaneously.

From the point of view of resistance to sudden damage, when the statically determinate structure has a connection broken, it becomes a geometric changeable system and therefore loses its load bearing capacity. But the statically indeterminate structure is different from it, when the redundant constraint is destroyed, the structure is still a geometric stable system and thus also has a certain load bearing capacity. Therefore, the statically indeterminate structure has a strong protective capacity. When choosing the form of structure in the design work, one should pay attention to this characteristic.

局部荷载在超静定结构中的影响范围，一般比在静定结构中的大，因为超静定结构内力分布的范围较广，其内力分布也比静定结构要均匀些，内力的峰值也要小些。

The influence of local load in the statically indeterminate structure is generally larger than that in the statically determinate structure because the internal force distribution of the statically indeterminate structure is wider and more uniform than that of the statically determinate structure, and the peak value of the internal force is smaller.

从结构刚度来看，在均布荷载作用下，简支梁的最大挠度为两端固定梁的5倍。这说明由于多余联系的存在，结构刚度有所提高，如果根据同样的容许应力和容许位移进行设计，显然超静定结构的设计截面将比静定结构的设计截面小得多，这无疑是比较经济的。

In terms of structural stiffness, the maximum deflection of the simply supported beam under uniform load is 5 times that of the fixed beam at both ends. This indicates that the structural stiffness is increased due to the existence of redundant links, and if the design is based on the same allowable stress and allowable displacement, it is obvious that the design section of the statically indeterminate structure will be much smaller than that of the statically determinate structure, which is undoubtedly more economical.

（2）在超静定结构中，由于温度改变、支座移动、制作误差、材料收缩等因素都可以引起内力。

(2) In statically indeterminate structure, internal forces can be caused by temperature changes, support movements, fabrication errors, material shrinkage, etc.

超静定结构的内力和支座反力除了与结构中杆的长度、荷载的大小有关外，还与各杆之间的刚度有关：超静定结构在荷载作用下的反力和内力，仅与各杆相对刚度有关；超静定结构在温度改变和支座位移时引起的内力，与各杆刚度的绝对值有关。这在前面的计算中可明显地看出：超静定结构可以通过调节各杆的相对刚度来调节内力；为了提高结构对支座位移和温度改变的抵抗能力，增大结构截面的尺寸，并不是有效的措施。在结构设计中也应注意这方面的特性。

The internal forces and support reaction forces of the statically indeterminate structure are not only related to the length of the rods in the structure and the magnitude of the load, but also to the stiffness between the rods: The reaction forces and internal forces of the statically indeterminate structure under load are only related to the relative stiffness of the rods; the internal forces of the statically indeterminate structure caused by temperature change and support displacement are related to the absolute value of the stiffness of the rods.

This is obvious in the previous calculation: The statically indeterminate structure can regulate the internal force by adjusting the relative stiffness of each rod; to improve the resistance of the structure to the support movement and temperature change, increasing the size of the structural section is not an effective measure. This aspect of the properties should also be noted in the design of the structure.

思 考 题

Questions

5.1 什么是超静定结构?它和静定结构有何区别?什么是超静定次数?如何确定超静定次数?撤除多余约束的方法有哪几种?

5.1 What is statically indeterminate structure? What is the difference between it and statically determinate structure? What is the statically indeterminate degree? How to determine the statically indeterminate degrees? What are the ways to remove redundant constraints?

5.2 为什么只用平衡条件不能确定超静定结构的反力和内力?

5.2 Why can't the reaction and internal forces of statically indeterminate structures be determined only by equilibrium conditions?

5.3 力法解超静定结构的思路是什么?什么是力法的基本结构和基本未知量?为什么首先要计算基本未知量?力法基本结构与原结构有何异同?基本体系与基本结构有何不同?

5.3 What is the idea of solving statically indeterminate structures by force method? What are the basic structure and basic unknowns of the force method? Why do we calculate the basic unknowns first? What are the similarities and differences between the basic structure of the force method and the original structure? What is the difference between the basic system and the basic structure?

5.4 力法典型方程的物理意义是什么?各系数和自由项的物理意义是什么?试从物理意义上说明,为什么主系数为必大于零的正值,而副系数可为正值、负值或零?

5.4 What is the physical meaning of the typical equations of the force method? What are the physical meanings of the coefficients and free terms? In the physical sense, why does the main coefficient have to be a positive value greater than zero, while the secondary coefficient can be positive, negative, or zero?

5.5 为什么在荷载作用下超静定结构的内力状态只与各杆刚度 EI（或 EA）的相对值有关，而与其绝对值无关？为什么静定结构的内力状态与 EI（或 EA）无关？

5.5 Why is the internal force state of a statically indeterminate structure under loads only related to the relative value of the stiffness EI (or EA) of each member, but independent of their absolute values? Why is the internal force state of statically determinate structures independent of EI (or EA)?

5.6 在超静定桁架、组合结构及厂房排架中，用撤去多余链杆的基本体系代替切开多余链杆的基本体系，这种算法是否正确？二者的力法方程有何异同？

5.6 In statically indeterminate trusses, composite structures and plant bents, is it correct to replace the basic system of cutting redundant links by the basic system of removing redundant links? What are the similarities and differences between their force method equations?

5.7 为什么用力法计算任何对称结构时，只要所取的基本结构是对称的，而基本未知量是对称力或反对称力，则力法方程自然地分成两组？

5.7 Why is it that when the force method is used to calculate any symmetric structure, as long as the basic structure is symmetric and the basic unknowns are symmetric or anti-symmetric forces, the force method equations are naturally divided into two groups?

5.8 为什么对称结构在对称或反对称荷载作用下可以取半边结构简化计算？

5.8 Why can a symmetrical structure be simplified with semi-structure under symmetrical or anti-symmetric loads?

5.9 能不能说两铰拱在竖向均布荷载 q 作用下的合理拱轴仍是抛物线［抛物线方程为 $y=\frac{4f}{l^2}x(l-x)$］？能不能说在均布荷载 q 作用下，$y=\frac{4f}{l^2}x(l-x)$ 形状的三铰拱与同样形状的两铰拱实际受力状态是一样的？为什么？

5.9 Is it ok that the reasonable arch axis of two hinged arch under the vertical uniformly distributed load q is still a parabola [The parabola equation is $y=\frac{4f}{l^2}x(l-x)$]? Is it ok that the three hinged arch with $y=\frac{4f}{l^2}x(l-x)$ shape is the same as the two hinged arch with that shape under the uniformly distributed load q? Why?

5.10 计算拱的位移时，M、F_Q、F_N的影响哪个重要？哪个次要？在什么情况下可以忽略F_Q和F_N的影响？

5.10 When calculating the displacement of an arch, which one is important among M, F_Q, and F_N? Which is secondary? Under what circumstances can the influence of F_Q and F_N be ignored?

5.11 什么叫弹性中心？怎样确定弹性中心？弹性中心法的好处是什么？

5.11 What is the elastic center? How to determine the elastic center? What are the benefits of the elastic center approach?

5.12 没有荷载就没有内力，这个结论在什么情况下适用？在什么情况下不适用？

5.12 There is no load, there is no internal force. Under what circumstances does this conclusion apply? Under what circumstances does it not apply?

5.13 计算超静定结构的内力时，在什么情况下只需给出EI（或EA）的相对值，在什么情况下需给出EI的绝对值？为什么？

5.13 When calculating the internal force of statically indeterminate structures, under what circumstances does only the relative value of EI (or EA) needs to be given, and under what circumstances the absolute value of EI needs to be given? Why?

5.14 计算超静定结构的位移与计算静定结构的位移，两者有何异同？为什么计算超静定结构的位移时，可以将所虚设的单位力施加于任一基本结构作为虚拟状态？

5.14 What are the similarities and differences between calculating the displacement of statically indeterminate structures and statically determinate structures? Why can the virtual unit force be applied to any basic structure as a virtual state when calculating the displacement of statically indeterminate structures?

5.15 说明变形条件式$\oint \frac{M}{EI}ds=0$的物理意义是什么？为什么用力法计算超静定结构EI的结果必须进行变形条件的校核？

5.15 Explain the physical meaning of the deformation conditional formula $\oint \frac{M}{EI}ds = 0$. Why is it necessary to check the deformation condition when calculating EI of a statically indeterminate structure by force method?

5.16 计算超静定结构的位移不能用各

5.16 The relative values of EI (or

杆 EI（或 EA）的相对值，而需用其绝对值；若用变形条件校核荷载产生的内力图时，为什么又可以用 EI 的相对值？用变形条件校核支座移动或温度变化产生的内力图时能否用 EI（或 EA）的相对值？为什么？

EA) of each rod are not used but the absolute values are used to calculate the displacement of statically determinate structure; If the deformation condition is used to check the internal force diagram generated by load, why can the relative value of EI be used? Can the relative value of EI (or EA) be used when checking the internal force diagram caused by support movement or temperature change with deformation conditions? Why?

习　题

Exercises

5.1　试确定图 5.51 所示结构的超静定次数。

5.1　Determine the statically indeterminate degrees of structures in Figure 5.51.

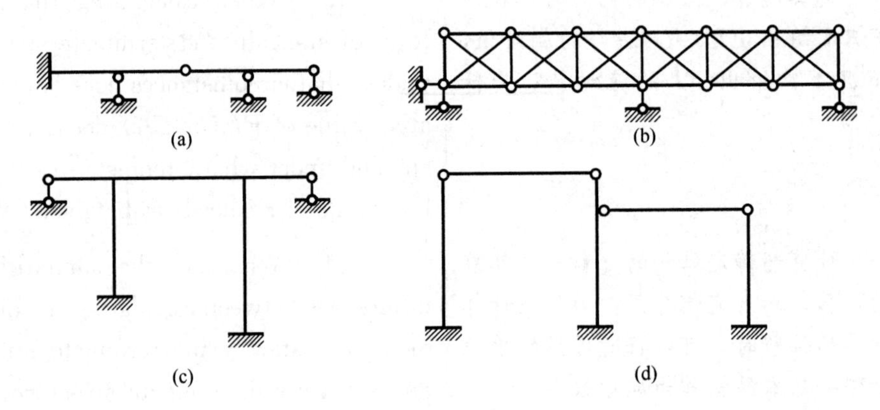

图 5.51　习题 5.1

Figure 5.51　Exercise 5.1

5.2　试确定图 5.52 所示结构的超静定次数。

5.2　Determine the statically indeterminate degrees of structures in Figure 5.52.

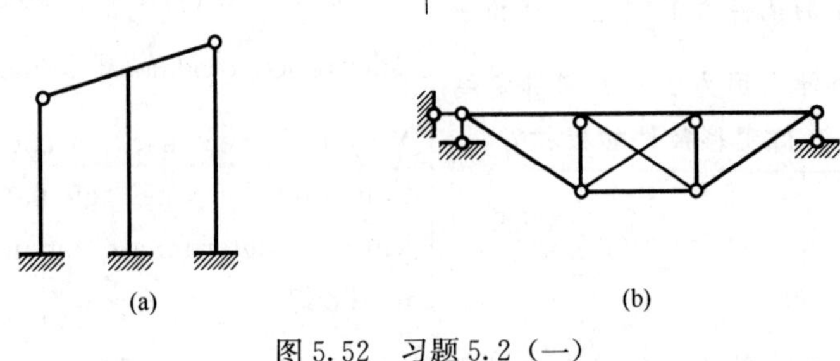

图 5.52　习题 5.2（一）

Figure 5.52　Exercise 5.2 (One)

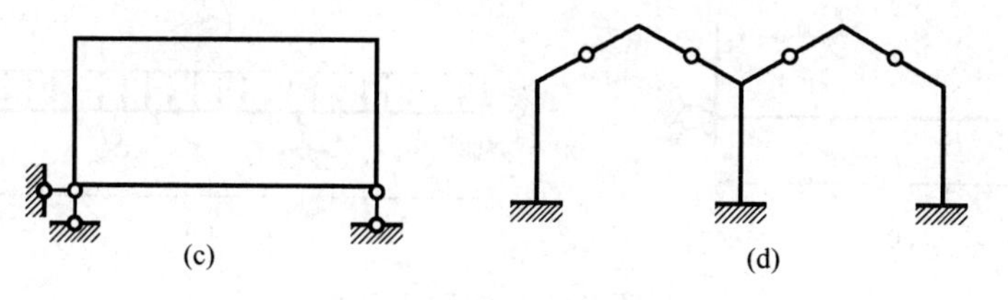

图 5.52 习题 5.2（二）

Figure 5.52 Exercise 5.2 (Two)

5.3 试确定图 5.53 所示结构的超静定次数。

5.3 Determine the statically indeterminate degrees of structures in Figure 5.53.

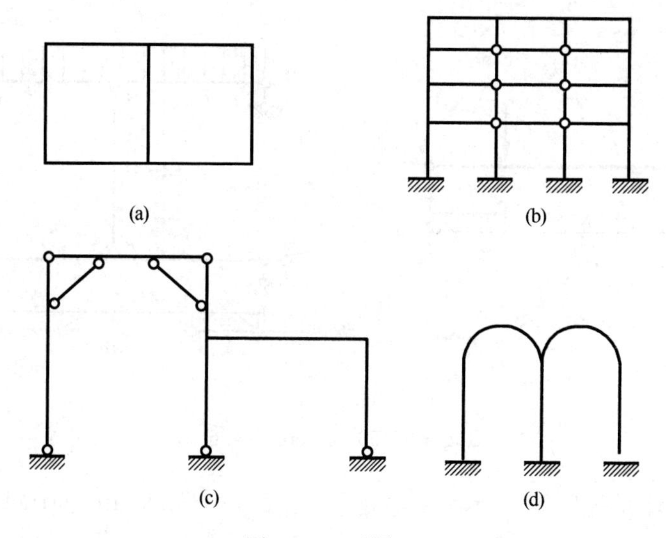

图 5.53 习题 5.3

Figure 5.53 Exercise 5.3

5.4 用力法计算图 5.54 所示结构，作 M，F_Q图。

5.4 Calculate structures in Figure 5.54 by force method, draw M, F_Q diagrams.

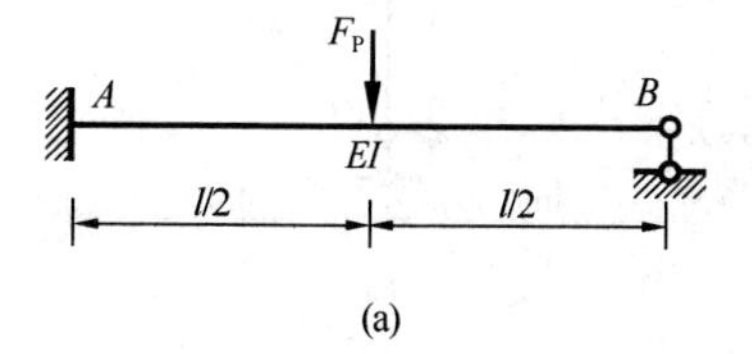

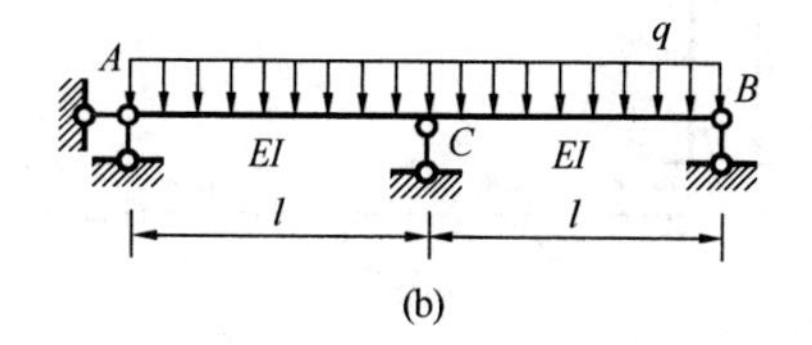

图 5.54 习题 5.4

Figure 5.54 Exercise 5.4

5.5 用力法计算图 5.55 所示结构，作 M，F_Q图，各图 EI=常数。

5.5 Calculate structures in Figure 5.55 by force method, draw M, F_Q diagrams, EI=constant.

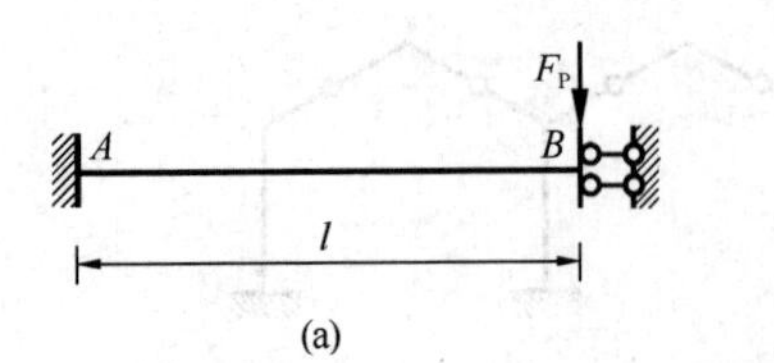

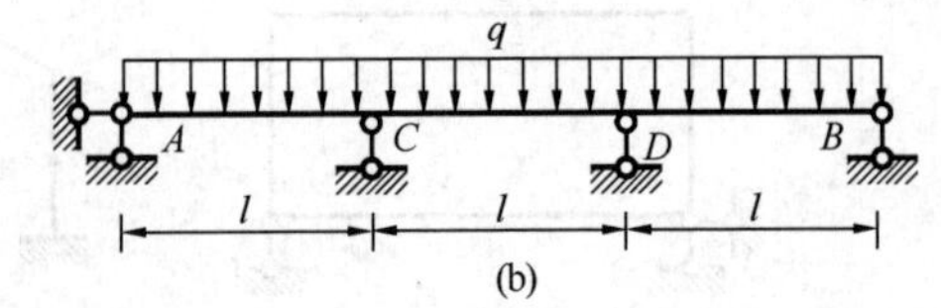

图 5.55　习题 5.5

Figure 5.55　Exercise 5.5

5.6　用力法计算图 5.56 所示结构，作 M，F_Q图，$I_1 = kI_2$。

5.6　Calculate structures in Figure 5.56 by force method, draw M, F_Q diagrams, $I_1 = kI_2$.

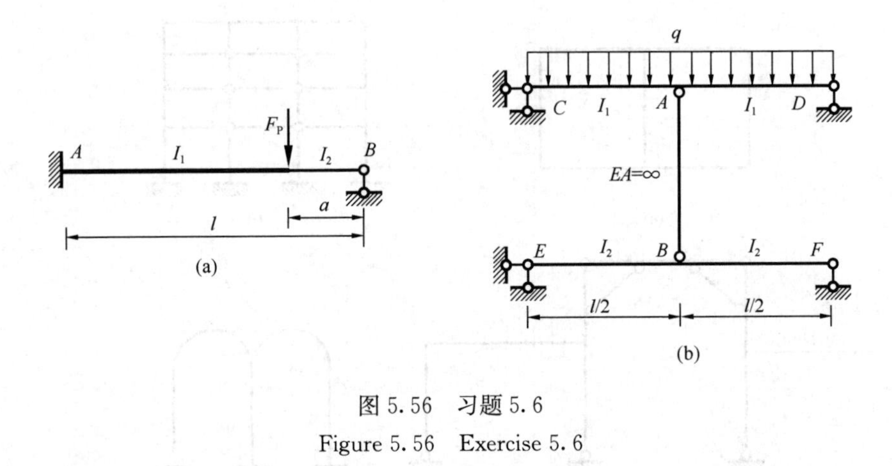

图 5.56　习题 5.6

Figure 5.56　Exercise 5.6

5.7　用力法计算图 5.57 所示刚架，作 M，F_Q，F_N图。

5.7　Calculate structures in Figure 5.57 by force method, draw M, F_Q, F_N diagrams.

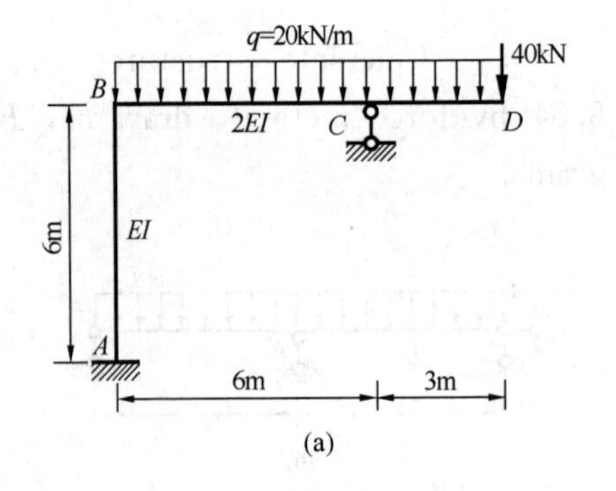

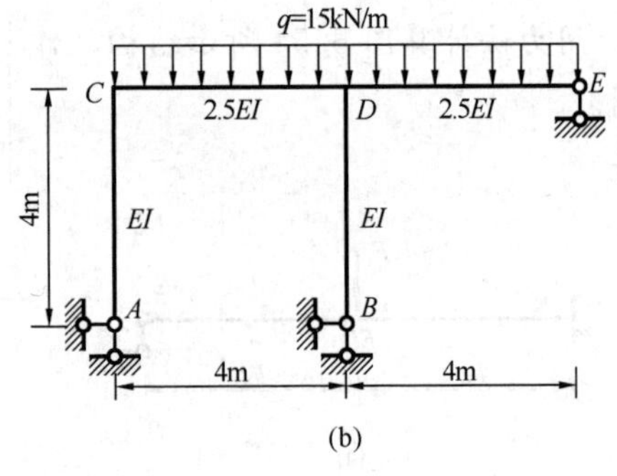

图 5.57　习题 5.7

Figure 5.57　Exercise 5.7

5.8　试用力法计算图 5.58 所示刚架，作 M，F_Q，F_N图。

5.8　Calculate rigid frames in Figure 5.58 by force method, draw M, F_Q, F_N diagrams.

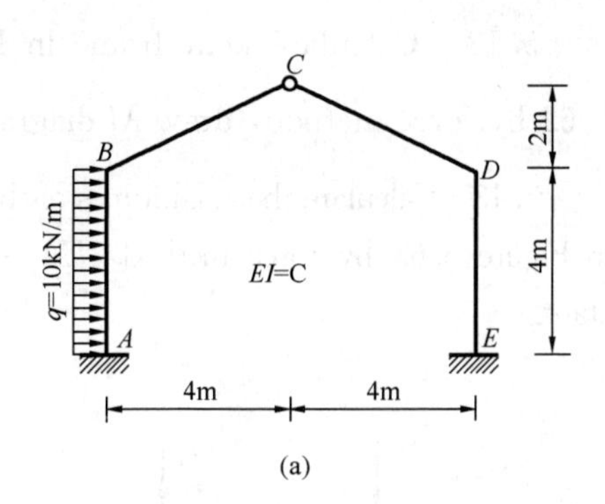

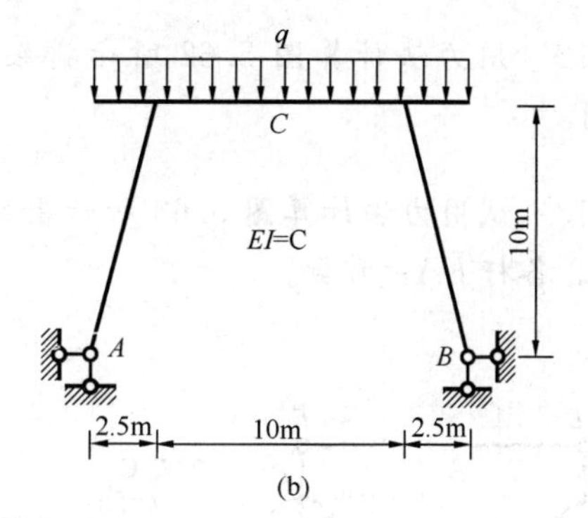

图 5.58 习题 5.8

Figure 5.58 Exercise 5.8

5.9 试用力法计算图 5.59 所示刚架，作 M，F_Q，F_N 图。

5.9 Calculate rigid frames in Figure 5.59 by force method, draw M, F_Q, F_N diagrams.

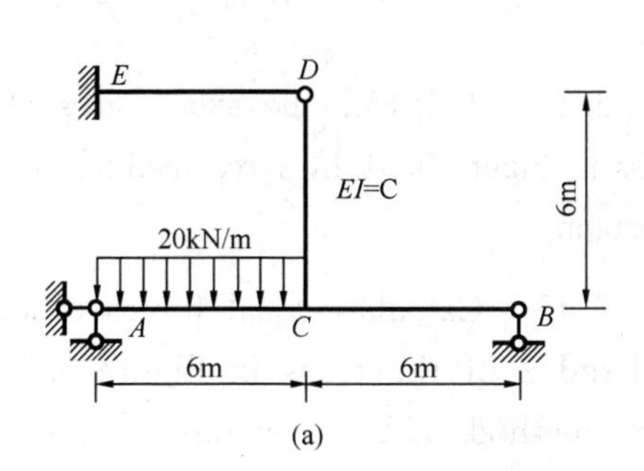

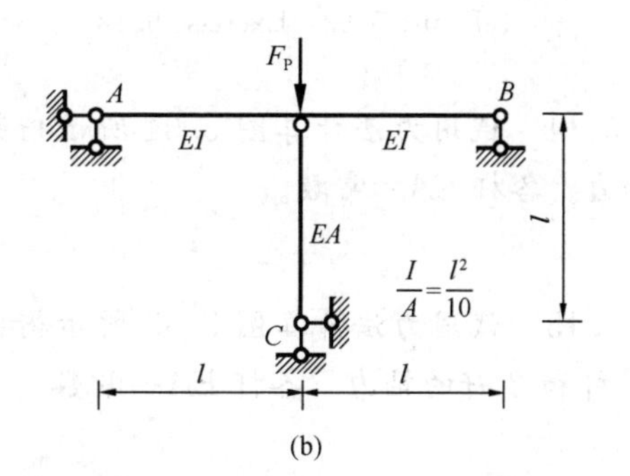

图 5.59 习题 5.9

Figure 5.59 Exercise 5.9

5.10 用力法计算图 5.60 所示排架，作 M 图。

5.10 Calculate bent frame in Figure 5.60 by force method, draw M diagram.

5.11 用力法计算图 5.61 所示排架，作 M 图。

5.11 Calculate bent frame in Figure 5.61 by force method, draw M diagram.

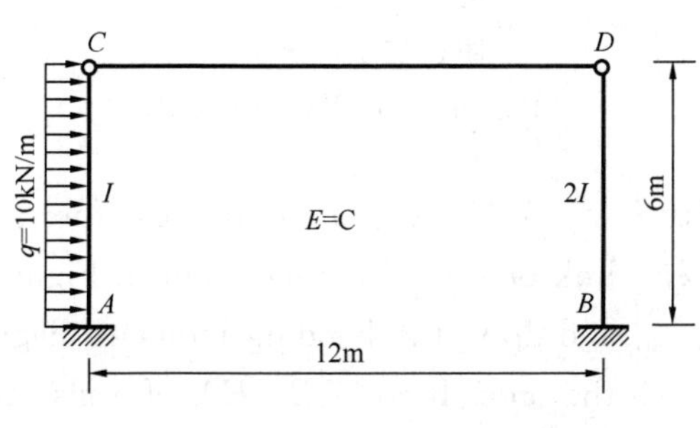

图 5.60 习题 5.10

Figure 5.60 Exercise 5.10

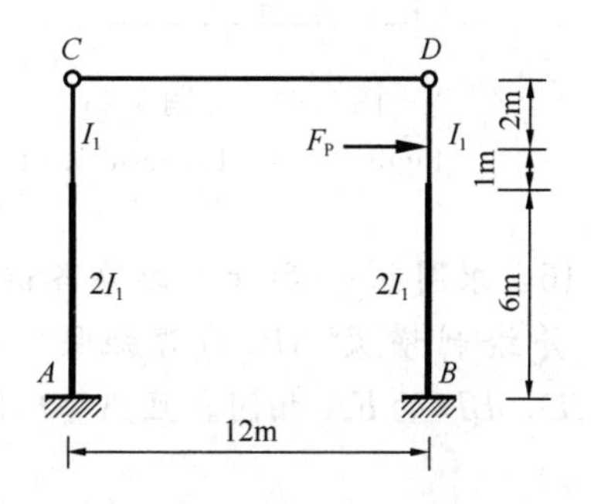

图 5.61 习题 5.11

Figure 5.61 Exercise 5.11

5.12 用力法计算图 5.62 所示排架，作 M 图。

5.12 Calculate bent frame in Figure 5.62 by force method, draw M diagram.

5.13 试用力法计算图 5.63 所示桁架的轴力。各杆 EA=常数。

5.13 Calculate the axial forces of the truss in Figure 5.63 by force method. EA = constant.

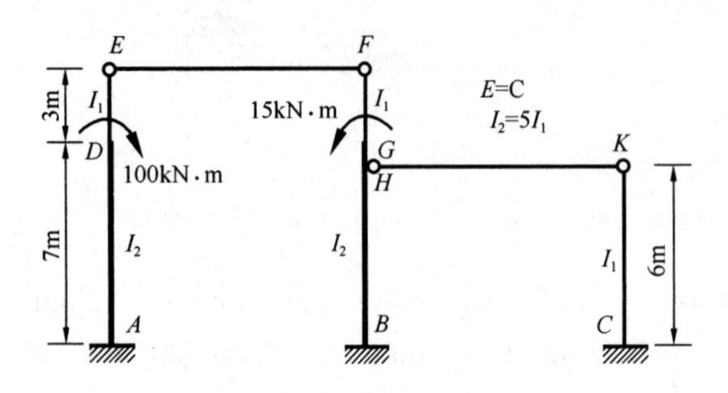

图 5.62 习题 5.12

Figure 5.62 Exercise 5.12

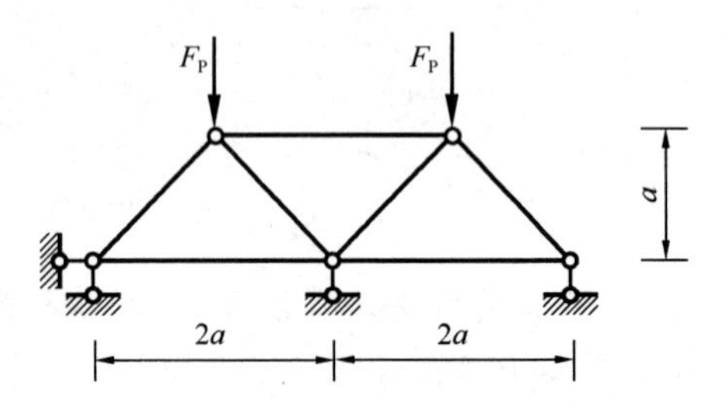

图 5.63 习题 5.13

Figure 5.63 Exercise 5.13

5.14 试用力法计算图 5.64 所示桁架的轴力。各杆 $EΛ$一常数。

5.14 Calculate the axial forces of the truss in Figure 5.64 by force method. EA = constant.

5.15 试用力法计算图 5.65 所示桁架中 1 杆和 2 杆的轴力。各杆 EA=常数。

5.15 Calculate axial forces of rod 1 and rod 2 of the truss in Figure 5.65 by force method. EA = constant.

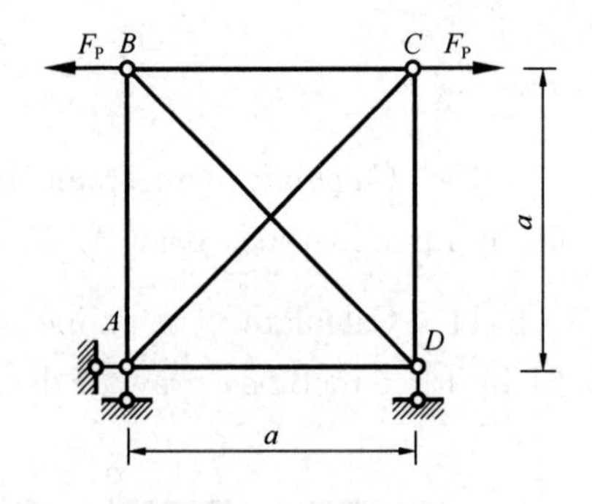

图 5.64 习题 5.14

Figure 5.64 Exercise 5.14

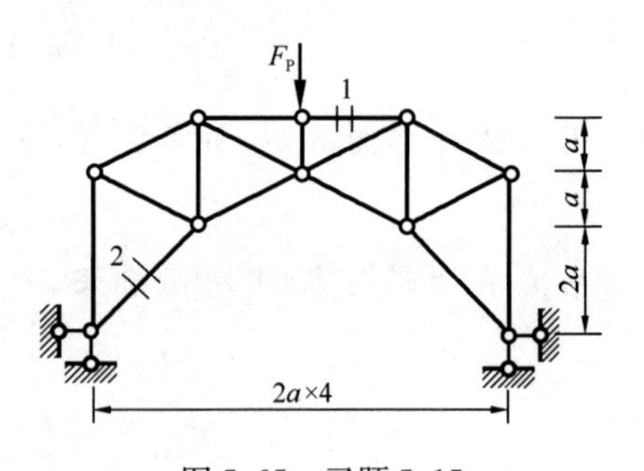

图 5.65 习题 5.15

Figure 5.65 Exercise 5.15

5.16 求图 5.66 所示加劲梁各链杆的轴力，并绘制横梁 AB 的弯矩图。设杆 AD、CD、BD 的 EA 相同，且 $A=I/16$。

5.16 Calculate the axial forces of each link of the stiffening beam in Figure 5.66, and draw the bending moment diagrams of the cross-beam AB. EA of rods AD, CD and BD are the same, $A=I/16$.

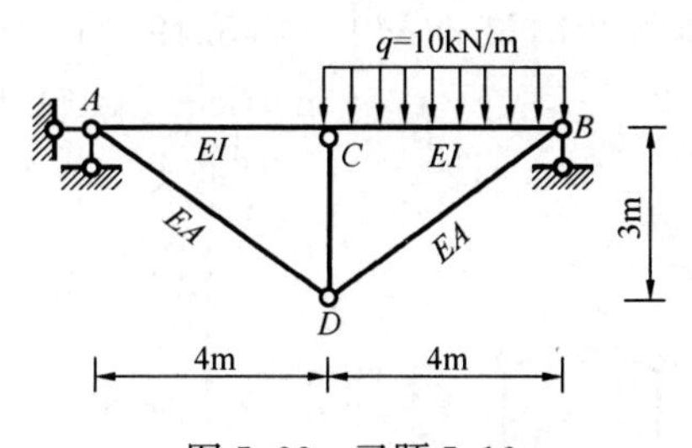

图 5.66 习题 5.16

Figure 5.66 Exercise 5.16

5.17 计算图 5.67 所示组合结构中各链杆的轴力，并绘制横梁的弯矩图。已知横梁的 $EI=1\times10^4\text{kN}\cdot\text{m}^2$，链杆的 $EA=15\times10^4\text{kN}$。

5.17 Calculate the axial forces of each link of the composite structure in Figure 5.67, and draw the bending moment diagram of the cross-beam. EI of the cross-beam is $EI=1\times10^4\text{kN}\cdot\text{m}^2$, EA of the chain rod is $EA=15\times10^4\text{kN}$.

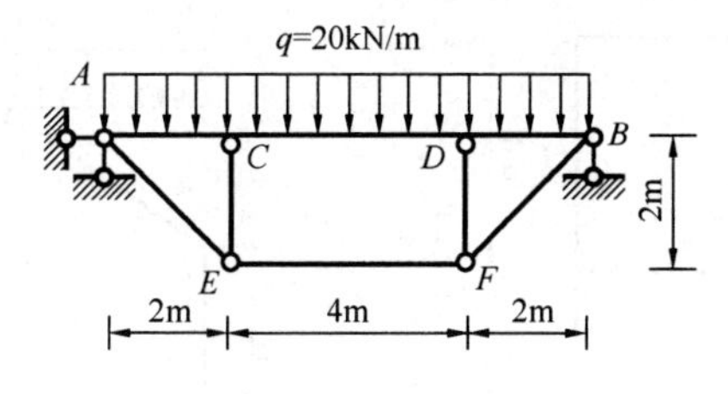

图 5.67 习题 5.17

Figure 5.67 Exercise 5.17

5.18 图 5.68 为一组合式吊车梁，上弦横梁截面的 $EI=1.4\times10^3\text{kN}\cdot\text{m}^2$，腹杆和下弦的 $EA=2.56\times10^5\text{kN}$。试计算各杆内力，作横梁的弯矩图。

5.18 Figure 5.68 shows a combined crane beam with $EI=1.4\times10^3\text{kN}\cdot\text{m}^2$ of the top chord beam, and $EA=2.56\times10^5\text{kN}$ of the web member and bottom chord. Try to calculate the internal forces of each rod and draw the bending moment diagram of the cross-beam.

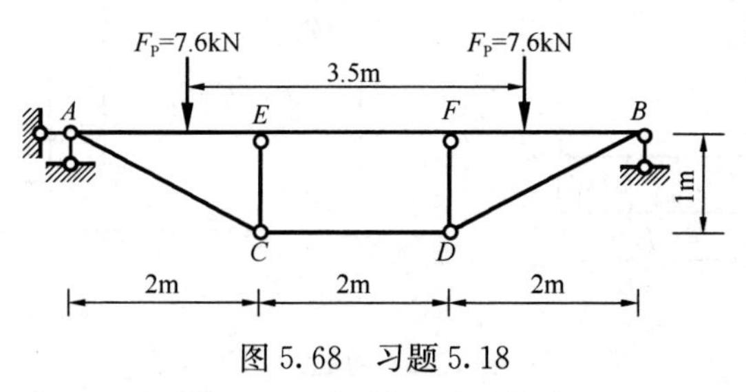

图 5.68 习题 5.18

Figure 5.68 Exercise 5.18

5.19　试作图5.69所示对称刚架的M图。

5.19　Try to draw M diagrams of symmetrical rigid frames in Figure 5.69.

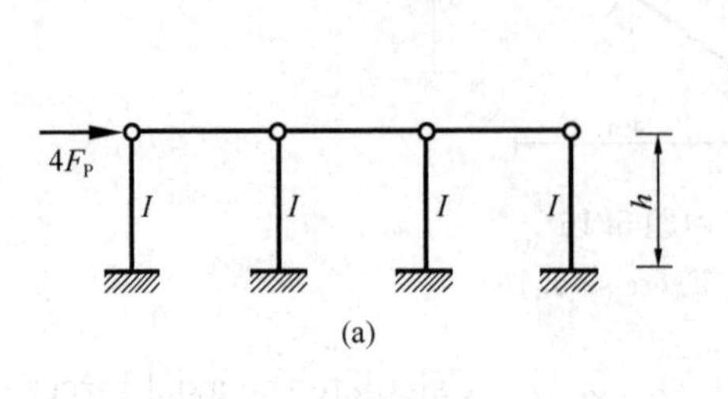

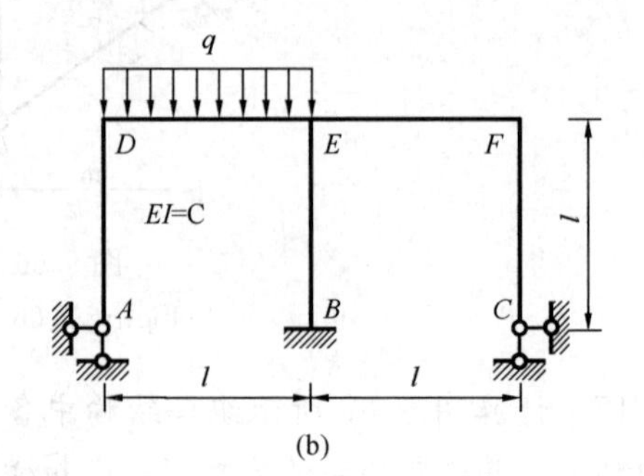

图5.69　习题5.19

Figure 5.69　Exercise 5.19

5.20　试作图5.70所示对称刚架的M图。

5.20　Try to draw M diagrams of symmetrical rigid frames in Figure 5.70.

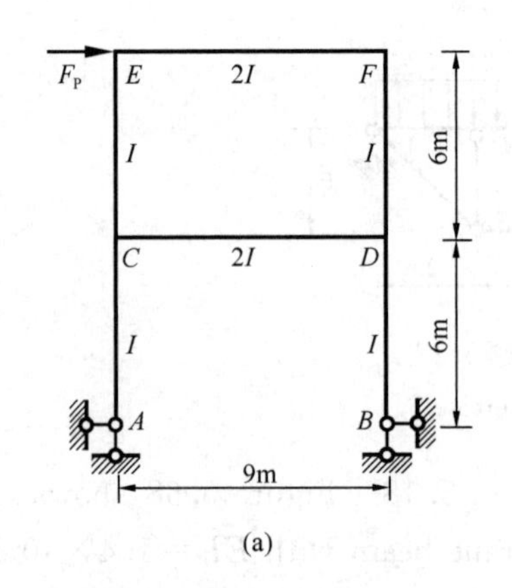

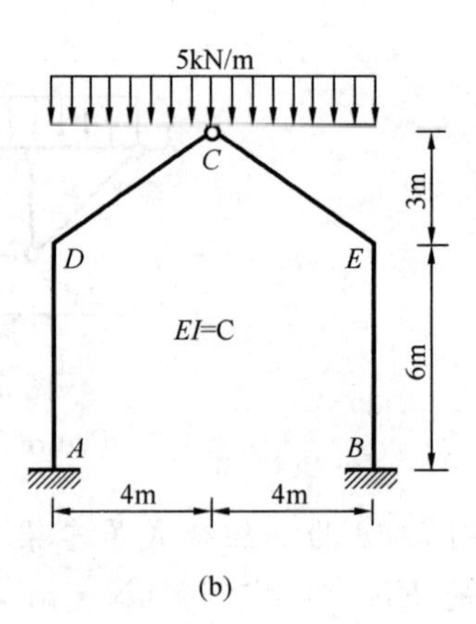

图5.70　习题5.20

Figure 5.70　Exercise 5.20

5.21　试作图5.71所示对称刚架的M图。

5.21　Try to draw M diagrams of symmetrical rigid frames in Figure 5.71.

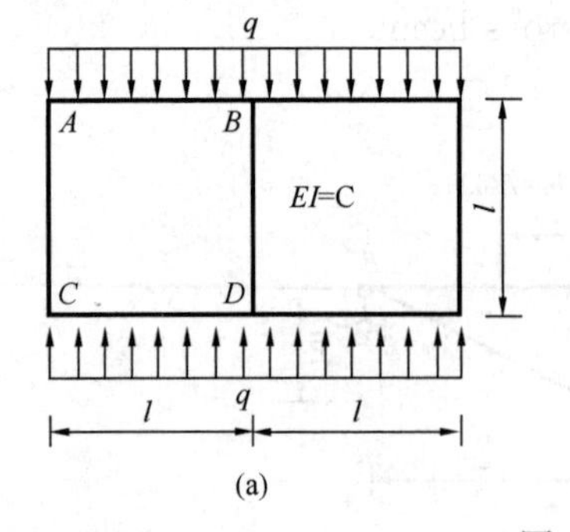

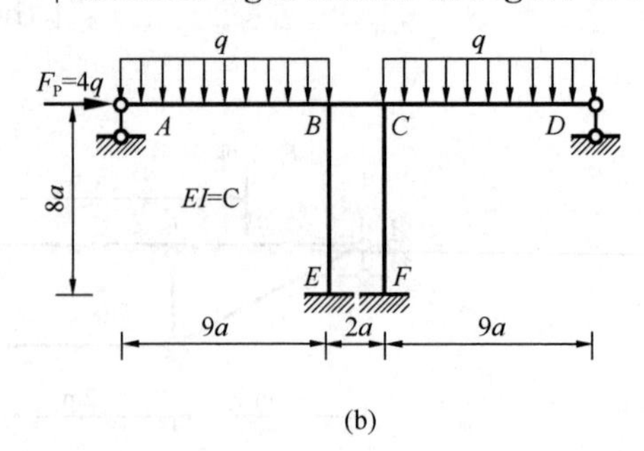

图5.71　习题5.21

Figure 5.71　Exercise 5.21

5.22 求作图 5.72 所示具有弹性支座结构的 M 图，图中弹性支座刚度 $k = 3EI/l^3$。

5.22 Draw M diagram of the structure with elastic support in Figure 5.72, and the stiffness of the elastic support is $k = 3EI/l^3$.

5.23 求作图 5.73 所示具有弹性支座结构的 M 图，图中弹性支座抗转动刚度 $k_\theta = EI/l$。

5.23 Draw M diagram of the structure with elastic support in Figure 5.73, and the resistant-rotation stiffness of the elastic support is $k_\theta = EI/l$.

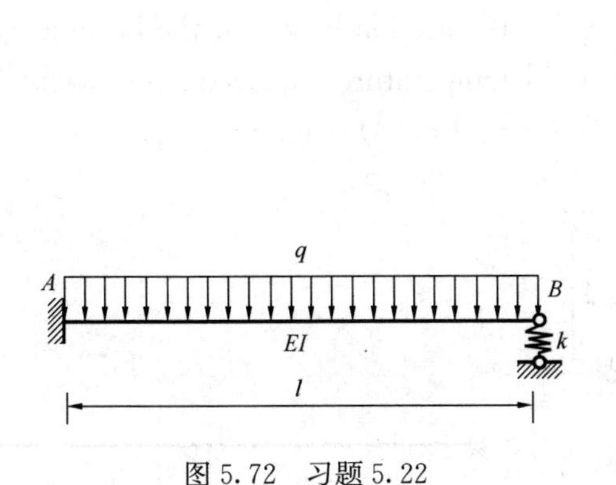

图 5.72 习题 5.22

Figure 5.72 Exercise 5.22

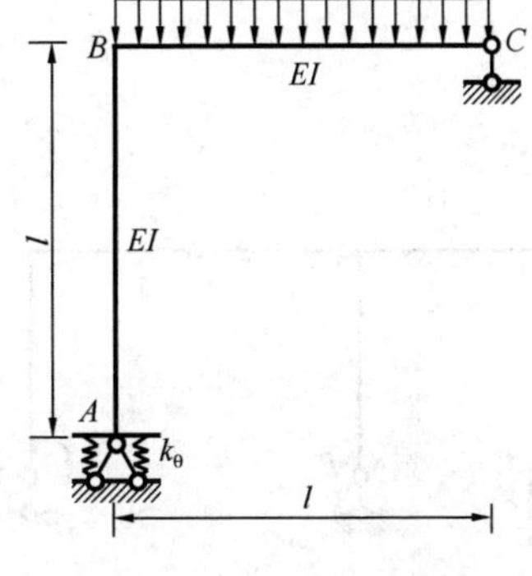

图 5.73 习题 5.23

Figure 5.73 Exercise 5.23

5.24 求作图 5.74 所示具有弹性支座结构的 M 图，图中弹性支座抗转动刚度 $k_\theta = EI/2$。

5.24 Draw M diagram of the structure with elastic support in Figure 5.74, and the resistant-rotation stiffness of the elastic support is $k_\theta = EI/2$.

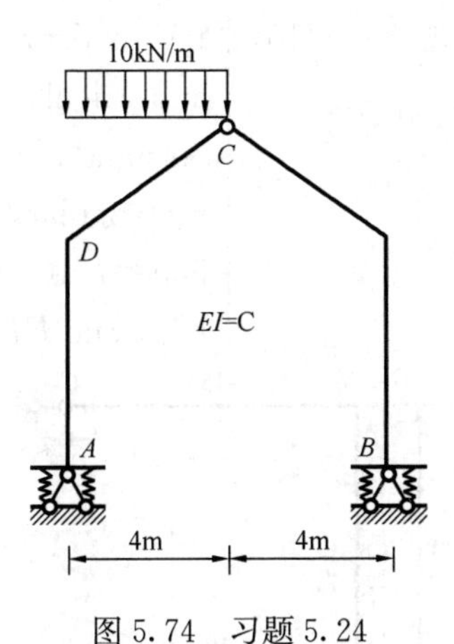

图 5.74 习题 5.24

Figure 5.74 Exercise 5.24

5.25 图 5.75 所示刚架支座 A 发生的水平位移和竖向位移分别为 $a=3\text{cm}$ 和 $b=1.5\text{cm}$，试求支座 C 截面的转角 ϕ_C 。

5.25 As shown in Figure 5.75, the horizontal displacement and vertical displacement of rigid frame support A are $a=3\text{cm}$ and $b=1.5\text{cm}$ respectively. Try to find out the angle ϕ_C in the section of support C.

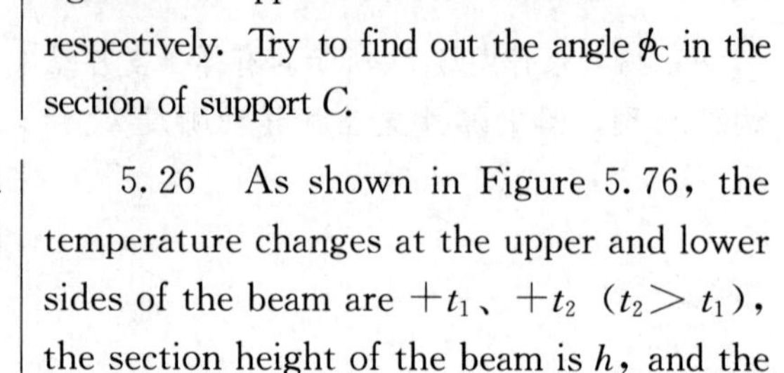

5.26 如图 5.76 所示，梁上、下侧温度变化分别为 $+t_1$、$+t_2$（$t_2>t_1$），梁截面高 h，温度膨胀系数 α，试作 M 图。

5.26 As shown in Figure 5.76, the temperature changes at the upper and lower sides of the beam are $+t_1$、$+t_2$ ($t_2>t_1$), the section height of the beam is h, and the temperature expansion coefficient is α, try to draw M diagram.

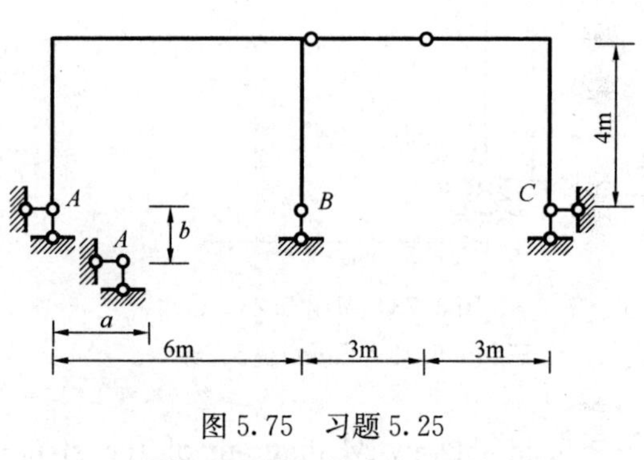

图 5.75 习题 5.25
Figure 5.75 Exercise 5.25

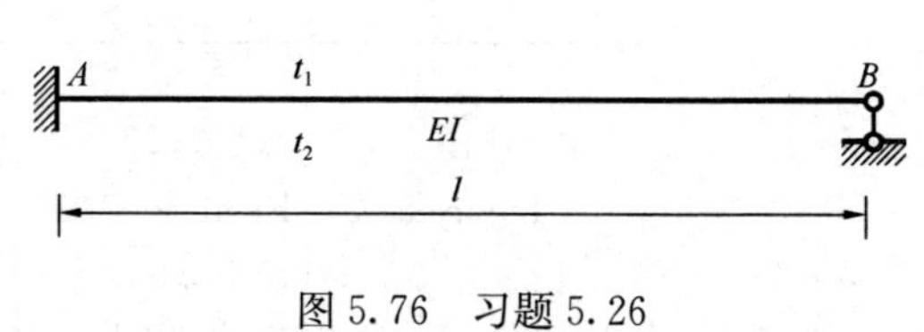

图 5.76 习题 5.26
Figure 5.76 Exercise 5.26

5.27 设结构的温度改变如图 5.77 所示，试绘制其弯矩图，并求 B 端的转角。设各杆截面为矩形，截面的高度 $h=l/10$，线膨胀系数为 α，EI 为常数。

5.27 Suppose the temperature change of the structure is shown in Figure 5.77, try to draw its bending moment diagram, and calculate the rotation angle of end B. Assume that the cross section of each rod is rectangular, the height of the cross section is $h=l/10$, the linear expansion coefficient is α, and EI is constant.

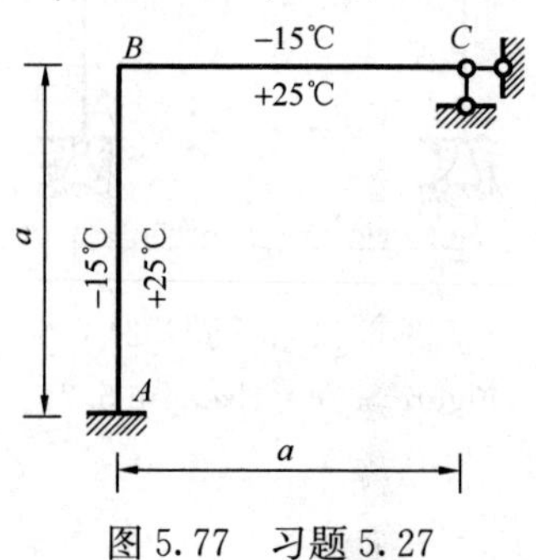

图 5.77 习题 5.27
Figure 5.77 Exercise 5.27

5.28 求作刚架因温度改变产生的弯矩图（图 5.78）。各杆截面为矩形，$h=l/10$，材料线膨胀系数为 α（E=常数）。

5.28 Draw bending moment diagram of rigid frame due to temperature change (Figure 5.78). The cross section of each rod is rectangular, $h=l/10$, and the linear expansion coefficient of the material is α (E = constant).

5.29 设图 5.79 所示梁 B 端下沉 c，试作梁的 M 图和 F_Q 图。

5.29 Suppose that a settlement c occurs at the end B of the beam in Figure 5.79, try to draw M diagram and F_Q diagram of the beam.

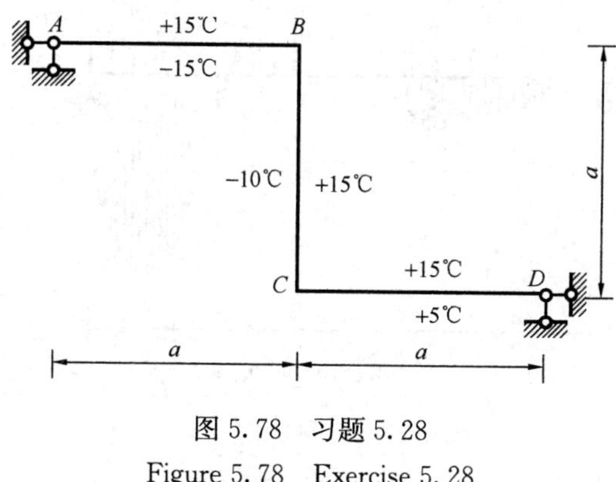

图 5.78 习题 5.28

Figure 5.78 Exercise 5.28

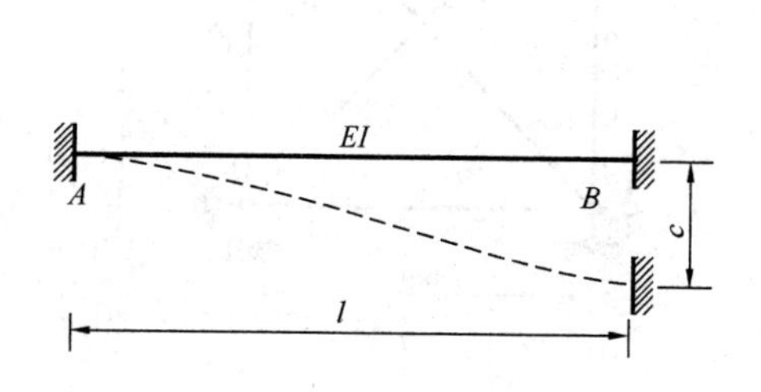

图 5.79 习题 5.29

Figure 5.79 Exercise 5.29

5.30 求图 5.80 所示刚架支座 A 发生转角位移 θ 时的弯矩图及 C 端的水平位移。

5.30 Calculate the bending moment diagram and horizontal displacement at end C when the angular displacement θ occurs at support A in Figure 5.80.

5.31 计算图 5.81 所示排架 C 点的水平位移 Δ_C。

5.31 Calculate the horizontal displacement Δ_C of the point C in bent frame in Figure 5.81。

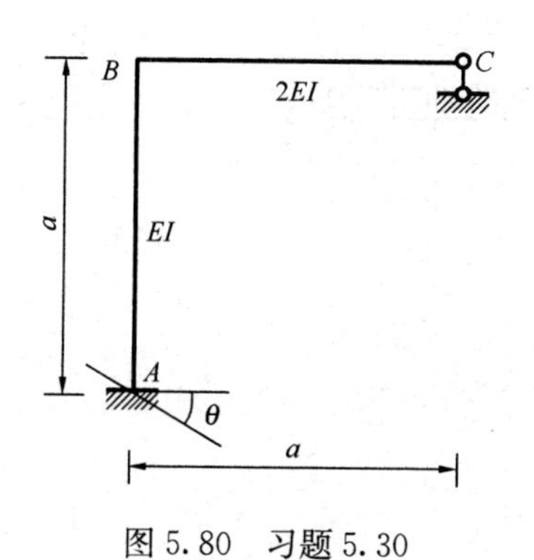

图 5.80 习题 5.30

Figure 5.80 Exercise 5.30

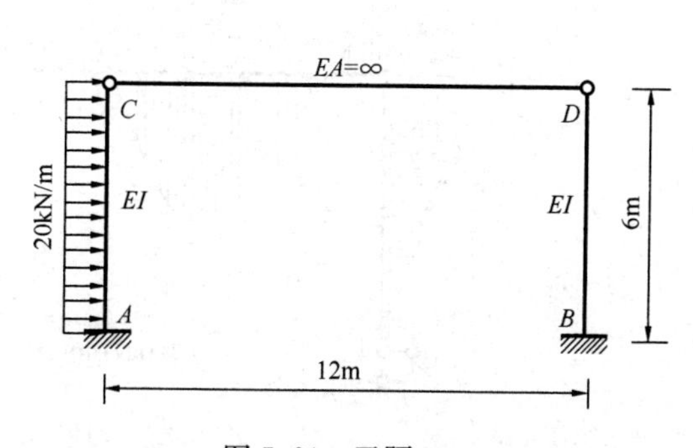

图 5.81 习题 5.31

Figure 5.81 Exercise 5.31

5.32 已知荷载作用下桁架各杆的内力如图5.82所示，试求节点D的水平位移。EI=常数。

5.32 The internal forces of each rod of the truss are shown in Figure 5.82. Try to calculate the horizontal displacement of node D. EI = constant.

5.33 图5.83所示连续梁EI=常数，已知其弯矩图（注意图中弯矩值均须乘以$qa^2/1000$）。据此计算截面C的转角。

5.33 The bending moment diagram of a continuous beam is shown in Figure 5.83, EI = constant (Note that the bending moment value has to be multiplied by $qa^2/1000$). Calculate the rotation angle of section C.

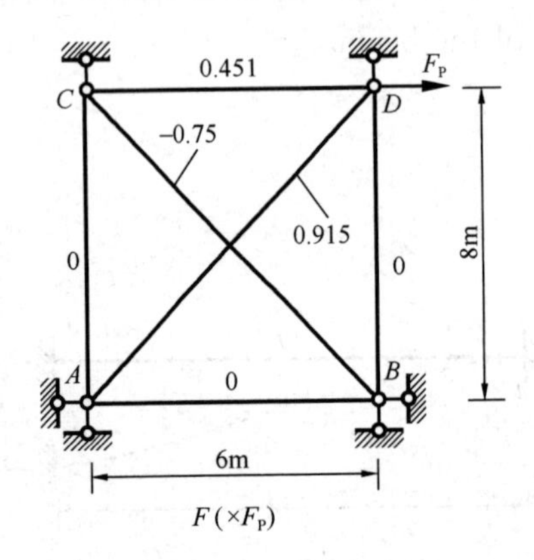

图5.82 习题5.32

Figure 5.82 Exercise 5.32

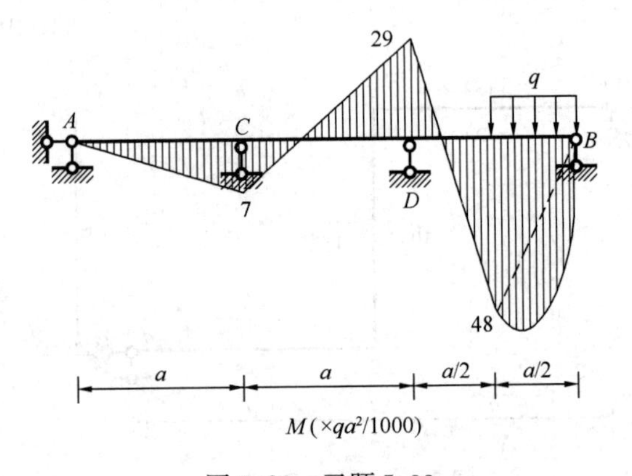

图5.83 习题5.33

Figure 5.83 Exercise 5.33

5.34 已知图5.84所示结构的M图（仅BD杆承受向下均布荷载，中点弯矩10.5kN·m），求C点竖向位移Δ_{CV}。各杆EI相同，杆长均为l=2m。

5.34 M diagram of the structure is shown in Figure 5.84 (Only BD rod bears the downward uniformly distributed load, and the midpoint bending moment is 10.5kN·m), calculate the vertical displacement Δ_{CV} of point C. EI of each rod are the same, and the rod length is l=2m.

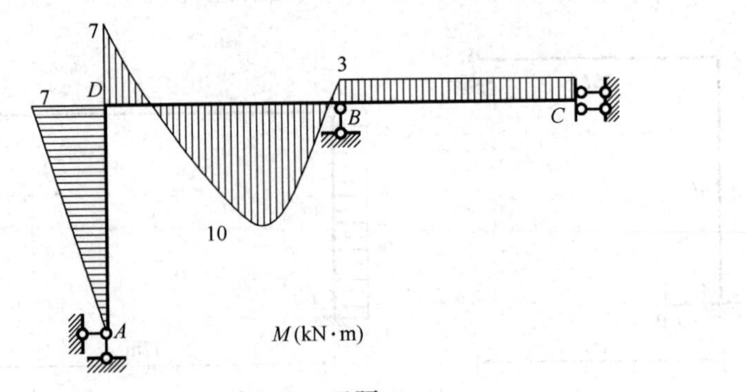

图5.84 习题5.34

Figure 5.84 Exercise 5.34

5.35 图 5.85 所示桁架，各杆长度均为 l，EA 相同。但杆 AB 制作时短了 Δ，将其拉伸（在弹性极限内）后进行装配。试求装配后杆 AB 的长度。

5.35 As shown in Figure 5.85, the length of each rod is l, and EA is the same. But during production rod AB shortens Δ, which is assembled after stretching (within the elastic limit). Try to find out the length of rod AB after assembly.

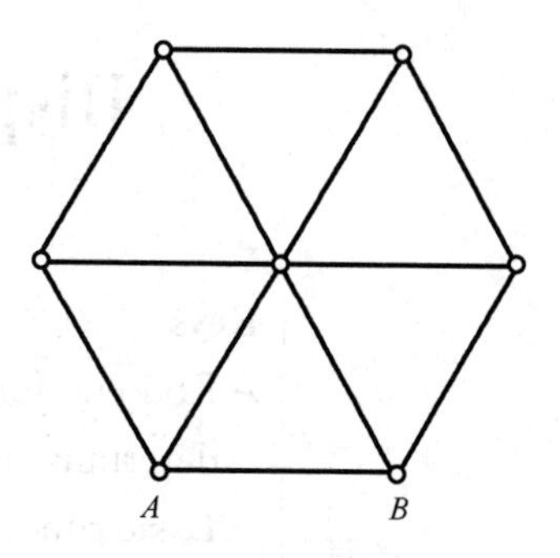

图 5.85 习题 5.35
Figure 5.85 Exercise 5.35

5.36 试判断图 5.86 所示超静定结构弯矩图形是否正确，并说明理由。

5.36 Try to judge whether statically indeterminate structures' bending moment diagrams are correct or not in Figure 5.86 and give the reasons.

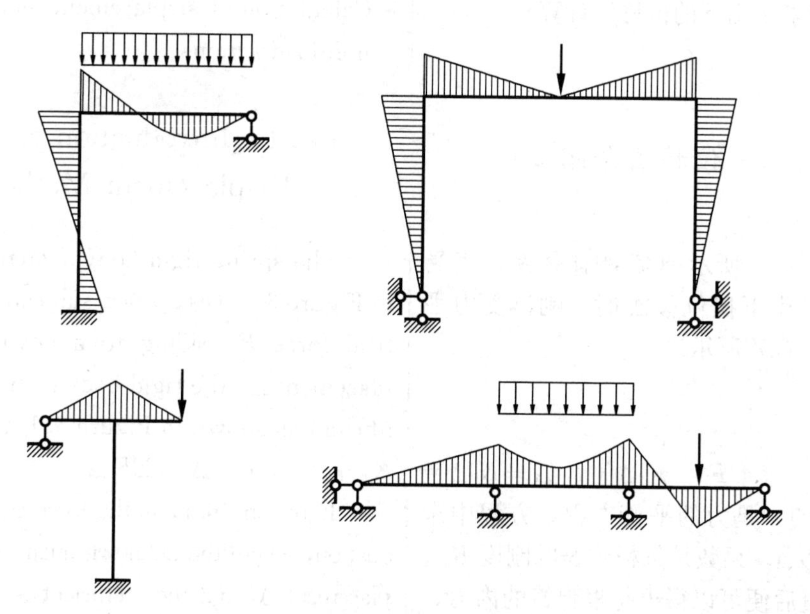

图 5.86 习题 5.36
Figure 5.86 Exercise 5.36

第6章 Chapter 6

位移法 Displacement Method

要点

- 单跨超静定梁的杆端力
- 位移法的基本概念
- 位移法基本未知量，即基本体系
- 无侧移刚架的计算
- 有侧移刚架的计算
- 非荷载因素作用下的位移法计算

Keys

- Rod-end forces in single-span statically indeterminate beams
- Basic concepts of the displacement method
- Fundamental unknown quantities, i. e. , fundamental system in displacement method
- Calculation of rigid frame without lateral displacements
- Calculation of rigid frame with lateral displacements
- Calculation of displacement method under non-load actions

6.1 位移法介绍

6.1 Introduction of the Displacement Method

图 6.1（a）所示弹簧-刚体体系，当受竖向力 F 产生下拉位移 Δ 时，刚体受力平衡如图 6.1（c）所示。

The spring-rigid body system is shown in Figure 6.1 (a), when subjected to a vertical force F leading to a downward displacement Δ, the rigid body is in force equilibrium as shown in Figure 6.1 (c).

$$F = K_1\Delta + K_2\Delta + K_3\Delta = (K_1 + K_2 + K_3)\Delta = \Sigma K_i\Delta \tag{6.1}$$

式（6.1）是力的平衡方程，方程中未知量是位移 Δ，系数是每根弹簧的刚度 K_i 。求出未知量后便可以得出每根弹簧的内力。

Equation (6.1) is the force equilibrium equation, where the unknown quantity is the displacement Δ, and the coefficient is the stiffness K_i of each spring. The internal force of each spring can be obtained after finding the unknown quantities.

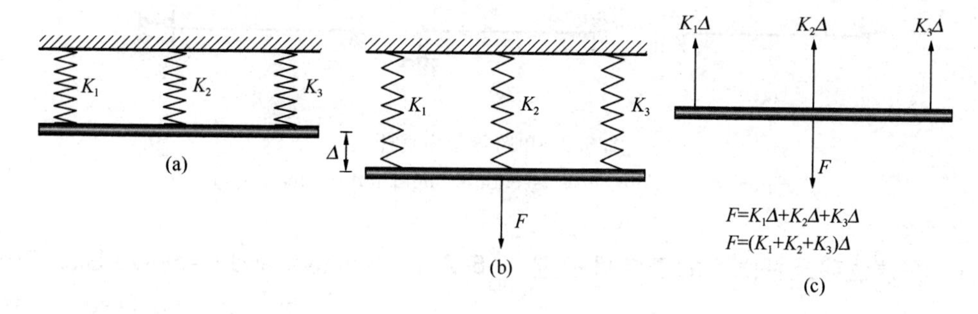

图 6.1 位移法

Figure 6.1 Displacement method

这一过程可分为以下几个步骤：

The process can be divided into the following steps:

（1）拆分单元：拆分出三个弹簧；

(1) Splitting of the units: Three springs are split;

（2）单元分析：获取每个弹簧的刚度；

(2) Unit analysis: Obtain the stiffnesses of each spring;

（3）整体分析：由位移未知量组合弹簧弹力，形成力的平衡方程。

(3) Overall analysis: The spring elasticity is combined from the displacement unknowns to form force equilibrium equations.

然后加以求解，这就是位移法的基本思想。

Then the equations are solved, which is the basic idea of the displacement method.

6.2 位移法的基本单元

6.2 Basic Units of the Displacement Method

单元杆端力与杆端位移之间的函数关系——杆件的转角位移方程，是位移法单元分析的基础。位移法中常见的等截面直杆包括以下三种，如图 6.2 所示。

The functional relationship between the force at the end and the displacement at the end of the rod—the equation for the angular displacement of the rod—is the basis for the unit analysis of the displacement method. Three common types of straight rods of an equal cross-section in the displacement method are shown in Figure 6.2.

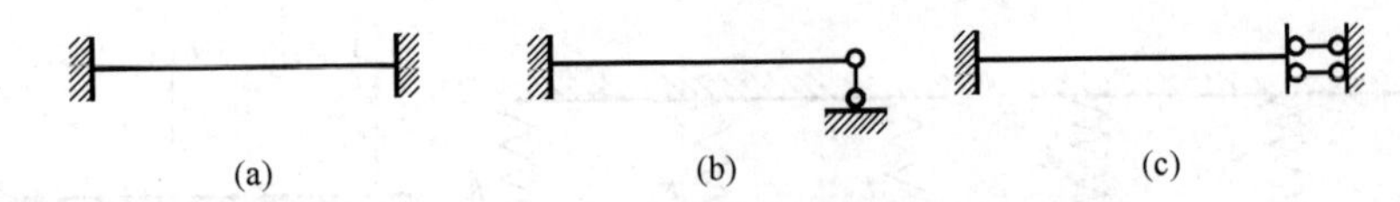

图 6.2 单跨等截面直杆

Figure 6.2 Equal cross-section straight rod of single span

6.2.1 杆端位移及杆端力的正负号规定

6.2.1 Positive and Negative Sign Provisions for Rod End Displacements and Rod End Forces

6.2.1.1 杆端内力的正负号

6.2.1.1 Positive and Negative Signs for Internal Forces at Rod Ends

杆端弯矩 M_{AB} 及 M_{BA} 以绕杆端顺时针转动为正（相应地，绕固端逆时针为正）；如图 6.3（a）所示（注：M_{AB} 下标的记法——A 近端；B 远端）。

Rod end bending moments M_{AB} and M_{BA} with clockwise rotation around the rod end are set positive (correspondingly, counter-clockwise around the solid end is set positive), as shown in Figure 6.3 (a) (Note: The notation of subscripts for M_{AB}—A represents proximal end and B represents distal end).

杆端剪力仍以绕截面顺时针转动为正，杆端轴力以拉力为正。

The shear force at the rod end is still positive with clockwise rotation around the section, and the axial force at the rod end is positive in tension.

6.2.1.2 杆端位移的正负号

6.2.1.2 Positive and Negative Signs for Rod End Displacements

杆端转角以顺时针为正，见图 6.3（b）；杆件弦转角如图 6.3（c）所示，定义见式（6.2）：

The rod end's rotation angle is positive clockwise, as shown in Figure 6.3 (b); the chord rotation angle of the rod is shown in Figure 6.3 (c) and is defined in Equation (6.2):

$$\beta=\frac{\Delta}{l} \tag{6.2}$$

侧移 Δ 以所引起的弦转角 β 顺时针为正，如图 6.3（c）所示。

The lateral displacement Δ is positive resulting chord rotation angle β in clockwise, as shown in Figure 6.3 (c).

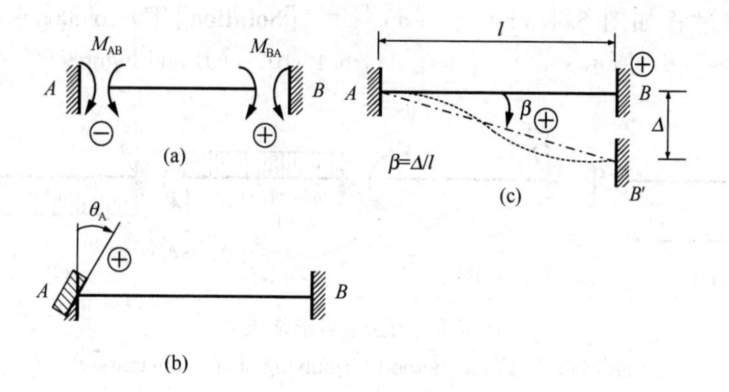

图 6.3 正负号规定

(a) 杆端弯矩；(b) 杆端转角；(c) 侧移

Figure 6.3 Specification of positive and negative signs

(a) Rod end bending moments; (b) Rod end rotation angles; (c) Lateral displacements

6.2.2 等截面直杆的形常数

6.2.2 Shape Constants for Straight Rods with Equal Section

令杆件的线刚度 $i=EI/l$，方便后面表达。单跨梁在单位杆端位移下的杆端内力，称为**形常数**。形常数通常可以由力法求解，下面用例 6.1 和例 6.2 加以说明。

Set the linear stiffness of the bar $i = EI/l$ convenient for later expression. The internal force at the rod end of a single-span beam by unit rod end displacement is called the **shape constant.** The shape constant can usually be solved by the force method and is illustrated below using Example 6.1 and Example 6.2.

【例 6.1】用力法求解图 6.4（a）、图 6.5（a）结构的形常数。

［**Example 6.1**］Use force method to solve the shape constants of the structures shown in Figure 6.4（a）and Figure 6.5（a）.

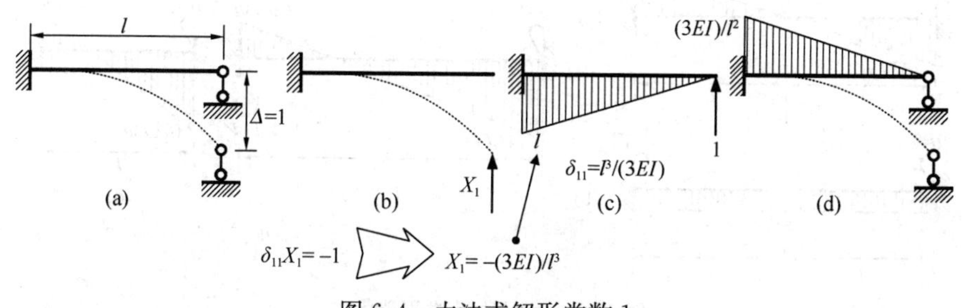

图 6.4 力法求解形常数 1

Figure 6.4 Force method for solving shape constants 1

【解】求解过程如图 6.4（b）～（d）和图 6.5（b）～（d）所示。

[Solution] The solution is shown in Figures 6.4 (b) - (d) and Figures 6.5 (b) - (d).

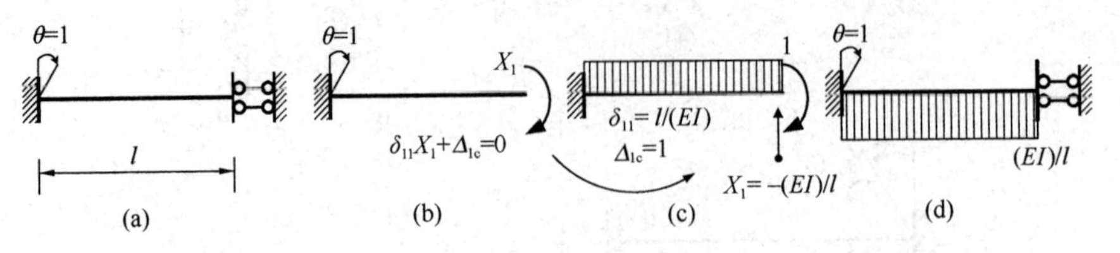

图 6.5　力法求解形常数 2

Figure 6.5　Force method for solving shape constants 2

【例 6.2】结合例 6.1 结果，用虚功原理求形常数。

[Example 6.2] Combine the results of Example 6.1 to find out the shape constants using the principle of virtual work.

【解】如图 6.6 所示。

[Solution] The solution is shown in Figure 6.6.

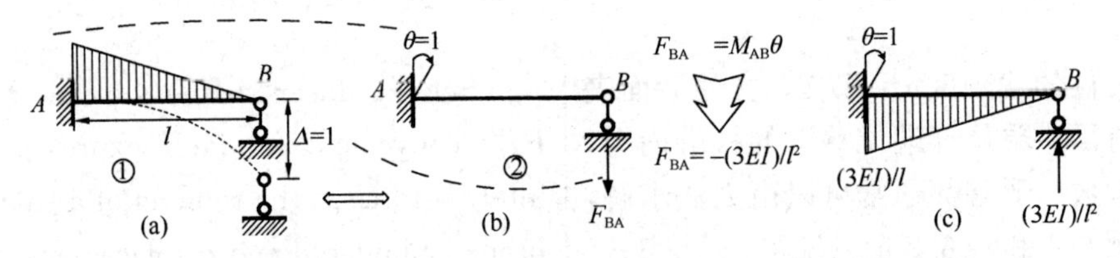

图 6.6　虚功原理求形常数

Figure 6.6　The principle of virtual work to find out the shape constants

【例 6.3】利用对称性，扩展例 6.1 和例 6.2 结果。

[Example 6.3] Using symmetry, extend the results of Example 6.1 and Example 6.2.

【解】如图 6.7 所示。

[Solution] The solution is shown in Figure 6.7.

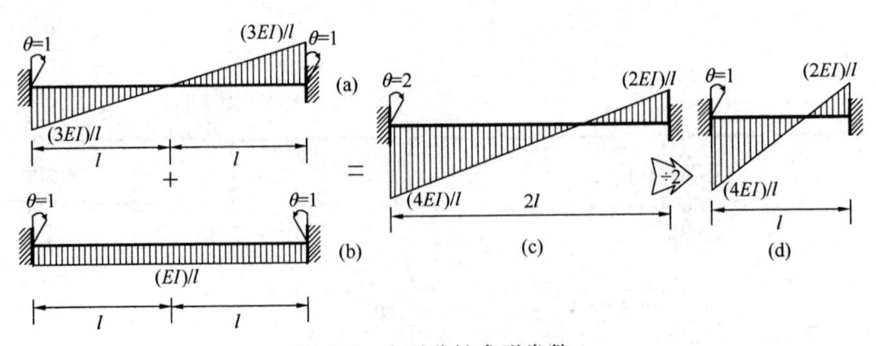

图 6.7　由对称性求形常数 1

Figure 6.7　Solving shape constants by using symmetry 1

【例 6.4】利用对称性，扩展例 6.1 结果。

[**Example 6.4**] Using symmetry, extend the results of Example 6.1.

【解】如图 6.8 所示。

[**Solution**] The solution is shown in Figure 6.8.

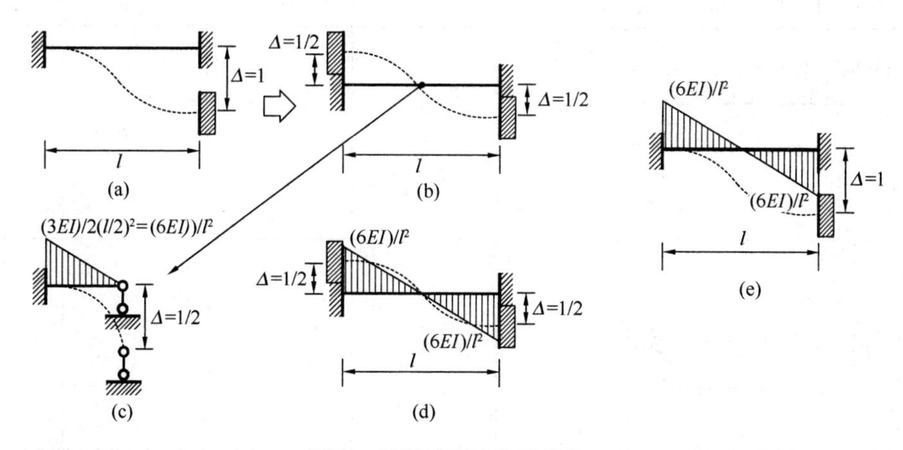

图 6.8 由对称性求形常数 2

Figure 6.8 Solving shape constants by using symmetry 2

至此，所有形常数推导完毕，汇总至表 6.1。

So far, all shape constants are derived and summarized in Table 6.1.

等截面直杆的形常数 **表 6.1**

Shape constants for straight rods with equal section **Table 6.1**

AB 杆件形常数，EI=C，杆长 l Shape constants of rod AB, EI = C, rod length l	杆端弯矩 Rod end bending moment		杆端剪力 Rod end shear force	
弯矩图 Bending moment diagram	M_{AB}	M_{BA}	F_{QAB}	F_{QBA}
$\theta=1$, $(2EI)/l$, $(4EI)/l$	$4i$	$2i$	$-6i/l$	$-6i/l$
$(6EI)/l^2$, $\Delta=1$, $(6EI)/l^2$	$-6i/l$	$-6i/l$	$12i/l^2$	$12i/l^2$
$\theta=1$, B, $(3EI)/l$	$3i$	0	$-3i/l$	$-3i/l$

续表

Continued

AB杆件形常数，EI=C，杆长l Shape constants of rod AB, EI = C, rod length l	杆端弯矩 Rod end bending moment		杆端剪力 Rod end shear force	
弯矩图 Bending moment diagram	M_{AB}	M_{BA}	F_{QAB}	F_{QBA}
(3EI)/l² Δ=1	$-3i/l$	0	$3i/l^2$	$3i/l^2$
θ=1 (EI)/l	i	$-i$	0	0

1. 形常数表6.1中杆端弯矩的系数可汇总口诀方便记忆：

1. The coefficients of the rod end bending moments in Table 6.1 of shape constants can be summarized by the following mnemonic for easy memory:

"近4，远2，侧—6；近角3，侧—3；近角i，远—i"

"near 4, far 2, side-6; near angle 3, side-3; near angle i, far-i"

2. 杆端剪力不用记忆，可以由弯矩图推导得到。

2. The shear force at the rod end does not need to be memorized and can be derived from the bending moment diagram.

3. 形常数仅与自身特性相关，与荷载无关。

3. The shape constants are only related to their properties and are not related to the loads.

6.2.3 等截面直杆的载常数

6.2.3 Load Constants for Straight Rods with Equal Section

由荷载等外界作用引起的杆端内力称为**载常数**。载常数的推导可以用力法，下面用简便方法推导。首先，两端固结等截面直杆的载常数可以由无铰封闭框弯矩图校核中的面积规律快速得出，如图6.9所示。

The internal force at the rod end caused by external actions is called the **load constant**. The derivation of the load constant can be done by force method, which is derived in the following simple way. First, the load constant for a straight rod with equal section of two solid ends can be quickly derived from the area law in the bending moment diagram calibration for a hingeless closed frame, as shown in Figure 6.9.

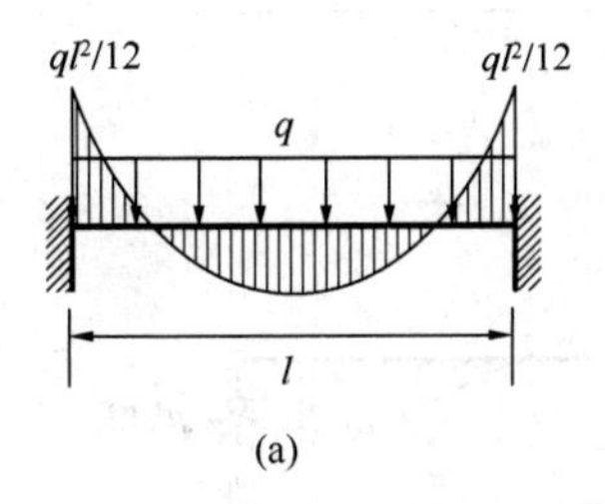

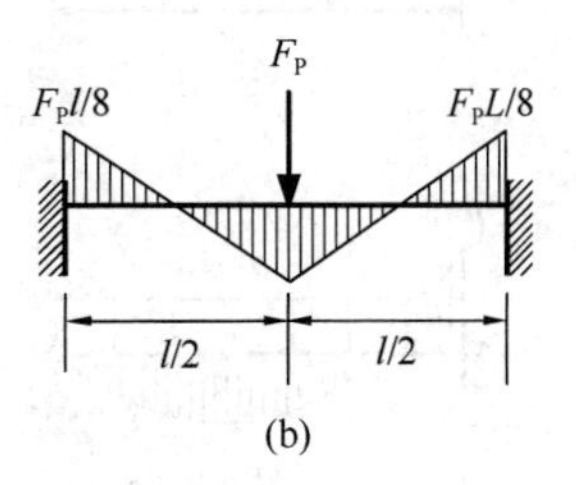

图 6.9 两端固结等截面直杆的载常数

Figure 6.9 Load constants for a straight rod with equal section of two solid ends

【例 6.5】 利用对称性，基于图 6.9，推导一端固结一端滑动等截面直杆的载常数。

［**Example 6.5**］Using symmetry, derive the load constants for a straight rod with equal section of one solid end and one sliding end, based on Figure 6.9.

【解】 如图 6.10 所示。

［**Solution**］The results can be shown in Figure 6.10.

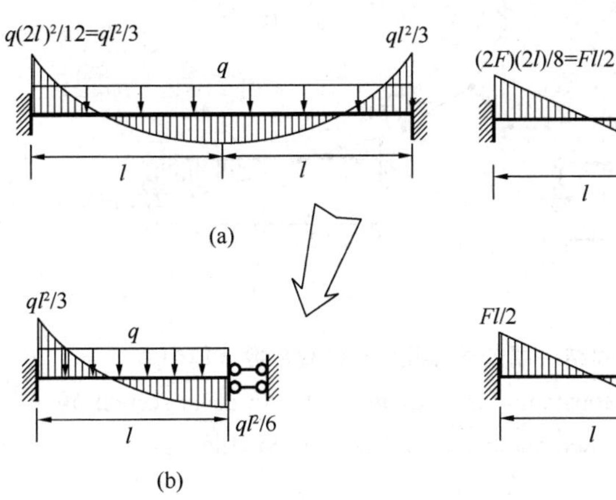

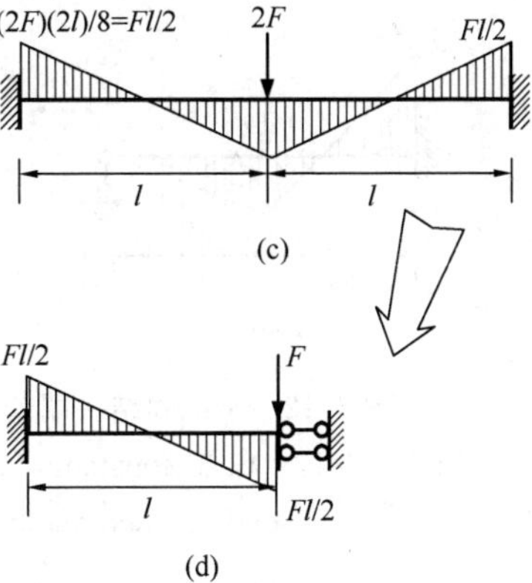

图 6.10 一端固结一端滑动等截面直杆的载常数

Figure 6.10 Load constants for a straight rod with equal section of one solid end and one sliding end

【例 6.6】 推导一端固结一端铰支等截面直杆的载常数（图 6.11、图 6.12）。

［**Example 6.6**］Derive the load constants for straight rods with equal section of one solid end and one hinge end (Figures 6.11 and 6.12).

【解】 如图 6.11、图 6.12 所示。

［**Solution**］The results can be shown in Figures 6.11 and 6.12.

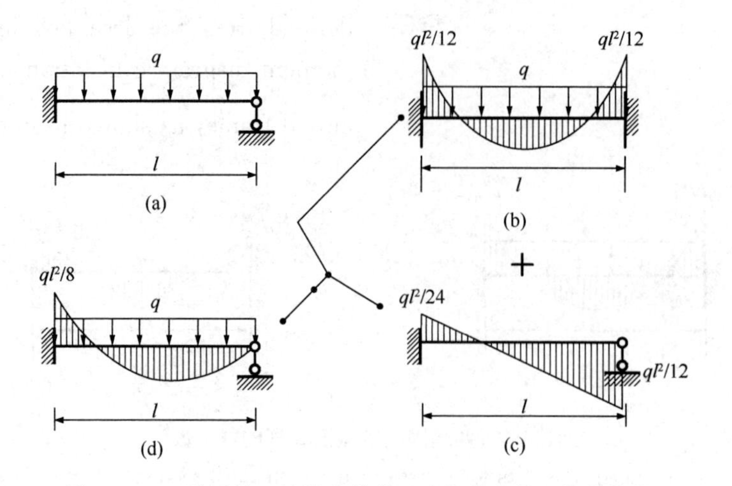

图 6.11　一端固结一端铰支等截面直杆载常数（均布荷载）

Figure 6.11　Load constants for a straight rod with equal section of one solid end and one hinge end (uniform loads)

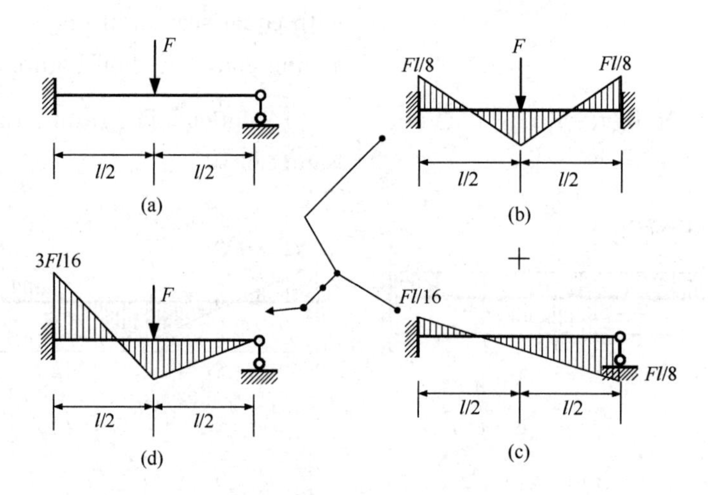

图 6.12　一端固结一端铰支等截面直杆载常数（集中力）

Figure 6.12　Load constants for a straight rod with equal section of one solid end and one hinge end (concentrated loads)

【例 6.7】利用对称性，求图 6.13（a）中两端固结等截面直杆在集中力偶作用下的载常数；然后在此基础上，求图 6.14（a）中一端固结一端铰支等截面直杆在集中力偶作用下的载常数。

[Example 6.7] Use symmetry, find out the load constants of the straight rod with equal section of two solid ends in Figure 6.13 (a) under the action of a concentrated couple; then, based on this, find out the load constants of the straight rod with equal section of one solid end and one hinge end in Figure 6.14(a) under the action of a concentrated couple.

【解】如图 6.13 和图 6.14 所示。

［**Solution**］The results are shown in Figures 6.13 and 6.14.

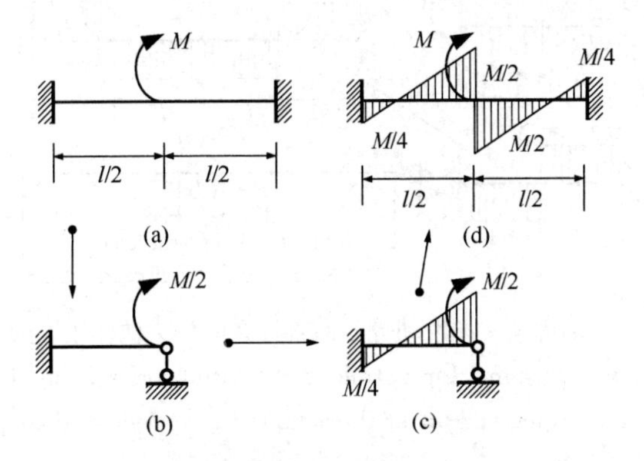

图 6.13 两端固结等截面直杆在集中力偶作用下的载常数

Figure 6.13 Load constants of a straight rod with equal section of two solid ends under a concentrated couple

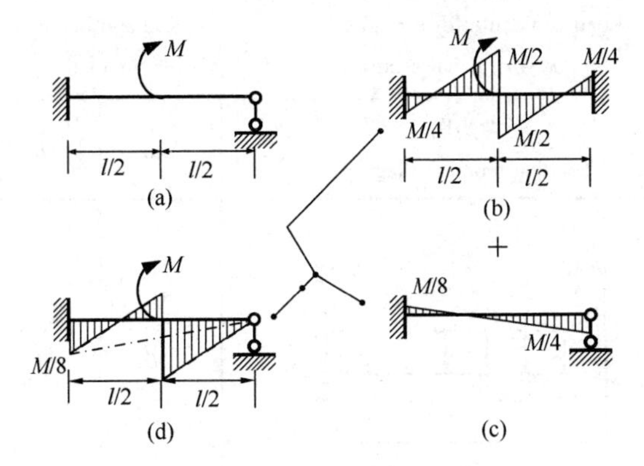

图 6.14 一端固结一端铰支等截面直杆在集中力偶作用下的载常数

Figure 6.14 Load constants of a straight rod with equal section of one solid end and one hinge end under a concentrated couple

【例 6.8】求图 6.15（a）中一端固结一端滑动的等截面直杆在集中力偶作用下的载常数。

［**Example 6.8**］Find out the load constants for a straight rod with equal section of one solid end and one sliding end in Figure 6.15 (a) under the action of a concentrated couple.

【解】如图 6.15 所示。

［**Solution**］The result is shown in Figure 6.15.

至此，常见载常数推导完毕，汇总至表 6.2 中。

So far, the common load constants have been derived and are summarized in Table 6.2.

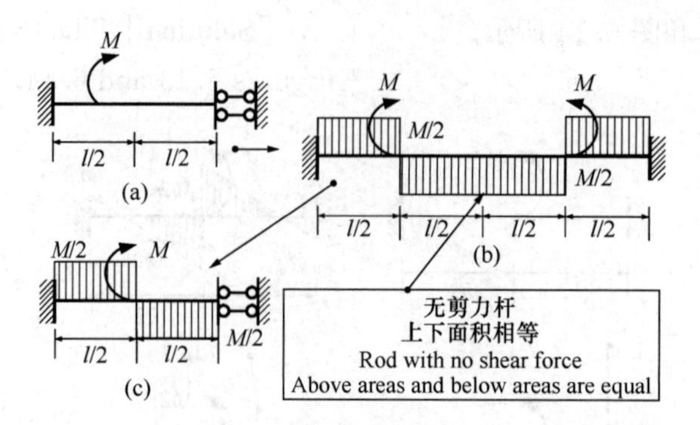

图 6.15 一端固结一端滑动等截面直杆在集中力偶作用下的载常数

Figure 6.15 Load constants for a straight rod with equal section of one solid end and one sliding end under the action of a concentrated couple

载常数表 **表 6.2**

Table of load constants **Table 6.2**

约束 Constraints	等截面直杆 AB 的载常数 Load constants for straight rod AB with equal section	杆端弯矩 Rod end bending moment		杆端剪力 Rod end shear force	
	弯矩图 Bending moment diagram	M_{AB}	M_{BA}	F_{QAB}	F_{QBA}
两端固结 Two solid ends	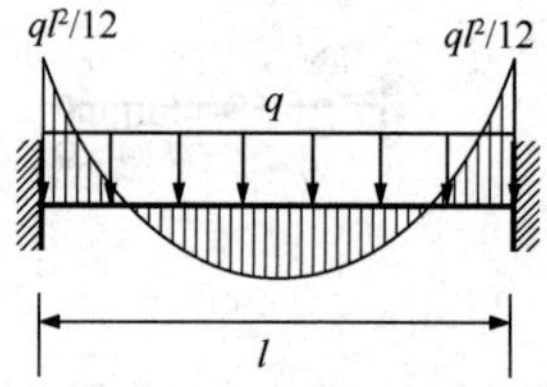 $ql^2/12$, q, $ql^2/12$, l	$-\frac{ql^2}{12}$	$\frac{ql^2}{12}$	$\frac{ql}{2}$	$-\frac{ql}{2}$
	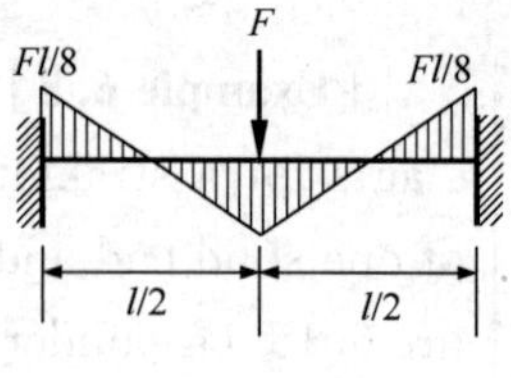 $Fl/8$, F, $Fl/8$, $l/2$, $l/2$	$-\frac{Fl}{8}$	$\frac{Fl}{8}$	$\frac{F}{2}$	$-\frac{F}{2}$
	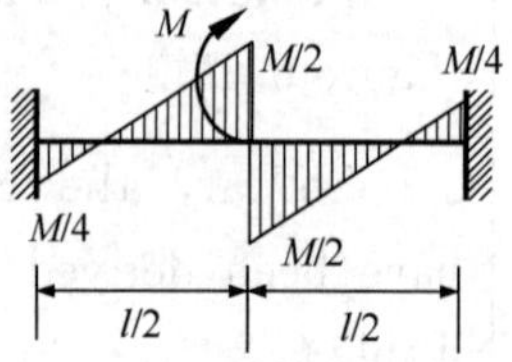 M, $M/2$, $M/4$, $M/4$, $M/2$, $l/2$, $l/2$	$\frac{M}{4}$	$\frac{M}{4}$	$-\frac{3M}{2l}$	$-\frac{3M}{2l}$

续表
Continued

约束 Constraints	等截面直杆 AB 的载常数 Load constants for straight rod AB with equal section	杆端弯矩 Rod end bending moment		杆端剪力 Rod end shear force	
	弯矩图 Bending moment diagram	M_{AB}	M_{BA}	F_{QAB}	F_{QBA}
一端固结一端滑动 One solid end and one sliding end	$ql^2/3$, q, l, $ql^2/6$	$-\frac{ql^2}{3}$	$-\frac{ql^2}{6}$	ql	0
	$Fl/2$, F, l, $Fl/2$	$-\frac{Fl}{2}$	$-\frac{Fl}{2}$	F	F
	$M/2$, M, $l/2$, $l/2$, $M/2$	$-\frac{M}{2}$	$-\frac{M}{2}$	0	0
一端固结一端铰支 One solid end and one hinge end	$ql^2/8$, q, l	$-\frac{ql^2}{8}$	0	$\frac{5ql}{8}$	$-\frac{3ql}{8}$
	$3Fl/16$, F, $l/2$, $l/2$	$-\frac{3Fl}{16}$	0	$\frac{11F}{16}$	$-\frac{5F}{16}$
	M, $M/8$, $l/2$, $l/2$	$\frac{M}{8}$	0	$-\frac{9M}{8l}$	$-\frac{9M}{8l}$

6.2.4 转角位移方程

6.2.4 Angular Displacement Equation

杆端内力和杆端位移之间的映射关系，称为杆件的转角位移方程。

The mapping relationship between the internal force and the displacement at the rod end is called angular displacement equation for the rod.

$$\{F\}=[K]\{\Delta\} \tag{6.3}$$

$$\{F\}=\{M_{AB} \quad M_{BA} \quad F_{QAB} \quad F_{QBA}\}^{T}$$

式中，[K] 可以参照形常数表得到。如两端固结等截面直杆的 [K] 为：

Where [K] can be obtained by referring to the shape constant table. For example, [K] for the straight rod with equal section of two solid ends is:

$$[K]=\begin{bmatrix} 4i & 2i & -\frac{6i}{l} \\ 2i & 4i & -\frac{6i}{l} \\ -\frac{6i}{l} & -\frac{6i}{l} & \frac{12i}{l^2} \\ -\frac{6i}{l} & -\frac{6i}{l} & \frac{12i}{l^2} \end{bmatrix}$$

杆端位移为：

Displacement for rod end:

$$\{\Delta\}=\{\theta_{A} \quad \theta_{B} \quad \Delta\}^{T}$$

如果杆件上存在荷载，则式（6.3）可写为：

If there is a load on the member, Equation (6.3) can be written as:

$$\{F\}=[K]\{\Delta\}+\{F\}^{F} \tag{6.4}$$

式中

Where

$$\{F\}^{F}=\{M_{AB}^{F} \quad M_{BA}^{F} \quad F_{QAB}^{F} \quad F_{QBA}^{F}\}^{T}$$

可以通过载常数表得到。回顾 [K] 可知，矩阵第3行和第4行存在线性关系，所以杆端剪力是不独立的。

can be obtained by referring to the table of load constants. Looking back [K], it shows a linear relationship between rows 3 and 4 of the matrix, so the shear forces at the rod ends are not independent.

6.3 位移法的基本概念

6.3 Basic Concepts of the Displacement Method

6.3.1 位移法的基本未知量

6.3.1 Basic Unknown Quantities for the Displacement Method

位移法的基本未知量是独立的节点角位移和线位移。此外，如果没有特别说明，一般忽略受弯杆件的轴向变形。

The basic unknown quantities for the displacement method are the independent nodal angular and linear displacements. In addition, if not explicitly stated, the axial deformation of the bent rod is generally ignored.

6.3.1.1 节点独立角位移

6.3.1.1 Independent Angular Displacements of Nodes

1. 由于在同一刚节点处，各杆端转角都是相等的，所以每个刚节点只有一个角位移。

1. There is only one angular displacement per rigid node since the rotation angles at each rod end are equal at the same rigid node.

2. 铰接点或铰支座处各杆端的转角，由于已有一端固结一端铰支的形常数和载常数，所以其转角不作为基本未知量。

2. Each rod end's rotation angle at the hinge junction or hinge support is not treated as a basic unknown quantity since there is already a shape constant and a load constant for one solid end and one hinge end.

3. 杆件的变截面处有一个独立角位移，其原因是位移法基本单元是等截面直杆。

3. The reason for having an independent angular displacement at the variable section of the rod is that the basic unit of the displacement method is a straight rod with equal section.

4. 固定支座无转角，抗转动弹簧支座有一个独立角位移。

4. The fixed support has no rotation, and the rotation-resistant spring support has an independent angular displacement.

6.3.1.2 节点独立线位移

6.3.1.2 Independent Linear Displacements of Nodes

1. 杆件的变截面处有一个独立线位移，

1. The reason for having an independ-

其原因是位移法基本单元是等截面直杆。

ent linear displacement at the variable section of the rod is that the basic unit of the displacement method is a straight rod with equal section.

2. 每个节点在平面内有两个线位移，节点独立线位移比较好操作的计算方法是铰化法，将于下节详述。

2. Each node has two linear displacements in the plane, and a better operational method for calculating the independent node linear displacements is the hinging method, which will be described in detail in the next section.

6.3.1.3 附加刚臂

6.3.1.3 Additional Rigid Arms

位移法的思想是将每个单元"拆"出来，如图 6.16（a）所示结构，是由两个单元组成，为了将其拆开，引入图 6.16（b）所示附加刚臂，附加刚臂具有以下特点：

The idea of the displacement method is to "split" each unit, as the structure shown in Figure 6.16 (a), which consists of two units, and to split them, the additional rigid arms are introduced as shown in Figure 6.16 (b), which has the following characteristics:

1. 能锁死转动，不限制线位移；

2. 能阻断弯矩传递。

1. Capable of locking rotation without limiting linear displacement;

2. Capable of interrupting the transmission of bending moments.

引入附加刚臂后，B 支座相当于固定端（没有角位移和线位移），①②两个单元拆出。拆出后，单元①②在荷载作用下的杆端弯矩可由载常数得到，见图 6.16（c）、图 6.16（d）。

After the introduction of additional rigid arms, support B is equivalent to the fixed end (no angular or linear displacement), and the two units ① and ② are split. After splitting, the bending moments at the rod ends of units ① and ② under the load can be obtained from the load constants, as shown in Figure 6.16 (c) and Figure 6.16 (d).

再令 B 节点和附加刚臂一起转动 θ（和原结构图 6.16a 一致），单元①②在 B 端 θ 角的作用下产生的杆端弯矩可由形常数得到，见图 6.16（e）、图 6.16（f）。

Then make the node B and additional rigid arm rotate together by θ (in agreement with the original structure in Figure 6.16a), the rod end bending moments of the units ① and ② generated by the rotation angle θ at the end B can be obtained from the shape constants, as shown in Figure 6.16 (e) and Figure 6.16 (f).

最后，所有杆端弯矩的反作用力汇集于附加刚臂，由于附加刚臂是虚拟的、假设添加的，因此，有$\sum M=0$，这样与原结构保持一致。

Finally, the reactions of all rod end moments converge on the additional rigid arms, since the rigid arm is virtual and assumed to be added, $\sum M=0$, which is consistent with the original structure.

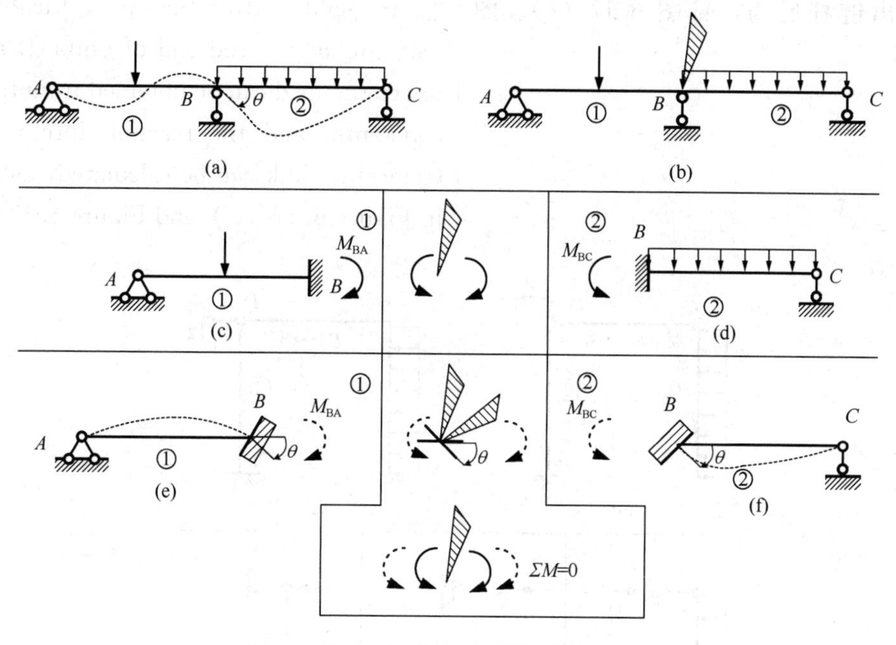

图 6.16 附加刚臂示意图

Figure 6.16 Schematic diagram of the additional rigid arm

6.3.1.4 附加链杆

和附加刚臂类似，为了拆出单元，引入附加链杆，如图 6.17 (a) 所示结构，是由①②两个单元组成，为了将其拆开，引入图 6.17 (b) 所示附加链杆，附加链杆具有以下特点：

6.3.1.4 Additional Connection Links

Similar to additional rigid arm, in order to split the unit, the additional connection link is introduced, as the structure shown in Figure 6.17 (a), which is composed of two units ① and ②, in order to split them, the additional connection link shown in Figure 6.17 (b) is introduced, and the additional connection link has the following characteristics:

1. 能限制线位移，不限制角位移；

1. Capable of limiting linear displacements without limiting angular displacements;

2. 能阻断链杆方向力的传递。

2. Capable of Blocking the transmission of forces in the direction of the connection link.

引入附加链杆后，C、D 支座没有线位移，①②两个单元拆出。拆出后，单元①②在荷载作用下的杆端弯矩可由载常数得到，并可算出链杆反力，见图 6.17（c）、图 6.17（d）。

Introducing the additional connection link, there is no linear displacement at the supports C and D, and the two units ① and ② are split. After the split, the bending moments at the rod end of units ① and ② under the load can be obtained from the load constants, and the reaction forces of the connection link can be calculated, as shown in Figure 6.17（c）and Figure 6.17（d).

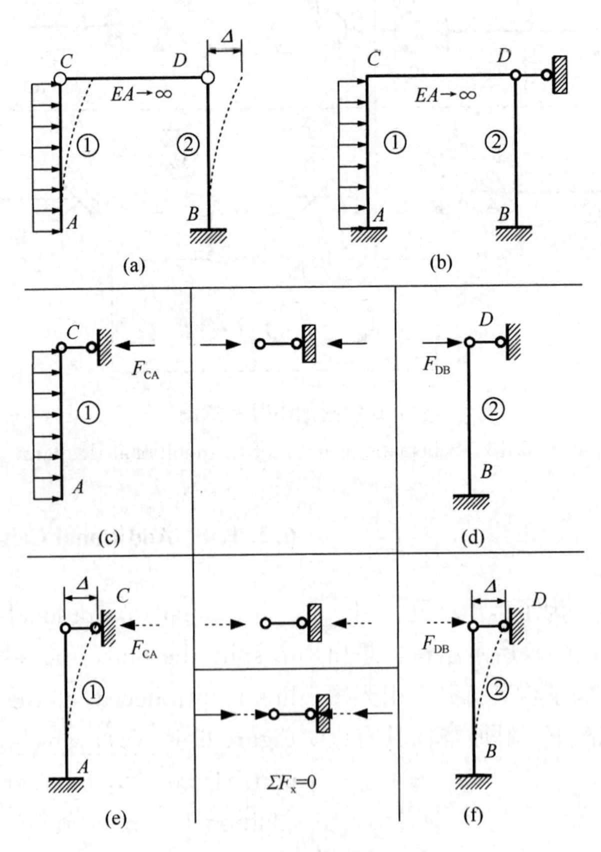

图 6.17　附加链杆示意图

Figure 6.17　Schematic diagram of the additional connection link

再令 C、D 节点和附加链杆一起平移 Δ（和原结构图 6.17a 一致），单元①②在 C、D 端侧移 Δ 下产生的杆端弯矩可由形常数得到，并可算出链杆反力，见图 6.17（e）、图 6.17（f）。

Then make the nodes C, D and the additional connection link together translate Δ (in agreement with the original structure in Figure 6.17a), the rod end bending moment of units ① and ② generated by the lateral shift Δ at the ends C and D can be obtained from the shape constants, and the link reaction force can be calculated as shown in Figure 6.17 (e) and Figure 6.17 (f).

最后，所有链杆反力的反作用力汇集于附加链杆，由于附加链杆是虚拟的、假设添加的，因此，有 $\sum F_x=0$，这样与原结构保持一致。

Finally, the reaction forces of all links' reactions converge on the additional connection link, and since the additional connection link is virtual and assumed to be added, $\sum F_x=0$, which remains consistent with the original structure.

6.3.2 位移法的基本结构

6.3.2 Basic Structure of the Displacement Method

在原结构的刚节点上加附加刚臂，在产生独立线位移的节点上增加附加链杆，所得到的组合体称为**基本结构**。确定基本结构是位移法解题的关键所在，确定基本结构的方法在于如何添加附加刚臂和附加链杆。

Adding additional rigid arms to the rigid nodes of the original structure and adding connection links to the nodes that produce independent linear displacements, the resulting combination of that is called the **basic structure.** Determining the basic structure is the key to the solution of the displacement method, and the method of determining the basic structure lies in how to add the additional rigid arms and additional connection links.

6.3.2.1 附加构件设置原则

6.3.2.1 Principle of Setting Additional Members

从前文分析可以，一个附加刚臂或附加链杆对应一个未知量。如何增设附加刚臂和附加链杆，使其既有必要又不多余？可遵从下面的附加构件设置原则：

From the previous analysis, an additional rigid arm or connection link corresponds to an unknown quantity. How to add additional rigid arms and connection links that are both necessary and not redundant? The following principles of additional member setting can be followed:

1. 附加刚臂所约束角位移对应杆件弯矩如果静定，则不加附加刚臂；

1. If the bending moment of the member that corresponds to the restrained angular displacement by additional rigid arm is stationary, no additional rigid arm is needed;

2. 附加链杆所约束线位移对应杆件剪力如果静定，则不加附加链杆。

2. If the shear force of the member that corresponds to the restrained linear displacement by additional connection link is stationary, no additional connection link is needed.

图 6.18 中 A 节点的区别即为加与不加附加刚臂的区别。

Figure 6.18 shows the differences between setting additional rigid arms and not setting additional rigid arms of node A.

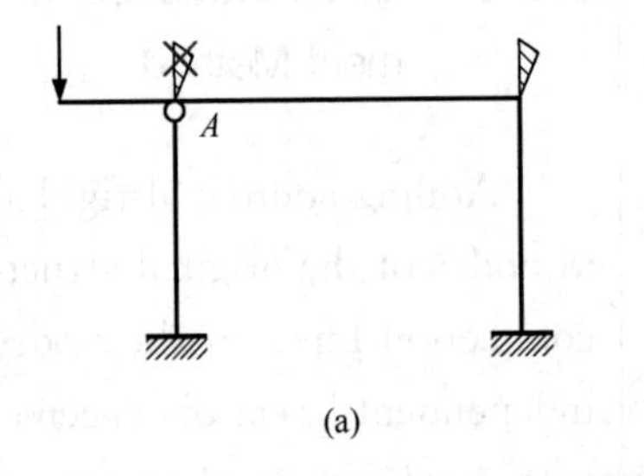

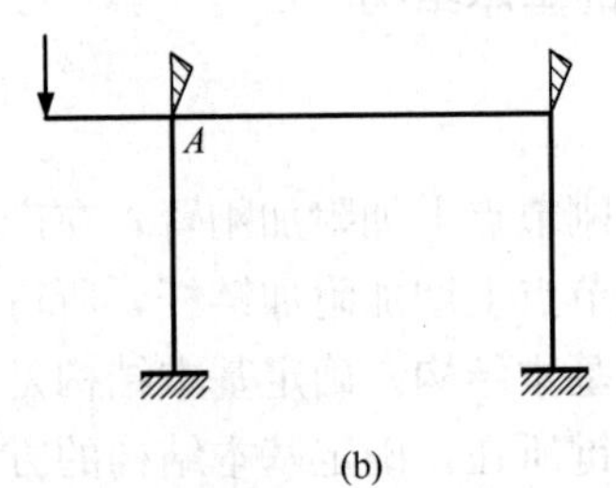

图 6.18 加或不加附加刚臂的情况举例

Figure 6.18 Example of setting additional rigid arm or not

图 6.19（b）是图 6.19（a）的基本结构，图 6.19（d）是图 6.19（c）的基本结构，注意两者在附加链杆添加的区别：其原因就在于 AB 杆的剪力是否能够由静力平衡条件得到，如可以，该杆称为剪力静定杆，相应的线位移处不加附加链杆。

Figure 6.19（b）is the basic structure of Figure 6.19（a），and Figure 6.19（d）is the basic structure of Figure 6.19（c），note the difference between them of the setting additional connection links：The reason for this lies in whether the static equilibrium condition can obtain the shear force of the rod AB，if so，the rod is a static shear bar，and no additional link is added at the point of corresponding linear displacements.

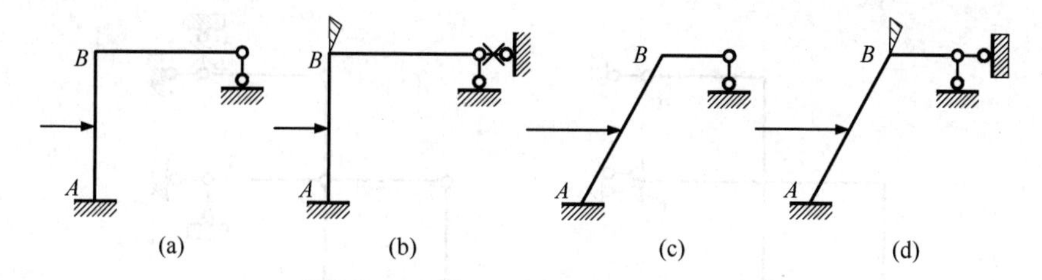

图 6.19 附加链杆设置举例

Figure 6.19 Example of additional connection link settings

6.3.2.2 用铰化法确定附加链杆

6.3.2.2 Determination of Additional Connection Links by Hinging Method

相比于附加刚臂，附加链杆的设置是难点，铰化法是非常具有操作性的一种做法。其步骤如下，参见图 6.20：

The setting of the additional connection link is the problematic part compared to the additional rigid arm, and the hinging method is a very operational practice. The steps are as follows and can be shown in Figure 6.20:

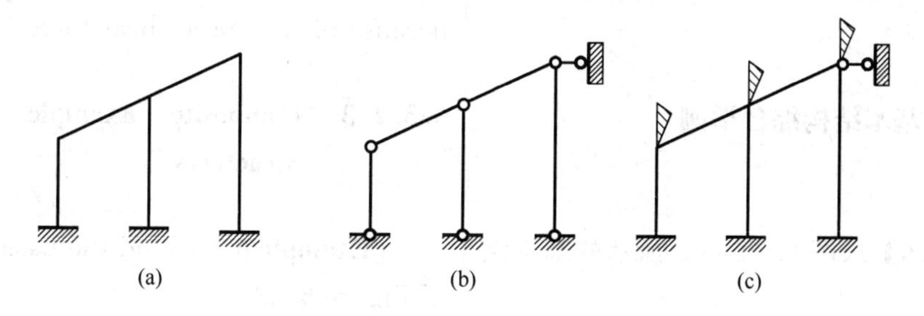

图 6.20 铰化法

(a) 原结构；(b) 铰化体系；(c) 基本结构

Figure 6.20 Hinging method

(a) Original structure; (b) Hinging system; (c) Basic structure

1. 将结构中所有刚节点（含组合节点）改为铰节点，固定支座改为铰支座，每一根杆件转化为刚性链杆；

1. Replace all rigid joints (including combined joints) in the structure by hinge joints, replace all fixed supports by hinge supports, and every rod is converted into a rigid link;

2. 对体系进行几何组成分析，增加最少的附加链杆使之几何不变。

2. The geometric composition of the system is analyzed, and a minimum number of additional connection links are added to realize geometry constant.

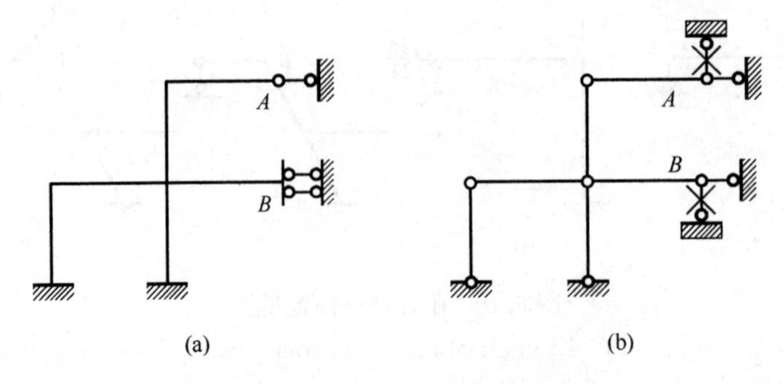

图 6.21　铰化法和附加构件设置原则

(a) 原结构；(b) 铰化体系

Figure 6.21　Hinging method and the principle of additional member setting

(a) Original structure; (b) Hinging system

值得注意的是，铰化法优先级在附加构件设置原则之后，如图 6.21（a）所示结构，按照铰化体系需要在 AB 处增设附加链杆，但由于此处剪力静定，故最终结果是不设置附加链杆。

It is worth noting that the principle of additional member setting takes priority of hinging method, as the case of the structure shown in Figure 6.21 (a), where additional connection links are required at AB according to the hinging system, but the final result is not setting additional connection link because of the static shear force.

6.3.2.3　基本结构综合举例

6.3.2.3　Composite Example of Basic Structures

【例 6.9】求图 6.22（a）所示的基本结构。

[Example 6.9] Find the basic structure of Figure 6.22 (a).

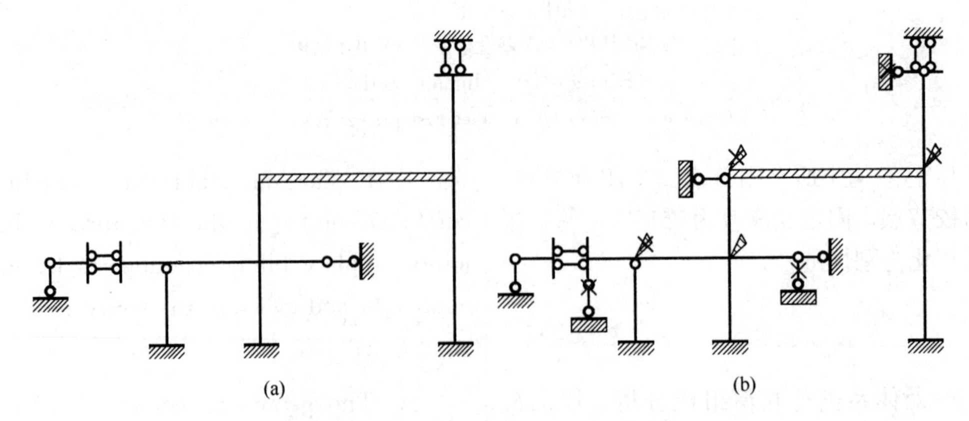

图 6.22　位移法基本结构综合举例

Figure 6.22　Composite example of the basic structure of the displacement method

【解】基本结构如图 6.22（b）所示，要注意的是，如果结构中含有刚性杆，由于刚性杆杆端（包括连接的其他杆）无转角，故此处不加附加刚臂。

[**Solution**] The basic structure is shown in Figure 6.22 (b). Note that if the structure contains rigid rods, no additional rigid arms are added since there is no rotation angle at the ends of the rigid rods (including the other rods connected).

6.4 位移法的基本体系与典型方程的建立

6.4 The Basic System of the Displacement Method and the Development of Typical Equations

从力法的解题思想可以归纳出超静定结构的求解过程：

The solution procedure for a statically indeterminate structure can be generalized from the solution ideas of the force method as follows:

取一基本体系，让基本体系与原结构变形一致（力法）或受力一致（位移法），然后根据对补充方程的求解得到基本未知量，从而达到求解目的。

The solution is achieved by taking a basic system, making the basic system consistent with the deformation of the original structure (force method) or consistent with the forces (displacement method), and then obtaining the basic unknown quantities based on the solution of the supplementary equations.

下面进行举例说明。

The following examples are given.

注：物理量的下标含义：第一个下标表示位置；第二个下标表示原因。

Note: The meanings of subscripts for physical quantities are: The first subscript indicates the position; The second subscript indicates the reason.

【例 6.10】用位移法求解图 6.23（a）弯矩图。

[**Example 6.10**] Use the displacement method to solve the bending moment diagram in Figure 6.23 (a).

【解】（1）添加附加刚臂、附加链杆（本题没有，不添加），形成位移法基本结构，如图 6.23（b）所示。

[**Solution**] (1) Add additional rigid arms and additional connection links (In this example, no additional connection links are added) to form the basic structure of the displacement method, as shown in Figure 6.23 (b).

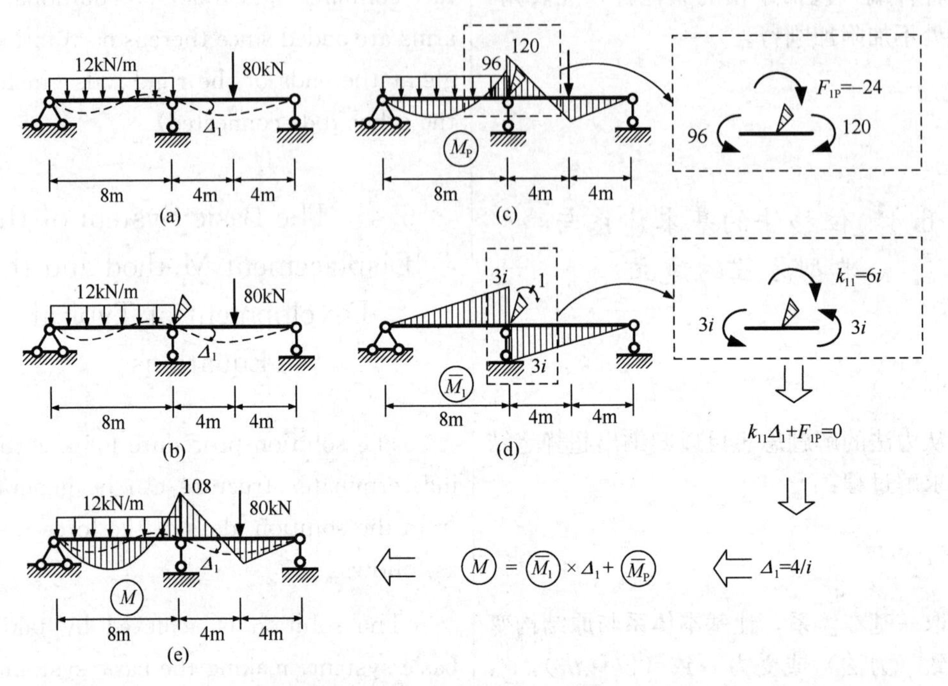

图 6.23　位移法基本过程（单位：kN·m）

Figure 6.23　Basic process of displacement method (Unit: kN·m)

（2）考虑荷载作用，利用载常数，得到M_P图；取附加刚臂为隔离体，根据平衡条件可得F_{1P}。

(2) Considering the load action, the M_P diagram is obtained by load constant; taking the additional rigid arm as the isolator, F_{1P} is obtained according to the equilibrium condition.

（3）连带附加刚臂发生单位转角$\delta_1=1$，利用形常数，得到$\overline{M}_1$图；取附加刚臂为隔离体，根据平衡条件可得k_{11}。

(3) The unit angle of rotation $\delta_1=1$ occurs with the attached rigid arm, using the shape constants, $\overline{M}_1$ is obtained; taking the additional rigid arm as an isolator, k_{11} is obtained according to the equilibrium condition.

（4）由于附加刚臂是虚拟的，所以其约束反力之和应为零，即：

(4) Since the additional rigid arm is virtual, the sum of its restraint reactions should be zero:

$$k_{11}\Delta_1+F_{1P}=0 \tag{6.5}$$

这就是位移法的基本方程。

This is the basic equation of the displacement method.

（5）解方程得到基本未知量 Δ_1，叠加可得弯矩图：

（5） Solve the equation to obtain the basic unknown quantity Δ_1 and superimpose to obtain the bending moment diagram:

$$M = \overline{M}_1 \times \Delta_1 + M_P$$

下面进一步拓展，如图 6.24（a）所示，有两个基本未知量，使用位移法进行求解，仍然是上述步骤，从图 6.24（b）的基本结构开始，依次画出弯矩图（图 6.24c～e），写出位移法的基本方程：

Now we extend further, as in Figure 6.24（a）, there are two fundamental unknowns, which are solved using the displacement method, still follow the steps described above, start with the basic structure of Figure 6.24（b）and draw the bending moment diagrams（Figures 6.24c-e）in turn, write the basic equations of the displacement method:

$$\left.\begin{aligned} k_{11}\,\Delta_1 + k_{12}\,\Delta_2 + F_{1P} = 0 \\ k_{21}\,\Delta_1 + k_{22}\,\Delta_2 + F_{2P} = 0 \end{aligned}\right\} \tag{6.6}$$

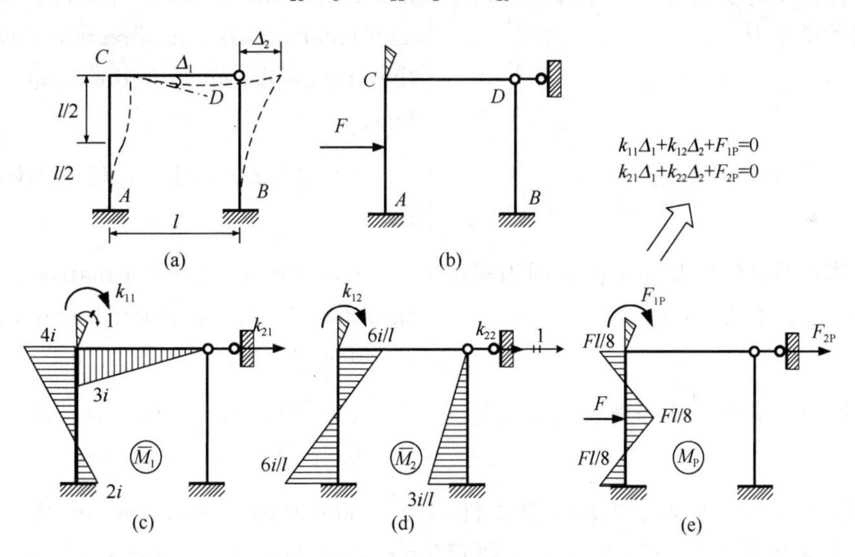

图 6.24 位移法求解 2 个基本未知量示意图

Figure 6.24 Schematic diagram of the displacement method to solve for the 2 basic unknowns

然后求解基本未知量，叠加弯矩图即可。

Then solve for the fundamental unknowns and superimpose the bending moment diagram.

将位移法的基本方程写成矩阵形式，有：

Write the basic equations of the displacement method in matrix form:

$$[K]\{\Delta\}+\{F\}=\{0\} \tag{6.7}$$

式中，$[K]$ 称为刚度系数矩阵；$\{F\}$ 为自由项。

Where $[K]$ is stiffness factor matrix; $\{F\}$ is free term.

一般的，对于 n 个基本未知量的位移法基本方程，可表示为：

In general, the basic equations of the displacement method for n fundamental unknowns can be expressed as:

$$\begin{bmatrix} k_{11} & \cdots & k_{1i} & \cdots & k_{1n} \\ \vdots & \ddots & \vdots & \ddots & \vdots \\ k_{i1} & \cdots & k_{ii} & \cdots & k_{in} \\ \vdots & \ddots & \vdots & \ddots & \vdots \\ k_{n1} & \cdots & k_{ni} & \cdots & k_{nn} \end{bmatrix} \begin{bmatrix} \Delta_1 \\ \vdots \\ \Delta_i \\ \vdots \\ \Delta_n \end{bmatrix} + \begin{bmatrix} F_{1P} \\ \vdots \\ F_{iP} \\ \vdots \\ F_{nP} \end{bmatrix} = \begin{bmatrix} 0 \\ \vdots \\ 0 \\ \vdots \\ 0 \end{bmatrix} \tag{6.8}$$

刚度系数矩阵具有以下特点：

The rigid factor matrix has the following characteristics:

（1）刚度系数 k_{ij} 的物理含义是 j 方向单位位移单独作用时所引起的 i 方向上的附加构件中的约束反力。

(1) The rigid factor k_{ij} has the physical meaning of the restraint reaction force in the additional member in direction i caused by the unit displacement in direction j acting alone.

（2）主系数 k_{ii} 恒为正。

(2) Main coefficient k_{ii} is always positive.

（3）刚度系数矩阵是对称的，因为根据反力互等定理，有 $k_{ij}=k_{ji}$。

(3) The rigid factor matrix is symmetric, since by the reaction forces reciprocity theorem, $k_{ij}=k_{ji}$.

（4）刚度系数与荷载等外界因素无关。

(4) The rigid factor is independent of the external factors including the load.

位移法典型方程是表示力的平衡条件，故又称为**刚度方程**，位移法又被称为**刚度法**。

The typical equation of the displacement method is to express the equilibrium conditions of the forces, so it is also called the **rigid equation** and the displacement method is also known as the **rigid method.**

综上所述，可归纳出位移法的计算步骤如下：

In summary, the calculation steps of the displacement method can be summarized as follows:

（1）确定基本未知量，形成位移法基本

(1) Determine the basic unknown

结构。

quantities that form the basic structure of the displacement method.

(2) 建立位移法典型方程。

(2) Establish the typical equations for the displacement method.

(3) 由形常数表和载常数表画出基本结构的单位弯矩图、荷载弯矩图 M_P。由平衡条件求系数和自由项。

(3) From the table of shape constants and load constants, draw the unit moment diagram and load moment diagram M_P of the basic structure. Find the coefficients and free terms from the equilibrium conditions.

(4) 解方程，求基本未知量。

(4) Solve the equation for the basic unknown quantities.

(5) 按叠加求最后弯矩图，进而由弯矩图求剪力，由剪力图求轴力。

(5) By superimposing, the final moment diagram is obtained, the shear force is then found from the moment diagram and the axial force is found from the shear force diagram.

(6) 校核。

(6) Check the results.

6.5 侧移刚架

6.5 Rigid Frame with Lateral Displacements

6.5.1 位移法求解侧移刚架

6.5.1 Displacement Method to Solve for Rigid Frame with Lateral Displacements

如果刚架有节点线位移，便称为有侧移刚架。图 6.24 (a) 就是有侧移刚架，前文已经建立其基本结构，下面仍以该题为例，说明刚度系数和自由项的求解，如图 6.25 所示。

If a rigid frame has nodal linear displacements, it is called a rigid frame with lateral displacements. Figure 6.24 (a) shows a rigid frame with lateral displacements, the basic structure of which has established in the previous section, taking this as an example again, the solution of the rigid factors and free terms is illustrated in Figure 6.25.

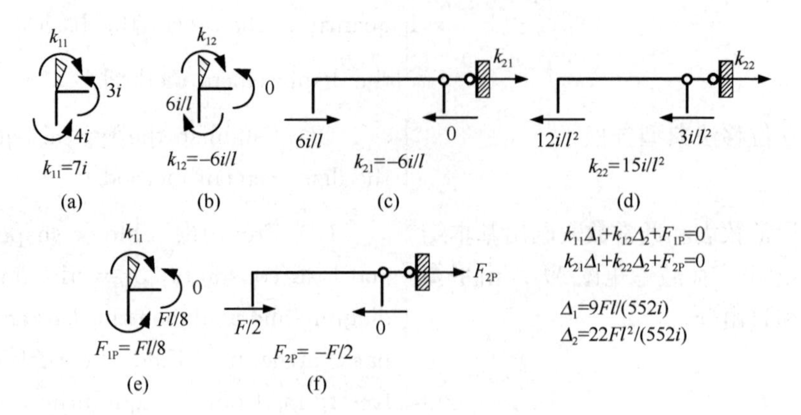

图 6.25　刚度系数和自由项的求解

Figure 6.25　Solution of stiffness factors and free terms

$$M=\{\overline{M}\}^{\mathrm{T}}\{\Delta\}+\{M_{\mathrm{P}}\} \tag{6.9}$$

由式（6.9）叠加弯矩图可得最终弯矩图如图 6.26 所示。

According to Equation (6.9), the final bending moment can be superimposed in Figure 6.26.

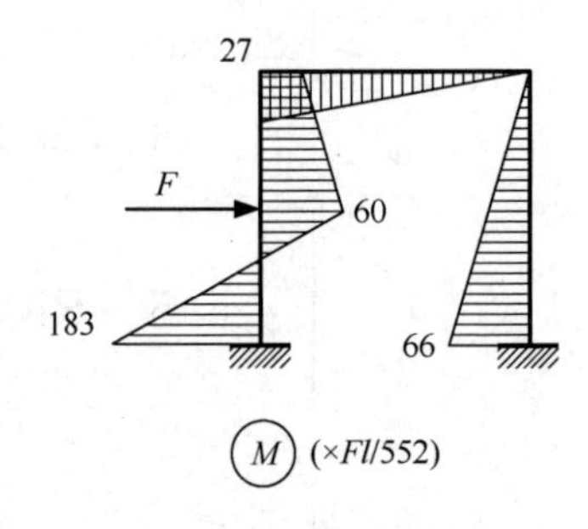

图 6.26　最终弯矩图

Figure 6.26　Final bending moment diagram

6.5.2　剪力分配法

6.5.2　Shear Force Distribution Method

由前文所述，当侧移刚架中含有 $EI\rightarrow\infty$ 横梁，如果横梁只发生平动，则其相连杆端没有角位移。

From the previous section, when a rigid frame with lateral displacements contains a cross-beam of $EI\rightarrow\infty$, there is no angular displacement at its connected rod ends if the cross-beam only undergoes translational movement.

如图 6.27（a）所示，由于刚性杆没有转动，所以结构只有一个基本未知量，即水平侧移。取基本结构如图 6.27（b）所示，列出位移法典型方程：

As in Figure 6.27 (a), there is only one basic unknown quantity in the structure, namely the horizontal lateral displacement since there is no rotation of the rigid rod. Take the basic structure as in Figure 6.27 (b) and list the typical equation for the displacement method:

$$k_{11}\,\Delta_1 + F_{1P} = 0$$

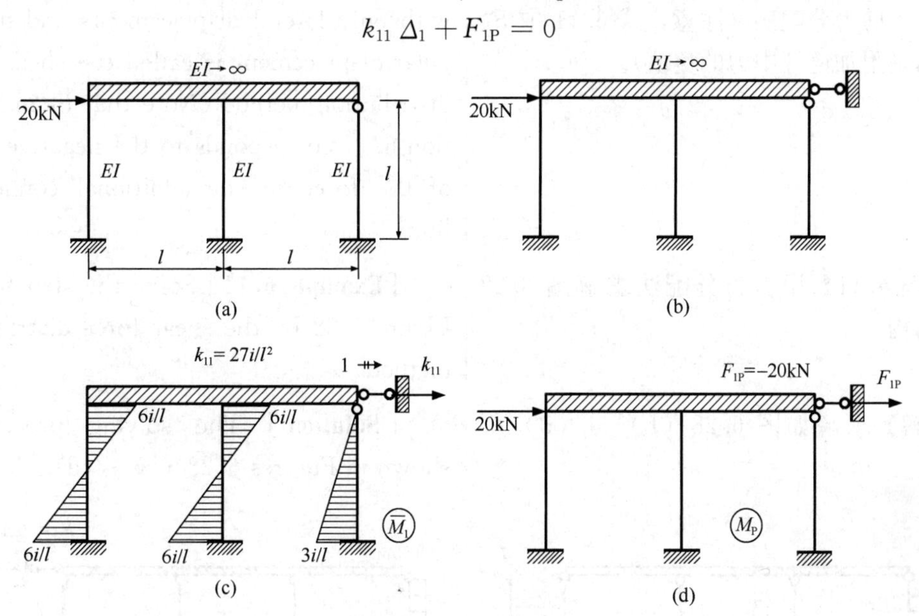

图 6.27 剪力分配法

Figure 6.27 Shear force distribution method

代入刚度系数和自由项，可求得基本未知量：

Substitute the stiffness coefficients and free terms, the fundamental unknowns can be obtained as:

$$\Delta_1 = \cfrac{20\,l^3}{\cfrac{27EI}{l^3}}$$

若杆件的杆端只有线位移，没有角位移，可定义其抗侧移刚度 J 为：

If the rod end has only linear displacement and no angular displacement, the lateral displacement resistance stiffness J can be defined as:

$$J = d\,\frac{EI}{l^3} \tag{6.10}$$

式中，d 为约束系数，两端固结杆件 $d=12$，一端固结一端铰支杆件 $d=3$。

Where d is the restraint factor, $d=12$ for the rod with two solid ends and $d=3$ for the rod with one solid end and one hinge end.

则每根杆的剪力可表示为：

The shear force of each rod can be expressed as:

$$F_{Qi} = \frac{J_i}{\sum J_i} F_P \tag{6.11}$$

这种对只有侧移没有转角的刚架计算的方法称为剪力分配法（注意，式6.11中F_P对应的是附加链杆中力的负值）。

This method of calculating rigid frames with only lateral displacements and no angular displacement is called the shear force distribution method (Note that F_P in Equation 6.11 corresponds to the negative value of the force in the additional connection link).

【例6.11】用剪力分配法求解图6.28所示结构。

[Example 6.11] Solve the structure in Figure 6.28 by the shear force distribution method.

【解】求解如图6.28（b）～（d）所示。

[Solution] The solving process is shown in Figures 6.28 (b) - (d).

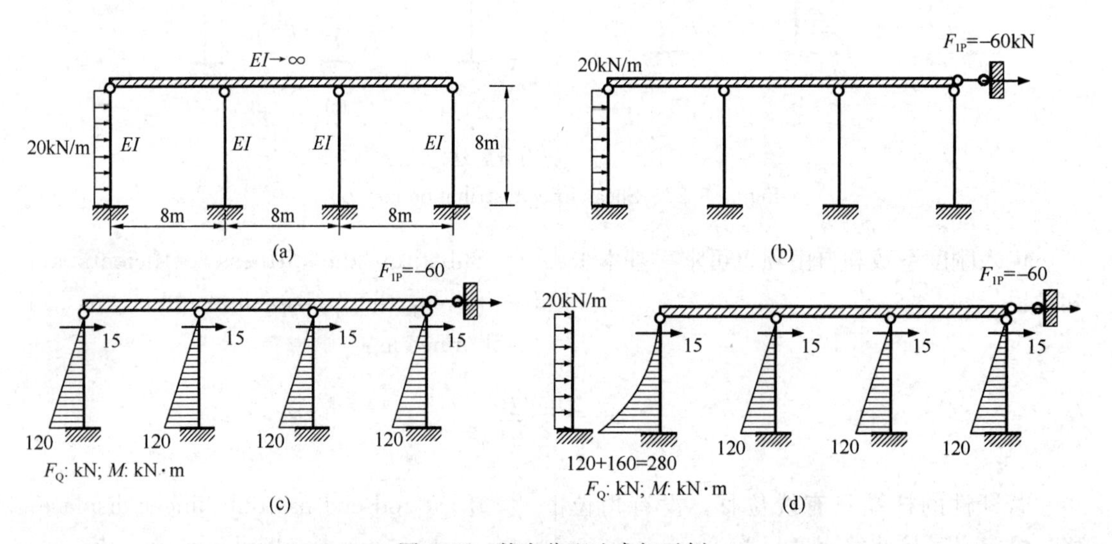

图6.28　剪力分配法求解示例

Figure 6.28　Example of shear force distribution method

6.5.3　侧移刚架中的刚性杆问题

6.5.3　Rigid Rod Problems in Rigid Frames with Lateral Displacements

侧移刚架含有刚性杆，如果刚性杆只有平移没有转动，则不会有杆端的角位移牵连，不用考虑所连杆件的杆端牵连角位移，如图6.29所示。注意，图6.29中刚性杆无论是否水平，均只有平动，没有转动。

There are rigid rods in the rigid frame with lateral displacements, if no angular displacement but only translational displacement occurs, there is no angular displacement implication of rod ends, i. e., implication angular displacement of the rod ends of the connecting rods are not considered, as shown in Figure 6.29. Note that the rigid rods in Figure 6.29 only have translational displacements and no rotational displacements whether they are horizontal or not.

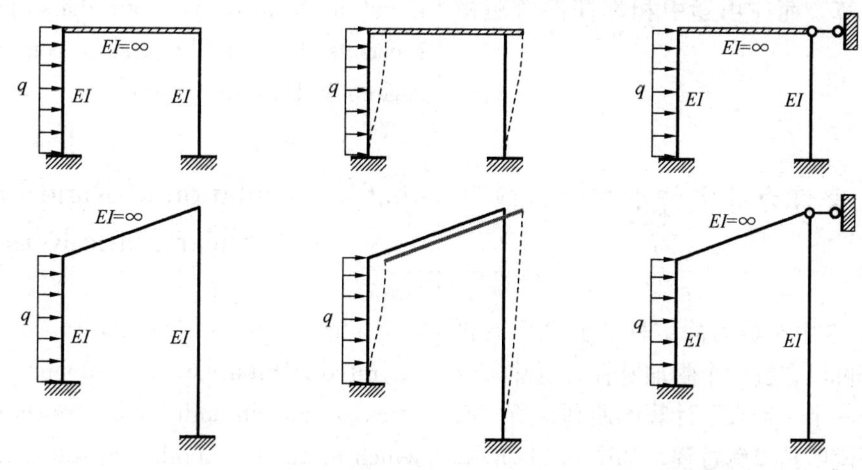

图 6.29 刚性杆平动

Figure 6.29 Rigid rod with translational displacements

【例 6.12】 求图 6.30 所示超静定梁的弯矩图。

[Example 6.12] Find out the bending moment diagram for the statically indeterminate beam shown in Figure 6.30.

【解】 求解过程如图 6.30 所示。

[Solution] The solving process can be shown in Figure 6.30.

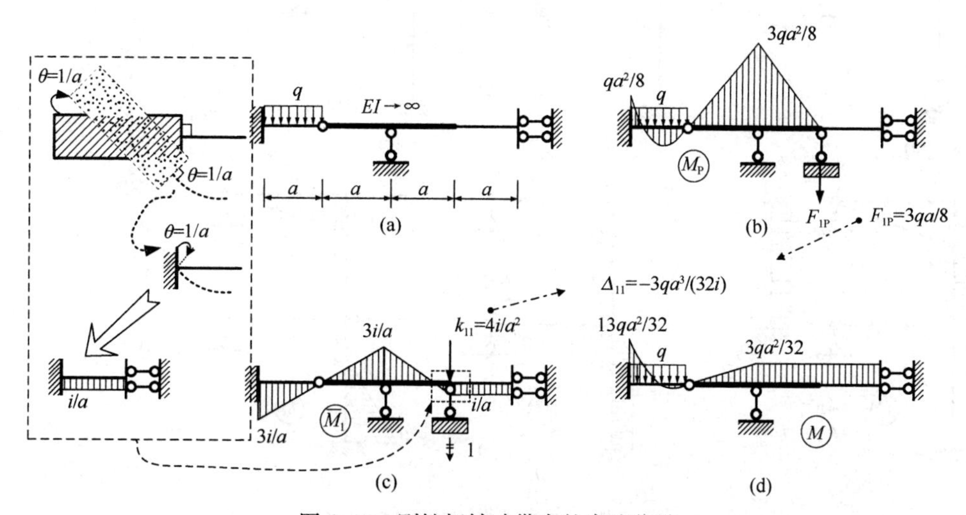

图 6.30 刚性杆转动带来的牵连位移

Figure 6.30 Implication displacement due to rigid rod rotation

【解题的关键点】

（1）刚性杆转动带来的牵连位移，如图6.30（c）所示；

（2）刚性杆自身弯矩图不能由形常数或载常数推算，而应由题中相邻杆件弯矩推算。

[Key points for solving the problem]

(1) Implication displacement due to rigid rod rotation, as shown in Figure 6.30 (c);

(2) The bending moment diagram of a rigid rod cannot be derived form the shape or load constants, but from the bending moments of the adjacent rods in the question

6.6　含剪力静定杆刚架的计算

6.6　Calculation of Rigid Frames with Shear Static Rods

图6.19中对剪力静定杆的概念已有说明，由于可以省去一个附加链杆，意味着求解次数少一个，带来了计算的便利，所以有必要举例说明其求解过程，如图6.31所示。

The concept of a shear static rod is illustrated in Figure 6.19, and since it is possible to omit an additional connection link, which means the number of solution can decrease by one, bringing ease of calculation, it is necessary to give an example of the solution process in Figure 6.31.

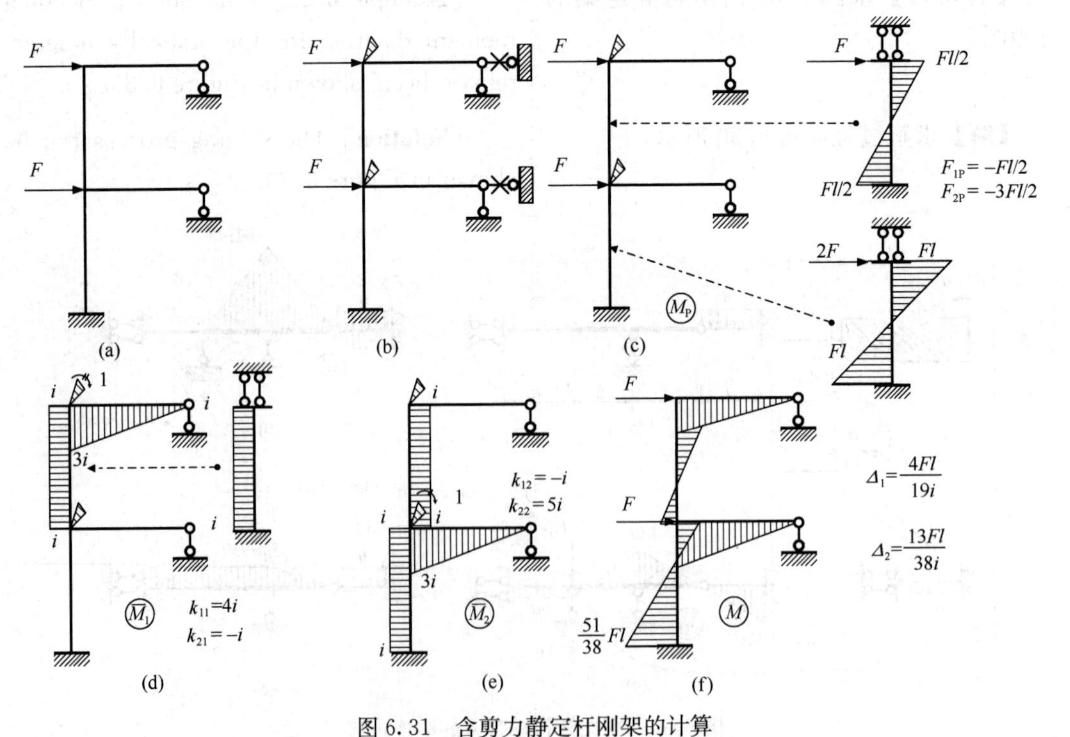

图6.31　含剪力静定杆刚架的计算

Figure 6.31　Calculation of a rigid frame with shear static rods

6.7 支座位移的位移法计算

6.7 Calculation of Displacement Method for Support Displacement

用位移法计算结构由于支座位移（移动或转动）引起的内力时，解题步骤与前述用位移法求解荷载作用下结构内力的计算步骤完全相同，区别仅在于典型方程中的自由项F_{iC}不是由荷载引起而是由支座位移引起的。

When calculating the internal forces of a structure due to support displacement (movement or rotation) by the displacement method, the solution procedure is the same as that described above for solving the internal forces of a structure under loads by the displacement method, with the difference being only that the free term F_{iC} in the typical equation is not caused by the load but by the support displacement.

$$[K]\{\Delta\}+\{F_C\}=\{0\} \tag{6.12}$$

【例 6.13】试作图 6.32（a）所示连续梁由于支座 B 下沉C_B时产生的弯矩图。

［Example 6.13］ Try to calculate the bending moment diagram of the continuous beam shown in Figur 6.32 (a) due to the settlement C_B of support B.

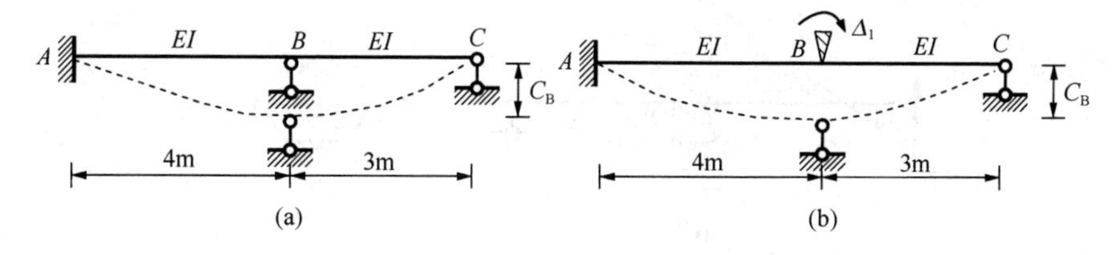

图 6.32 支座移动下的位移法计算

（a）原结构；（b）基本体系

Figure 6.32 Calculation of displacement method under support movement

(a) Original structure; (b) Basic system

【解】（1）确定基本未知量。此连续梁只有一个刚节点 B，只有节点 B 的角位移Δ_1。

［Solution］ (1) Determine the basic unknown quantity. This continuous beam has only one rigid node B with only the angular displacement Δ_1 .

（2）取基本体系如图 6.32（b）所示。

(2) Take the basic system as shown in Figure 6.32 (b).

（3）建立位移法方程：

(3) Establish the equation for the displacement method:

$$k_{11}\Delta_1+F_{1C}=0$$

(4) 计算系数和自由项。

(4) Calculate the coefficients and free terms.

由形常数表画出 $\overline{M}_1$ 和 M_C 图，如图 6.33 (a) 和 (c) 所示。

From the table of shape constants, draw $\overline{M}_1$ and M_C as shown in Figures 6.33 (a) and (c).

$$k_{11} = EI + EI = 2EI$$

由图 6.33 (d) 所示刚节点 B 的力矩平衡条件可得：

From the moment balance conditionof the rigid node B shown in Figure 6.33 (d):

$$F_{1C} = -\frac{1}{24}EIC_B$$

(5) 解方程求基本未知量：

(5) Solve equations to find out basic unknown quantity:

$$\Delta_1 = -\frac{F_{1C}}{k_{11}} = \frac{1}{48}C_B$$

(6) 根据叠加原理 $M = \overline{M}_1\Delta_1 + M_C$ 求杆端弯矩，并作出 M 图如图 6.33 (e) 所示。

(6) Find out the rod end bending moment according to the superposition principle $M = \overline{M}_1\Delta_1 + M_C$ and make M diagram as shown in Figure 6.33 (e).

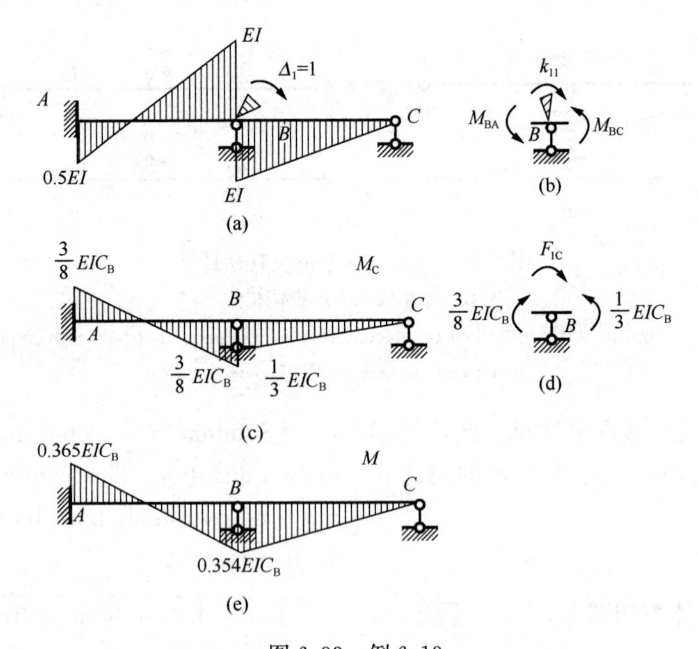

图 6.33　例 6.13

Figure 6.33　Example 6.13

思考题

6.1 为什么位移法中铰节点或铰支座处的角位移可以不选作基本未知量？

6.2 为什么位移法中滑动支座处的线位移可以不选作基本未知量？

6.3 试述位移法基本未知量中节点角位移数目的确定方法。

6.4 试述位移法基本未知量中独立节点线位移数目的确定方法。

6.5 位移法两种计算方法（直接平衡法和基本体系法）的位移法方程是否相同？

6.6 试述位移法典型方程中系数和自由项的物理意义。

6.7 位移法典型方程建立的依据是什么？

6.8 试说明力法和位移法在基本未知量、基本体系、基本方程三方面的不同之处。

Questions

6.1 Why can the hinge support or angular displacement at the hinge joint not be selected as the basic unknown quantities in the displacement method?

6.2 Why can the linear displacement at the sliding support not be selected as the basic unknown quantities in the displacement method?

6.3 Try to describe the determination of the number of nodal angular displacement in the basic unknown quantities of displacement method.

6.4 Try to describe the determination of the number of nodal independent linear displacements in the basic unknowns of displacement method.

6.5 Are the displacement method equations of the two methods (direct balance method and basic system method) the same?

6.6 Try to explain the physical meaning of coefficients and free terms in typical equations of displacement method.

6.7 What is the basis for establishing the typical equations of the displacement method?

6.8 Try to explain the differences between the force method and the displacement method in three aspects: basic unknown quantities, basic system, and basic equations.

习 题

Exercises

6.1 试确定图6.34所示结构用位移法计算时独立节点角位移和独立节点线位移的数目。

6.1 Determine the number of independent nodal angular displacements and independent nodal linear displacements for the calculations of the structures in Figure 6.34 by displacement method.

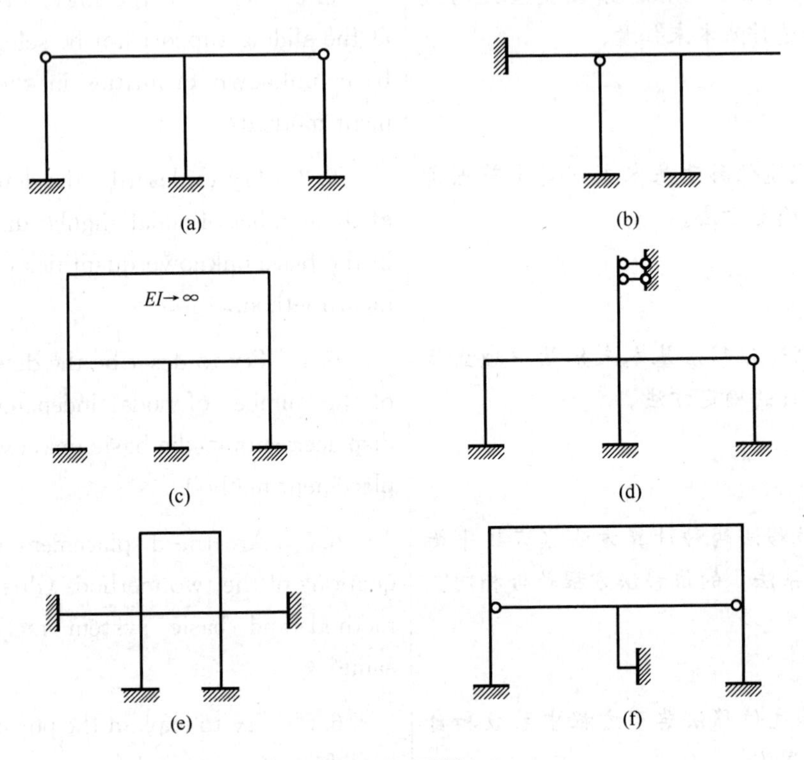

图6.34 习题6.1

Figure 6.34 Exercise 6.1

6.2 试用位移法计算图6.35所示连续梁，并绘制弯矩图和剪力图。各杆 EI 相同且为常数。

6.2 Use displacement method to calculate the continuous beams in Figure 6.35, and draw bending moment diagrams and shear force diagrams. EI of each rod is the same and constant.

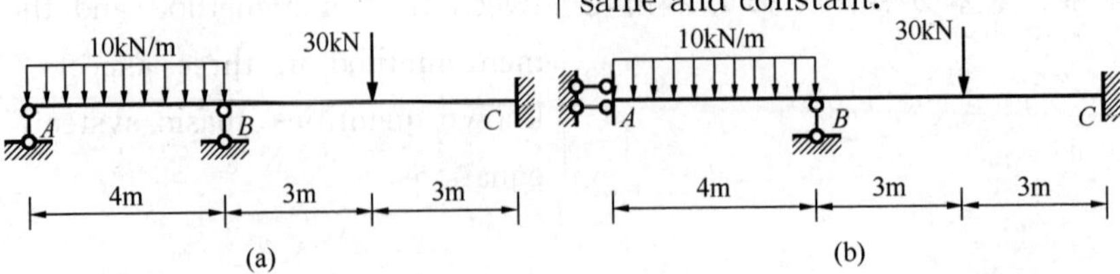

图6.35 习题6.2

Figure 6.35 Exercise 6.2

6.3 试用位移法计算图 6.36 所示刚架，并绘制内力图。各杆 EI 相同且为常数。

6.3 Calculate the rigid frames by displacement method in Figure 6.36 and draw the internal force diagrams. EI of each rod is the same and constant.

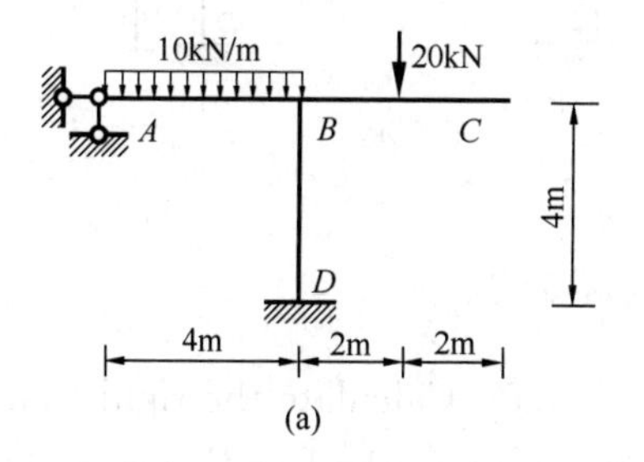

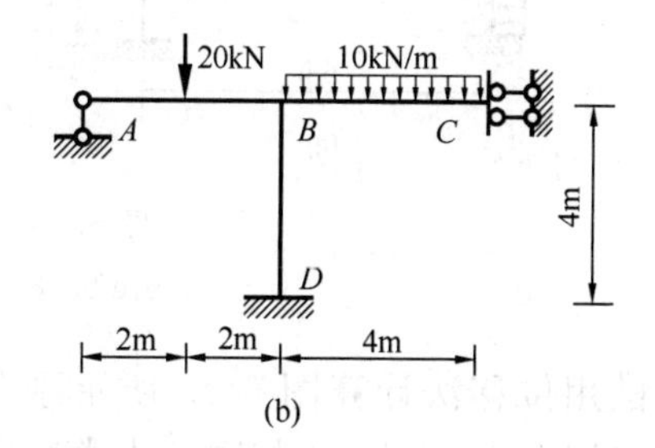

图 6.36 习题 6.3

Figure 6.36 Exercise 6.3

6.4 试用位移法计算图 6.37 所示刚架，并绘制弯矩图。各杆 EI 相同且为常数。

6.4 Calculate the rigid frames by displacement method in Figure 6.37 and draw the internal force diagrams. EI of each rod is the same and constant.

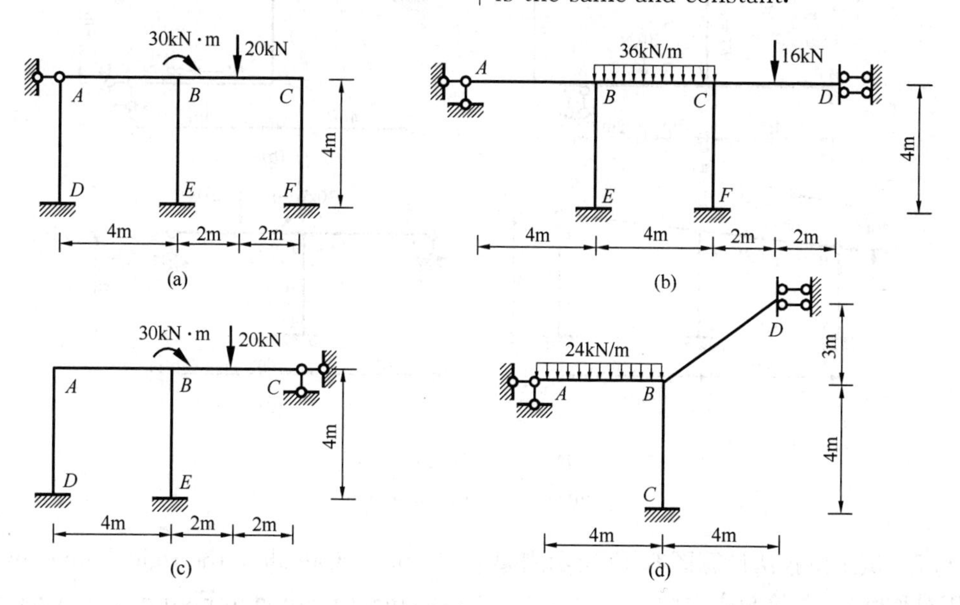

图 6.37 习题 6.4

Figure 6.37 Exercise 6.4

6.5 试用位移法计算图 6.38 所示结构，并绘制内力图。各杆 EI 相同且为常数（注明者除外）。

6.5 Calculate the structures by displacement method in Figure 6.38 and draw the internal force diagrams. EI of each rod is the same and constant (except for specified conditions).

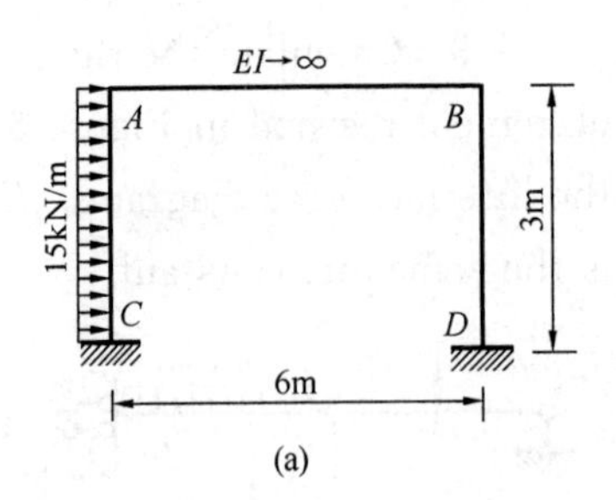

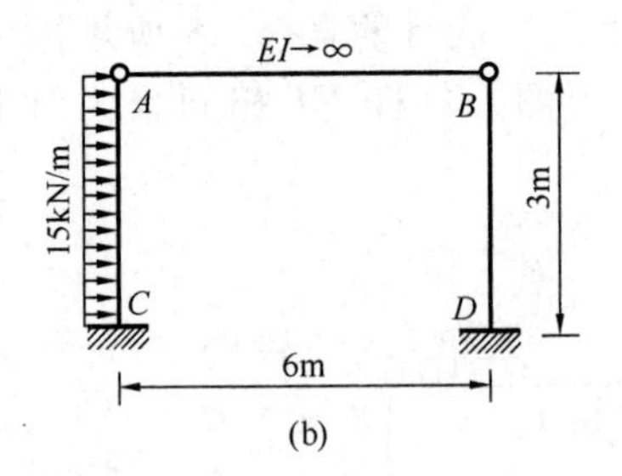

图 6.38 习题 6.5

Figure 6.38 Exercise 6.5

6.6 试用位移法计算图 6.39 所示刚架，并绘制弯矩图。各杆 EI 相同且为常数（注明者除外）。

6.6 Calculate the rigid frames by displacement method in Figure 6.39 and draw the internal force diagrams. EI of each rod is the same and constant (except for specified conditions).

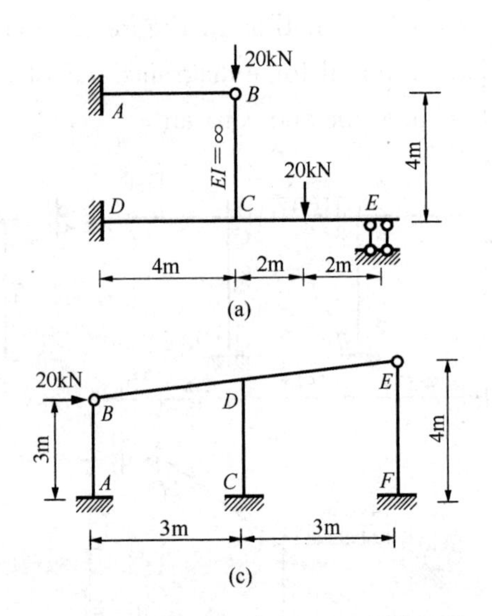

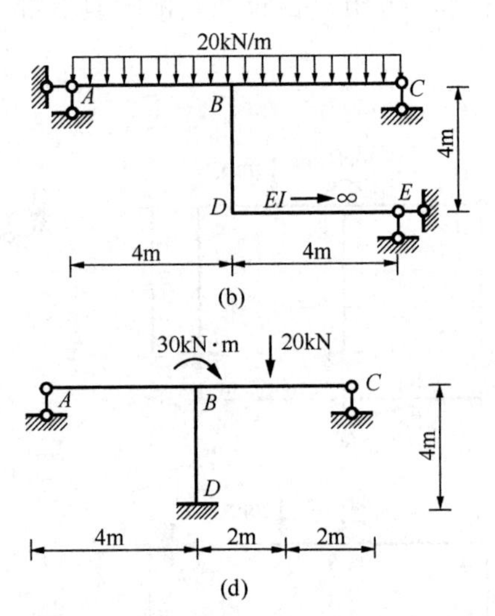

图 6.39 习题 6.6

Figure 6.39 Exercise 6.6

6.7 试用位移法计算图 6.40 所示刚架（利用对称性），并绘制内力图。

6.7 Calculate the rigid frames by displacement method in Figure 6.40 (use symmetry) and draw the internal force diagrams.

6.8 试用位移法计算图 6.41 所示刚架（利用对称性），并绘制弯矩图（各杆刚度均为 EI）。

6.8 Calculate the rigid frames by displacement method in Figure 6.41 (use symmetry) and draw the internal force diagrams (The stiffness of each rod is EI).

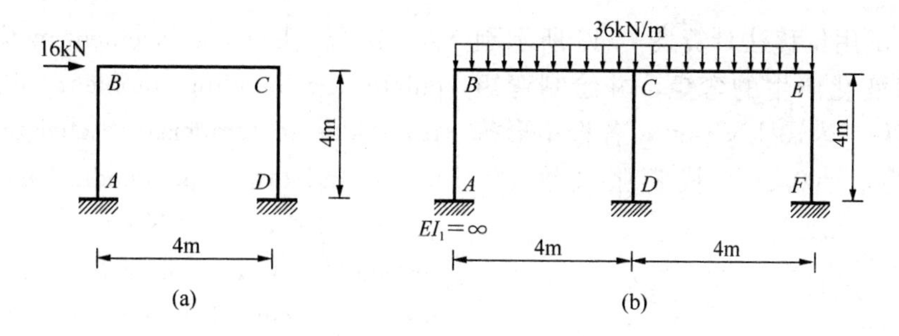

图 6.40 习题 6.7

Figure 6.40 Exercise 6.7

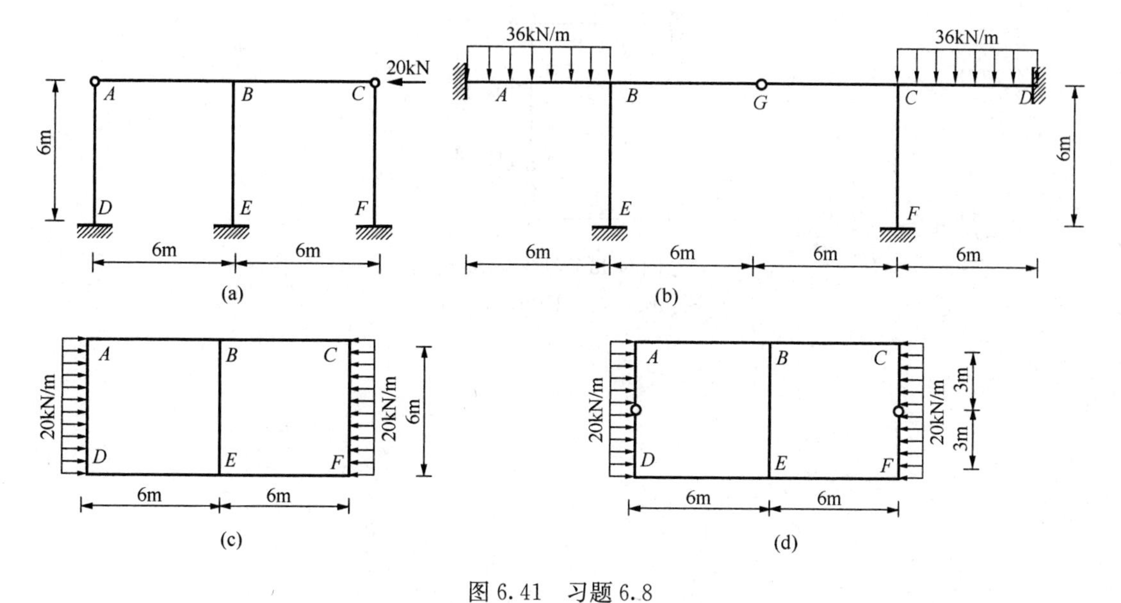

图 6.41 习题 6.8

Figure 6.41 Exercise 6.8

6.9 试用位移法计算图 6.42 所示超静定梁和刚架由于支座位移产生的弯矩，并作弯矩图。各杆 EI 相同且 $EI=1.4\times10^5\text{kN}\cdot\text{m}^2$。

6.9 Use displacement method to calculate the bending moments of statically indeterminate beam and rigid frame due to support displacements in Figure 6.42, and draw the bending moment diagrams. EI of each rod is the same and $EI=1.4\times10^5\text{kN}\cdot\text{m}^2$.

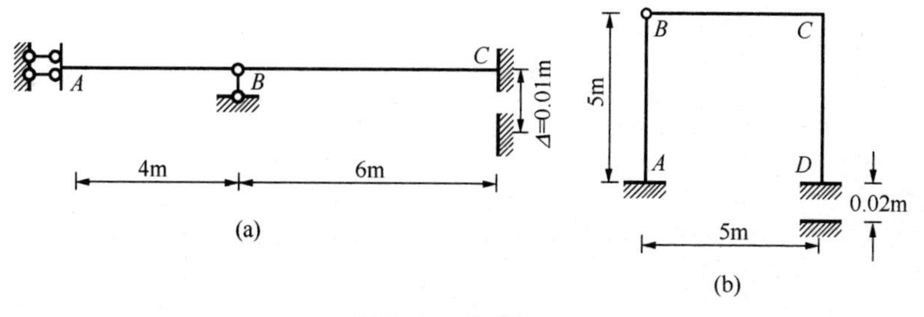

图 6.42 习题 6.9

Figure 6.42 Exercise 6.9

6.10　试用位移法计算图6.43所示刚架由于温度变化产生的弯矩，并绘制弯矩图。已知$EI=1\times10^4\text{kN}\cdot\text{m}^2$，各杆矩形截面高度均为$h=0.5\text{m}$，线膨胀系数$\alpha=0.00001$。

6.10　Use displacement method to calculate the bending moment of the rigid frame due to temperature change in Figure 6.43, and draw the bending moment diagram. $EI=1\times10^4\text{kN}\cdot\text{m}^2$, The height of rectangular section of each rod is $h=0.5\text{m}$, the coefficient of linear expansion is $\alpha=0.00001$.

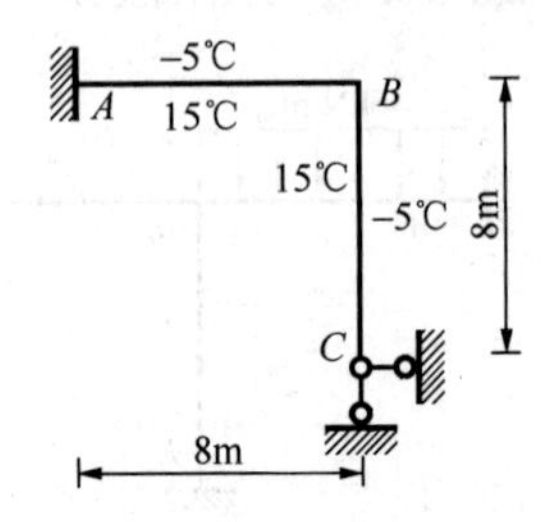

图6.43　习题6.10

Figure 6.43　Exercise 6.10

第 7 章
Chapter 7

渐近法
The Progressive Method

要点

- 力矩分配法
- 多节点力矩分配
- 无剪力分配法

Keys

- Moment distribution method
- Multi-node moment distribution
- Shear-free distribution method

7.1 渐近法概述

7.1 Overview of the Progressive Method

20 世纪 30 年代以来，为避免求解大规模联立的方程组，陆续出现了一些适合手算的方法，如力矩分配法、无剪力分配法等，其特点在于随着计算轮次的增加，计算精度逐渐提高，从而快速向精确解收敛，因此这些方法统称渐近法。

Since the 1930s, to avoid solving large-scale simultaneous equations, some methods suitable for hand calculations have appeared one after another, such as the moment distribution method and the shear-free distribution method, which are characterized by a gradual increase in computational accuracy as the number of computational rounds increases, leading to a rapid convergence to the exact solution, so these methods are collectively called progressive methods.

7.2 力矩分配法的基本概念

7.2 Basic Concepts of Moment Distribution Method

力矩分配法由位移法演变而来，故其节点角位移、杆端力的方向规定均与位移法一致。力矩分配法适于计算连续梁和无节点线

Moment distribution method is evolved from the displacement method, so its nodal angular displacement and direction of rod

位移刚架。

end force are consistent with the displacement method. Moment distribution method is suitable for calculating the continuous beams and rigid frames with no nodal linear displacements.

7.2.1 转动刚度

7.2.1 Rotational Stiffness

转动刚度表示杆端对转动的抵抗能力。杆端的转动刚度用 S 表示，对于等截面直杆 AB，转动刚度 S_{AB} 表示 A 端（近端）产生单位转角时需要施加的力矩 M_{AB}。其不仅与杆件线刚度 $i=EI/l$ 有关，也与 B 端的约束情况有关。常见的转动刚度见表 7.1。

Rotational stiffness represents the resistance of the rod end to rotation. The rotational stiffness of a rod end is denoted by S. For a straight rod with equal section AB, the rotational stiffness S_{AB} represents the moment M_{AB} required to produce a unit angle of rotation at end A (proximal end), which is not only related to the rod linear stiffness $i=EI/l$, but also related to the restraint at end B. Common rotational stiffnesses are shown in Table 7.1.

等截面直杆的转动刚度和传递系数 **表 7.1**

Rotational stiffness and transfer coefficients of straight rods with equal section **Table 7.1**

远端约束情况 Distant constraints	S_{AB}	S_{BA}	C_{AB}	C_{BA}
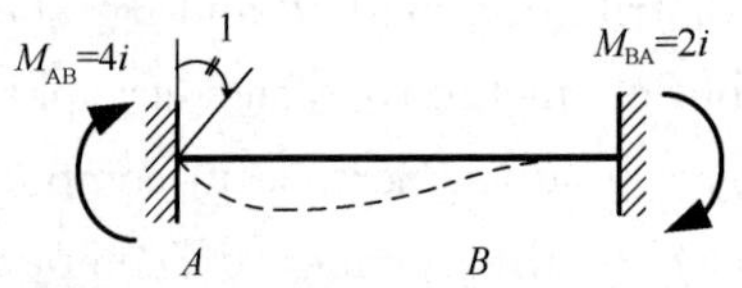 $M_{AB}=4i$, 1, $M_{BA}=2i$, A, B	$4i$	$4i$	0.5	0.5
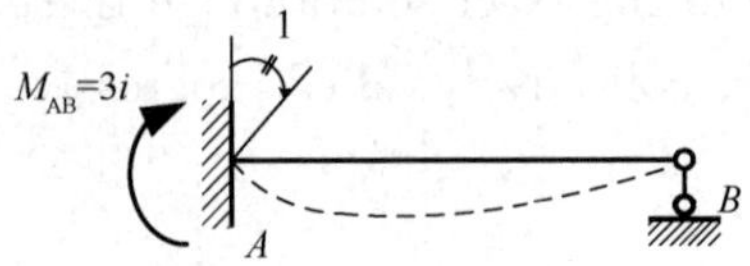 $M_{AB}=3i$, 1, A, B	$3i$	$4i$	0	0.5
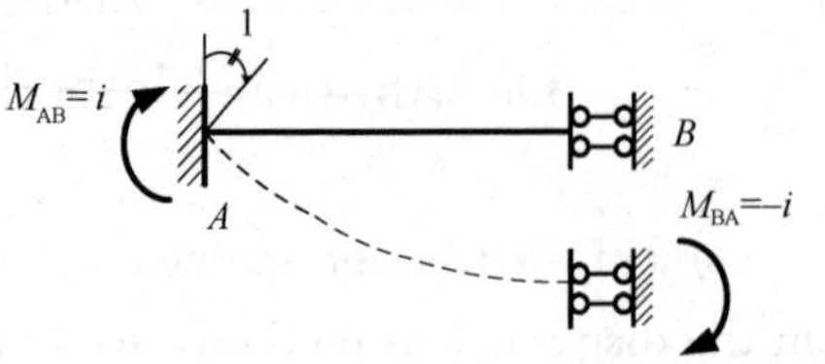 $M_{AB}=i$, 1, A, B, $M_{BA}=-i$	i	i	−1	−1

这里要说明的是，转动刚度可仅考虑远端的约束情况（近端加上附加刚臂），如图 7.1 所示。

It should be noted here that the rotational stiffness can be considered only for the distant end constraint case (proximal end adds additional stiffness arms), as shown in Figure 7.1.

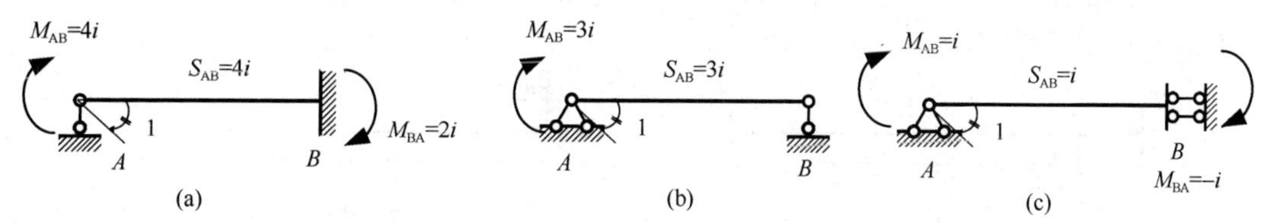

图 7.1 几种转动刚度的情况

Figure 7.1 Several cases of rotational stiffness

【例 7.1】 利用概念求解转动刚度 S_{AB}。

[Example 7.1] Use the concept to solve the rotational stiffness S_{AB}.

【解】 图 7.2（a）中，B 端自由，A 端转角不会受到抵抗，故 $S_{AB}=0$。

[Solution] In Figure 7.2 (a), end B is free, and the rotation angle of end A will not be resisted, $S_{AB}=0$.

图 7.2（b）中，可按图 7.3（a）（位移法思路：单位转角需要的弯矩）或图 7.3（b）（力法思路：单位弯矩产生的转角，即柔度，再取倒数）得到，结果是一样的，$S_{AB}=i$。

In Figure 7.2 (b), the results can be obtained as in Figure 7.3 (a) (displacement method idea: bending moment required per unit rotation angle) or in Figure 7.3 (b) (force method idea: rotation angle generated per unit bending moment, i. e., flexibility, and then take the inverse), and the results are the same, $S_{AB}=i$.

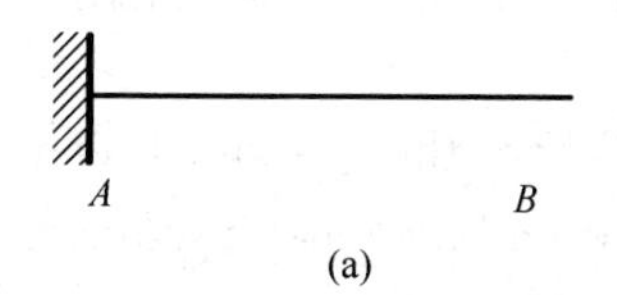

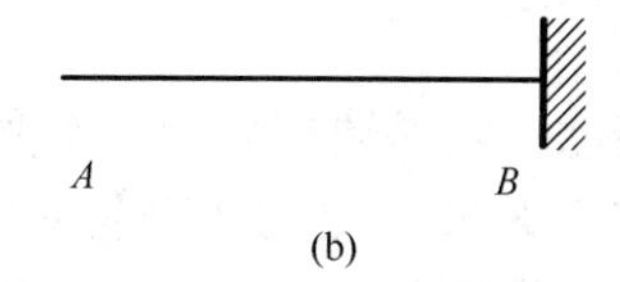

图 7.2 转动刚度的概念

Figure 7.2 Concept of rotational stiffness

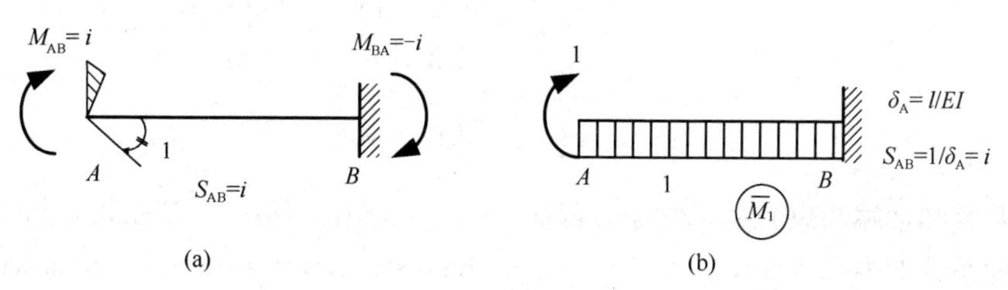

图 7.3 转动刚度的求解

Figure 7.3 Solving for rotational stiffness

【例 7.2】利用概念求解转动刚度 S_{AB}。已知 $EI=C$，令 $i=EI/l$。

[**Example 7.2**] Solving for the rotational stiffness S_{AB} using the concept. $EI=C$, $i=EI/l$.

【解】求解过程如图 7.4 所示。

[**Solution**] The solving process is shown in Figure 7.4.

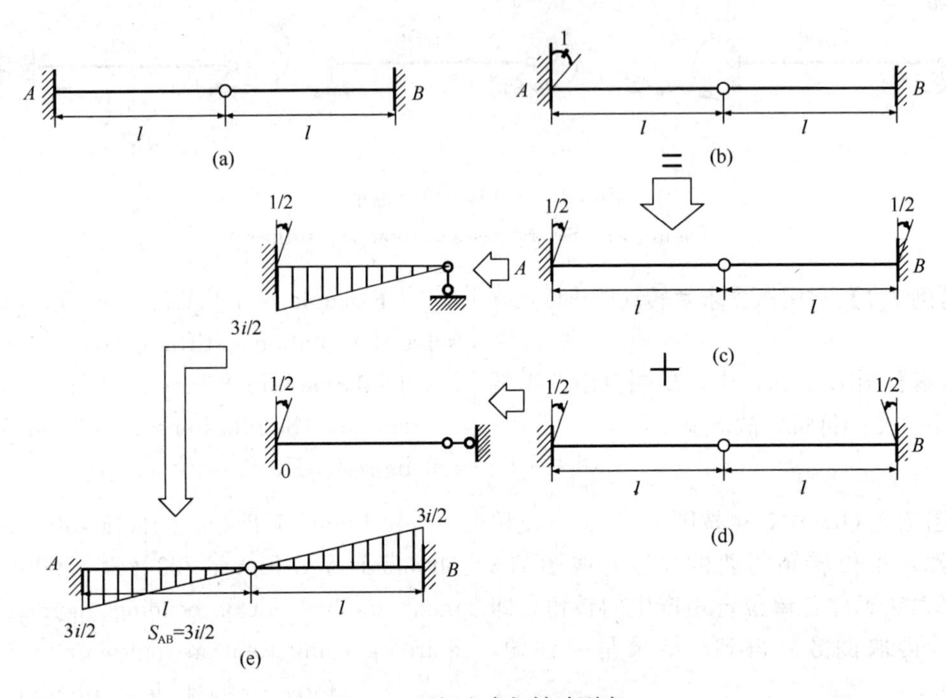

图 7.4 用概念求解转动刚度

Figure 7.4 Solving for rotational stiffness by concept

7.2.2 传递系数

7.2.2 Transfer Coefficient

当近端转动时，远端也可能产生弯矩，这好比是近端弯矩按一定的比例传到了远端，故此，定义传递系数如下：

When the proximal end rotates, the distant end may also produce a bending moment, like the proximal bending moment is transmitted to the distant end in a particular proportion. Therefore, the transfer coefficient is defined as:

$$C_{AB}=\frac{M_{BA}}{M_{AB}} \tag{7.1}$$

传递系数和远端约束有关，常见等截面直杆的传递系数如表 7.1 所示。

The transfer coefficients are related to the distant restraint and are shown in Table 7.1 for common straight rods with equal section.

7.2.3 分配系数

图 7.5 所示各等截面直杆线刚度均为 i，A 节点上作用一个集中力偶 M，使之产生转角 θ，那么杆端弯矩和 M 在 A 节点的力矩平衡方程可以写为：

7.2.3 Distribution Factor

The linear stiffness of each straight rod with equal section in Figure 7.5 is i. Concentrated couple M is acted on node A to produce an angle of rotation θ, the moment at the rod end and the moment balance equation of M at the node A can be written as:

$$M = M_{AB} + M_{AC} + M_{AD}$$

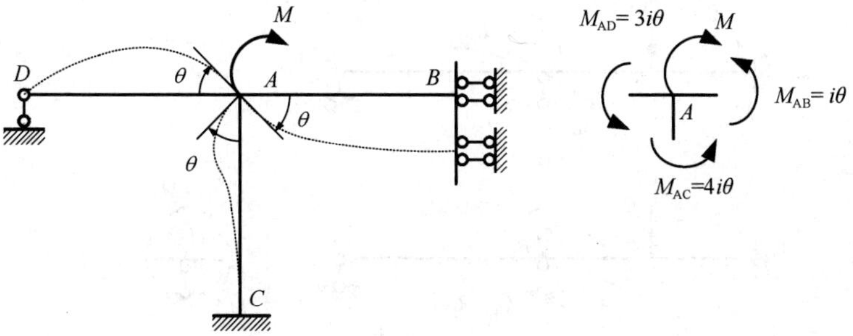

图 7.5 分配系数

Figure 7.5 Distribution factor

由转动刚度的概念，可以写为：

By the concept of rotational stiffness, it can be written as:

$$M = S_{AB}\theta_A + S_{AD}\theta_A + S_{AC}\theta_A$$

$$\theta_A = \frac{M}{\sum_A S}$$

$$M_{AB} = \frac{S_{AB}}{\sum_A S} M$$

$$M_{AC} = \frac{S_{AC}}{\sum_A S} M$$

$$M_{AD} = \frac{S_{AD}}{\sum_A S} M$$

引入分配系数：

Introduce distribution factor:

$$\mu_{Aj} = \frac{S_{Aj}}{\sum_A S} \tag{7.2}$$

则分配弯矩：

Then the distributed bending moment is:

$$M_{Aj} = \mu_{Aj} M \tag{7.3}$$

$$\sum_{A} \mu_{Aj} = 1 \tag{7.4}$$

7.3 单节点力矩分配

7.3 Moment Distribution of Single Node

用力矩分配法计算任意荷载作用下具有一个节点角位移的结构。以图 7.6（a）所示的结构说明其计算步骤。

The moment distribution method is used to calculate structures with one nodal angular displacement under arbitrary loads. The calculation procedure is illustrated with the structure shown in Figure 7.6（a）.

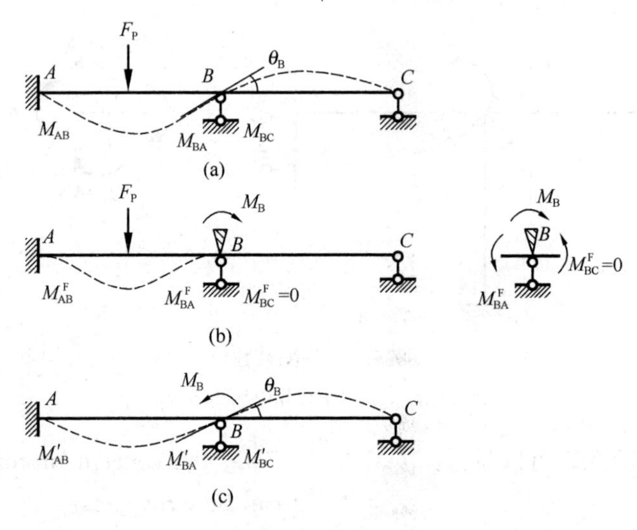

图 7.6 力矩分配法

Figure 7.6 Moment distribution method

由于力矩分配法是以位移法为理论基础的，故仿照位移法思路进行分析。

Since the moment distribution method is theoretically based on the displacement method, the analysis is modeled after the displacement method.

1. 固定节点，求约束力矩。设想在作为位移法基本未知量的节点角位移（θ_B）处，附加一个阻止其转动的约束——刚臂，即在节点 B 上加刚臂，阻止节点 B 的转动，然后将荷载加上去，如图 7.6（b）所示。此时连续梁 ABC 已分为两个单跨梁 AB 和 BC，各杆端产生固端弯矩，而节点上有约束力矩，它暂时由刚臂承担。利用节点 B 的力矩平衡条件，可求出节点 B 的**约束力**

1. Fix the node and find out the constraint moment. Suppose that at the angular displacement (θ_B) of the node, which is the basic unknown quantity of the displacement method, a restraint that prevent its rotation is attached, i. e., a rigid arm is added to the node B to prevent the rotation of the node B, and then the load is added, as shown in Figure 7.6（b）. At this point,

矩M_B（也称**不平衡力矩**）。从图 7.6（b）可以看出，杆 BC 的固端弯矩 $M_{BC}^{F}=0$，杆 BA 的固端弯矩为 M_{BA}^{F}。由 $\Sigma M_B=0$ 知，节点 B 的约束力矩 $M_B=M_{BA}^{F}+M_{BC}^{F}=M_{BA}^{F}$。约束力矩等于固端弯矩之和，以顺时针转向为正。

the continuous beam ABC has been divided into two single-span beams AB and BC, of which each rod end produces a solid end bending moment, while there is a restraining moment at the node, which is temporarily borne by the rigid arm. Using the moment balance condition of node B, the **restrained moment M_B** (also called **unbalanced moment**) of node B can be found. From Figure 7.6 (b), the solid end bending moment of rod BC is $M_{BC}^{F}=0$, the solid bending moment of rod BA is M_{BA}^{F}. According to $\Sigma M_B=0$, the constraint moment of nodal B is $M_B=M_{BA}^{F}+M_{BC}^{F}=M_{BA}^{F}$. The constraint moment equals the sum of the solid end bending moments, with clockwise rotation as positive.

2. 放松节点，求分配力矩和传递力矩。连续梁的节点 B 本来没有限制转动的约束，也不存在约束力矩 M_B。为了使其恢复到原来的状态，如图 7.4（a）所示，使节点 B 处的约束力矩由 M_B 回复到零，我们取消刚臂，放松节点 B 处的约束，让节点 B 转动，这相当于在节点 B 处新加一个等值、反向的力偶（$-M_B$），如图 7.6（c）所示，于是不平衡力矩被消除而节点获得平衡。此反号的不平衡力矩将按转动刚度大小的比例分配给各近端，于是各近端得到分配力矩，同时各自向其远端进行传递，各远端得到传递力矩。

2. Relax the node and find out the distributed moments and transfer moments. The node B of the continuous beam has no restraint to limit the rotation, and there is no restraint moment M_B. In order to restore it to its original state as shown in Figure 7.4 (a), so that the restraint moment at node B returns from M_B to zero, we remove the rigid arm, relax the restraint at node B, and let node B rotate, which is equivalent to adding a new equal and opposite force couple ($-M_B$) at node B, as shown in Figure 7.6 (c), the unbalanced moment is eliminated, and the node is balanced. The unbalanced moment of inverse sign will be distributed to each proximal end in proportion according to the rotational stiffness, so each proximal end will get the distributed moment, while each will transmit to its distant end, and each distant end will get the transfer moment.

3. 将图 7.6（b）、（c）所示两种情况叠加，就得到 7.6（a）所示情况，即将以上两步求出的各杆端弯矩加在一起，就得到实际的杆端弯矩。例如 $M^F_{BA}+M'_{BA}=M_{BA}$。

3. Superimpose the two cases in Figures 7.6（b） and （c） and obtain the case in Figure 7.6（a），that is，the bending moments from the above two steps of the rod end are added together to obtain the actual rod end bending moments，such as $M^F_{BA}+M'_{BA}=M_{BA}$.

现结合上例把一般荷载作用下单节点力矩分配法的计算步骤简述如下：

Combined with the above example，the calculation steps of the single node moment distribution method under general loads are briefly described as follows：

1. 在刚节点 B 上加入刚臂，把连续梁分为单跨梁，求出各杆端产生的固端弯矩，由节点 B 各杆固端弯矩之和求出约束力矩 M_B。

1. Add a rigid arm to the rigid node B，divide the continuous beam into single-span beams，find the solid end bending moment at each rod end，and find the restraint moment M_B by the sum of the solid end bending moments of each rod at node B.

2. 去掉约束，即相当于在节点 B 新加一个力偶（$-M_B$），求出各杆在 B 端的分配力矩和远端的传递力矩。

2. Remove the constraint，which is equivalent to adding a new moment （$-M_B$） at node B，and find the distribution moments of each rod at end B and the transmission moments at distant ends.

3. 叠加各杆端的力矩就得到实际的杆端弯矩。

3. Superimpose the moments under each rod end to get the actual rod end bending moment.

【例 7.3】 用力矩分配法求解图 7.7（a），EI=C。

［**Example 7.3**］ Solve the structure shown in Figure 7.7（a） by the moment distribution method，EI=C.

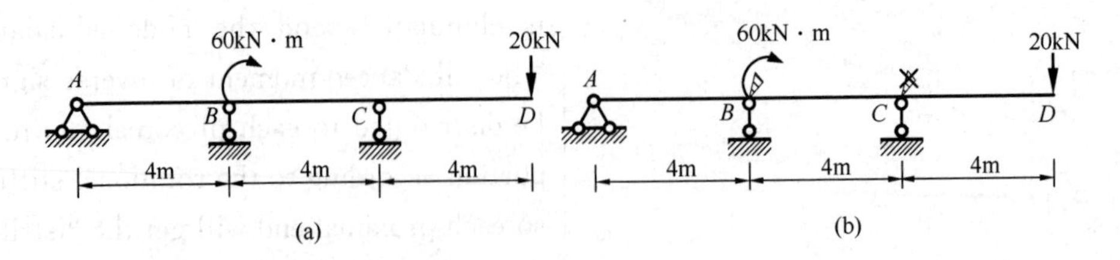

图 7.7 单节点力矩分配

Figure 7.7 The moment distribution of single node

【解】（1）建议按照位移法的思路首先

［**Solution**］（1） It is recommended that

明确基本结构，即如何添加附加刚臂。从图7.7（b）中可以看出，C节点由于弯矩静定，所以不需要添加附加刚臂，所以本题只有B节点进行力矩分配。

the basic structure be clarified first according to the displacement method, i. e., how to add the additional rigid arm. From Figure 7.7 (b), we can see that the node C does not need to add additional rigid arm because the moment is static, so only node B is conducted for moment distribution in this example.

（2）求出初始弯矩，如图7.8所示；注意B节点的固端弯矩的正负（以顺时针为负，和位移法规定一致）。

(2) Find out the initial bending moment, as shown in Figure 7.8; pay attention to the positive and negative bending moments of the solid end of node B (clockwise as negative, which is consistent with the displacement method).

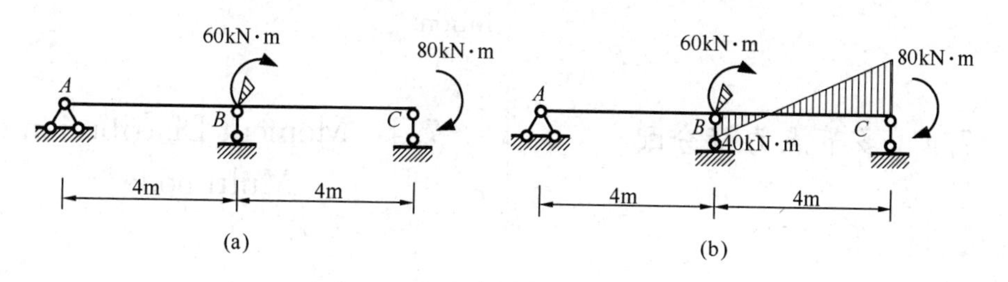

图7.8 初始弯矩

Figure 7.8 Initial bending moment

（3）计算转动刚度、分配系数和传递系数并填表（表7.2）。

(3) Calculate the rotational stiffness, distribution coefficients, and transfer coefficients and fill the table (Table 7.2).

单节点力矩分配表（kN·m） **表7.2**

Table of the moment distribution of single node (kN·m) **Table 7.2**

<table>
<tr><td>节点 Node</td><td>A</td><td colspan="3">B</td><td>C</td></tr>
<tr><td>杆端 Rod end</td><td>AB</td><td colspan="2">BA</td><td>BC</td><td>CA</td></tr>
<tr><td>分配系数
Distribution coefficient</td><td>—</td><td colspan="2">0.5</td><td>0.5</td><td>—</td></tr>
<tr><td>固端弯矩
Solid end bending moment</td><td>0</td><td>0</td><td>−60</td><td>40</td><td>80</td></tr>
<tr><td>分配传递
Distribution transfer</td><td>0</td><td colspan="2">← 10</td><td>10 →</td><td>0</td></tr>
<tr><td>最后弯矩
Final bending moment</td><td>0</td><td colspan="2">10</td><td>50</td><td>80</td></tr>
</table>

（4）画出最终弯矩图，如图7.9所示。

(4) Draw the final bending moment diagram, as shown in Figure 7.9.

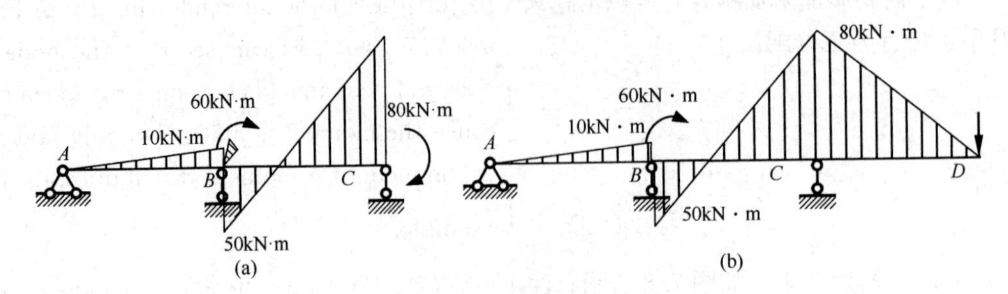

图7.9　最终弯矩图

Figure 7.9　Final bending moment diagram

总结可知，单节点力矩分配一次即可完成，且结果为精确解。

To sum up, it can be seen that the moment distribution of single node can be done only one time and the result is an exact solution.

7.4　多节点力矩分配

7.4　Moment Distribution of Multi-node

对于具有多个节点转角但无节点线位移的结构，可采用逐个节点轮流放松的办法，即每次只放松一个节点，其他节点仍暂时锁住，这样把各节点的约束力矩轮流地进行分配、传递，直到各节点的约束力矩可以略去为止，最后根据叠加原理求得结构的各杆端弯矩。

For the structure with multiple nodal rotation angles but no nodal linear displacement, the nodes can be relaxed in sequence, i.e., for each time only one node is relaxed, and the other nodes are temporarily locked, so that the restraint moment of each node is distributed and transferred in turn, until the restraint moment of each node can be omitted, and finally according to the principle of superposition the bending moment of each rod end of the structure can be obtained.

【例7.4】用力矩分配法求解图7.10所示刚架弯矩图。

[Example 7.4] Solve the bending moment of the rigid frame in Figure 7.10 by moment distribution method.

【解】（1）确定力矩分配节点为B、C；

[Solution] (1) Determine the moment distribution nodes B and C;

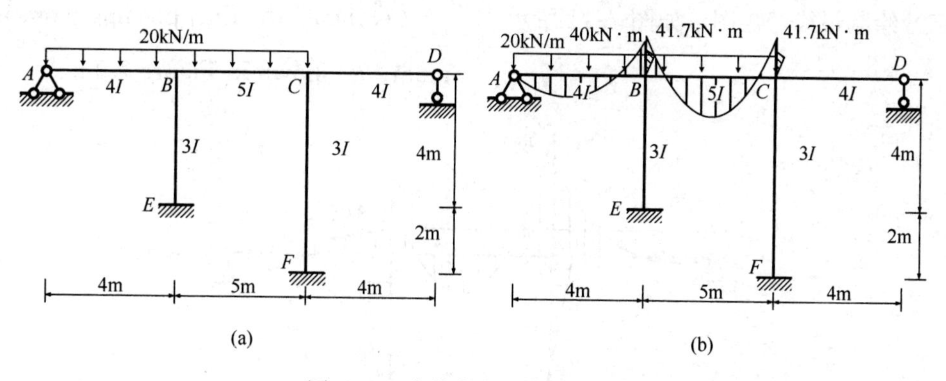

图 7.10 多节点力矩分配法举例

Figure 7.10 Example of moment distribution method of multi-node

（2）计算转动刚度、分配系数和传递系数；

（2）Calculate the rotational stiffness, distribution coefficients, and transfer coefficients;

（3）计算初始弯矩；

（3）Calculate the initial bending moment;

（4）填表、进行力矩分配（从不平衡力矩大的节点开始）（表 7.3）；

（4）Fill in the table and perform moment distribution (from the node with large unbalanced moment) (Table 7.3);

多节点力矩分配表（kN · m） **表 7.3**

Table of moment distribution of multi-node（kN · m） **Table 7.3**

节点 Node		A	B			C			D	E	F
杆端 Rod end		AB	BA	BE	BC	CB	CF	CD	DC	EB	FC
分配系数 Distribution coefficient			0.3	0.3	0.4	0.445	0.222	0.333			
固端弯矩 Solid end bending moment		0	40	0	−41.7	41.7	0	0	0	0	0
分配传递 Distribution transfer	Ⅰ				−9.3	←−18.5	−9.3	−13.9→	0		−4.7
		0	←3.3	3.3	4.4→	2.2				1.6	
	Ⅱ				−0.5	←−1.0	−0.5	−0.7→	0		−0.2
		0	←0.15	0.15	0.2					0.1	
结果 Final result		0	43.4	3.5	−46.9	24.4	−9.8	−14.6	0	1.7	−4.9

（5）绘制最终弯矩图，如图 7.11 所示。

(5) Draw the final bending moment diagram, as shown in Figure 7.11.

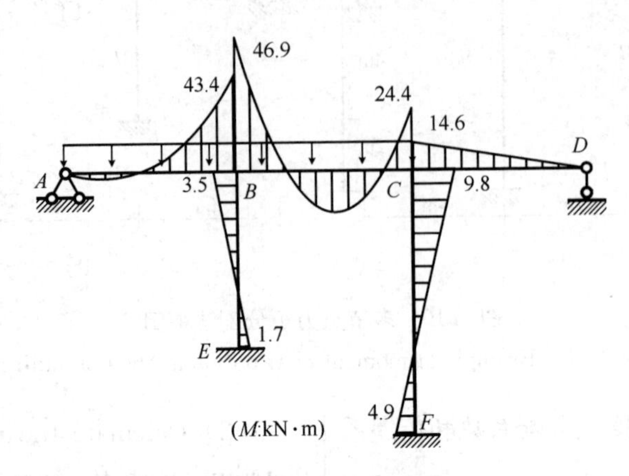

图 7.11 最终弯矩图

Figure 7.11 Final bending moment diagram

7.5 无剪力分配法

7.5 Distribution Method with No Shear Force

力矩分配法只能求解无侧移刚架，不能直接计算有侧移刚架。但对于某些特殊的有侧移刚架，可以用与力矩分配法类似的无剪力分配法进行计算。

The moment distribution method can only solve for rigid frames with no lateral displacement, and cannot directly calculate rigid frames with lateral displacements. However, for some special rigid frames with lateral displacements, the distribution method with no shear force can be conducted, which is similar to the moment distribution method.

7.5.1 无剪力分配法的适用条件

7.5.1 Applicable Conditions of the Distribution Method with No Shear Force

无剪力分配法并不能直接用于所有的有侧移刚架，其适用条件为：刚架中除了无侧移杆外，其余杆件都是剪力静定杆，如图 7.12 所示。

The distribution method with no shear force is not directly applicable to all rigid frames with lateral displacements. The applicable condition is that all the rods in the rigid frame are shear static rods except for the rods with no lateral displacements, as shown in Figure 7.12.

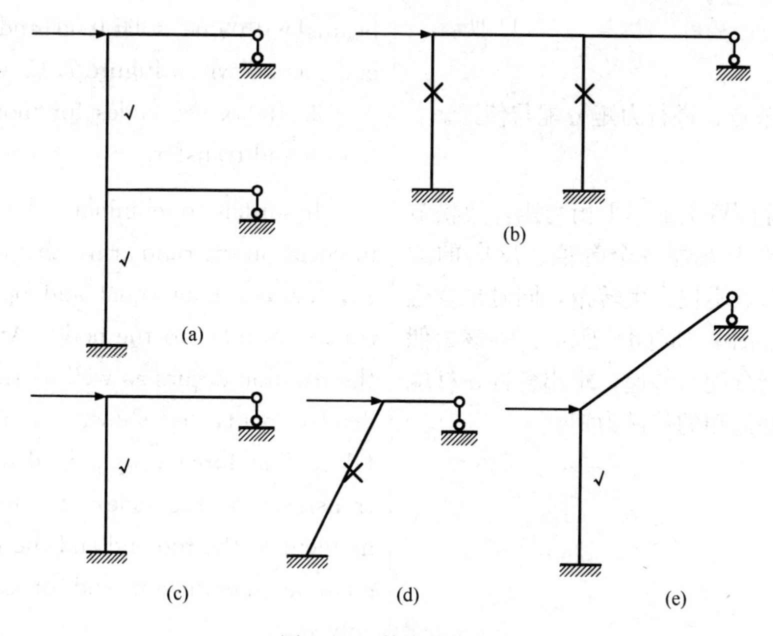

图 7.12 剪力静定杆

Figure 7.12 Shear static rod

7.5.2 无剪力分配法的计算方法和步骤

7.5.2 Calculation Methods and Steps of the Distribution Method with No Shear Force

采用无剪力分配法计算时，计算过程和力矩分配法相似，分为两步：

When using the distribution method with no shear force, the calculation process is similar to the moment distribution method and is divided into two steps:

1. 固定节点，求约束力矩。

1. Fix the node, find out the constraint moment.

添加附加刚臂，如图 7.13（a）所示。这样，杆 *AB*、*BC* 的上端虽不能转动但仍可自由地水平滑动，故可看作下端固定上端滑动的梁；杆 *CD*、*BE* 虽可发生水平位移，但各点水平位移相同，对其内力没有任何影响，故仍可看作一端固定，一端铰支的梁，如图 7.13（d）所示。

Add additional rigid arms, as shown in Figure 7.13（a）. In this way, the upper ends of rods *AB* and *BC* can not rotate but still slide freely and horizontally, so they can be regarded as beams with fixed lower ends and sliding upper ends; rods *CD* and *BE* can have horizontal displacements, but the horizontal displacement at each point is the same, which does not affect their internal forces, so they can still be regarded as beams with one solid end and one hinge end, as shown in Figure 7.13 (d).

2. 放松节点，进行力矩分配与传递。

2. Relax the nodes for moment distribution and transfer.

为了消除刚臂上的不平衡力矩，放松节点，即在节点上新加一个等值、反向的力偶。此时，节点不仅发生转角，同时也发生水平位移，如图7.13（d）所示。将该力偶在节点处进行分配与传递，求出各杆在杆端的分配力矩和远端的传递力矩。

In order to eliminate the unbalanced moment on the rigid arm, the node is relaxed, i.e., a new equal and opposite force couple is added to the node. At this point, the rotation occurs as well as the horizontal displacement, as shown in Figure 7.13 (d). The force couple is distributed and transferred at the node, and the distributed moment at the rod end and the transfer moment at the distant end of each rod are solved.

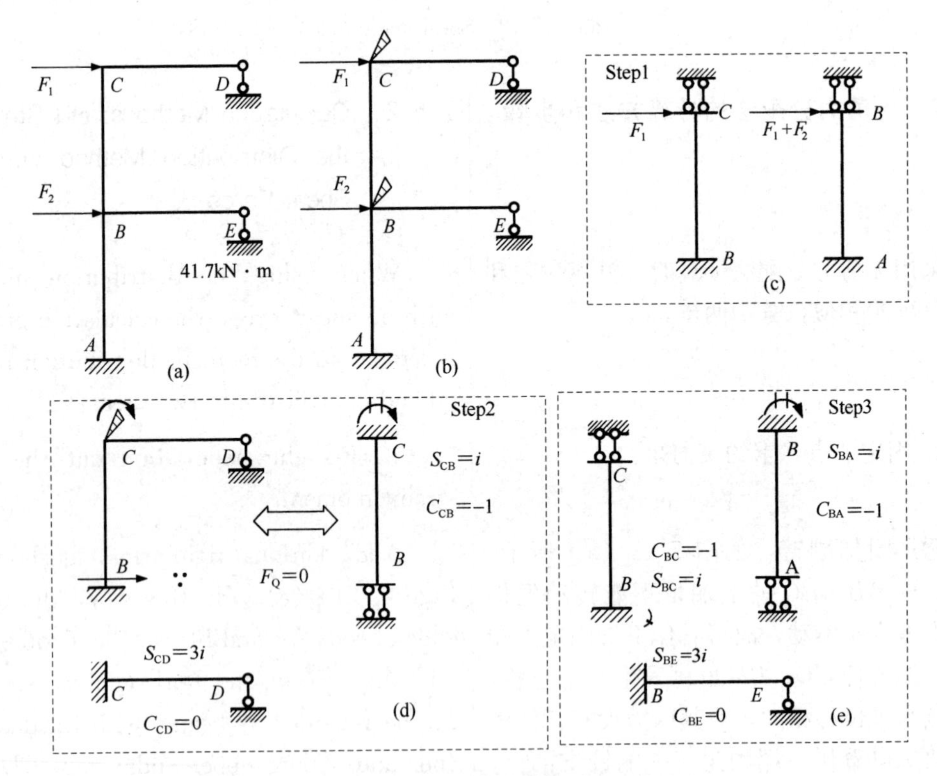

图7.13 无剪力分配法

Figure 7.13 Distribution method with no shear force

以B节点为例。因为此时B端没有剪力（图7.13d），所以等效为滑动支座，所

Take node B as an example. Since there is no shear force at end B (Figure

以这种方法称为"无剪力力矩分配法"，简称无剪力分配法。

7.13d), it is equivalent to a sliding support, so this method is called the "moment distribution method with no shear force", which is short of distribution method with no shear forces.

简言之，对无剪力杆杆端计算时：

In short, for the calculation of rod ends with no shear force:

1. 计算不平衡力矩时，近端修正为滑动支座；

1. Proximal end is corrected for sliding support when calculating unbalanced moments;

2. 计算转动刚度、分配系数、传递系数时，远端修正为滑动支座。

2. The distant end is corrected to sliding support when calculating the rotational stiffness, distribution coefficient, and transfer coefficient.

【例 7.5】试用无剪力分配法作图 7.14 所示刚架的 M 图。

[Example 7.5] Try to use the distribution method with no shear force to solve for M diagram of the rigid frame in Figure 7.14.

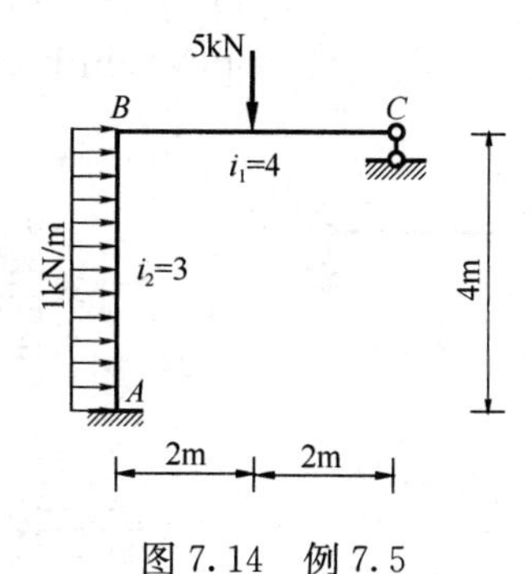

图 7.14 例 7.5

Figure 7.14 Example 7.5

【解】刚架中杆 BC 为无侧移杆，杆 AB 为剪力静定杆，可用无剪力分配法计算。

[Solution] The rod BC in the rigid frame has no lateral displacements, and the rod AB is a shear static rod, the structure can be calculated by the distribution method with no shear force.

（1）计算分配系数

(1) Calculate distribution coefficient

注意各柱端的转动刚度应等于其柱的线刚度：

Note that the rotational stiffness of each column end should be equal to its linear stiffness:

$$S_{BA} = i_2 = 3$$

$$S_{BC} = 3i_1 = 12$$

$$\mu_{BA} = \frac{3}{3+12} = 0.2$$

$$\mu_{BC} = \frac{12}{3+12} = 0.8$$

（2）计算固端弯矩

(2) Calculate the solid end bending moment

$$M^F_{AB} = -\frac{ql^2}{3} = -\frac{1\times4^2}{3} = -5.33\text{kN}\cdot\text{m}$$

$$M^F_{BA} = -\frac{ql^2}{6} = -\frac{1\times4^2}{6} = -2.67\text{kN}\cdot\text{m}$$

$$M^F_{BC} = -\frac{3F_Pl}{16} = -\frac{3\times5\times4}{16} = -3.75\text{kN}\cdot\text{m}$$

（3）放松节点 B，进行力矩的分配、传递

(3) Relax node B for moment distribution and transfer

注意杆 BA 的传递系数为－1。无剪力分配法的计算过程如图 7.15（a）所示。M 图如图 7.15（b）所示。

Note that the transfer coefficient of the rod BA is－1. The calculation procedure of distribution method with no shear force is shown in Figure 7.15 (a). M diagram is shown in Figure 7.15 (b).

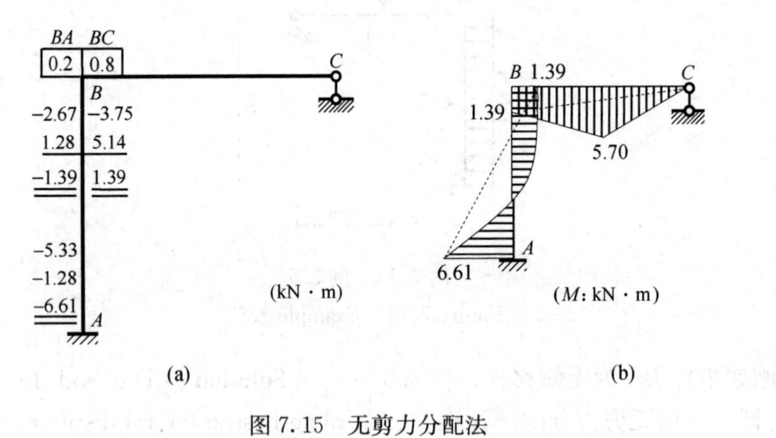

图 7.15 无剪力分配法

Figure 7.15 Distribution method with no shear force

思考题

Questions

7.1 什么是转动刚度？分配系数和转动刚度有何关系？为什么每一刚节点处各杆端的分配系数之和等于1？

7.1 What is rotational stiffness? What is the relationship between distribution coefficient and rotational stiffness?

Why is the sum of the distribution coefficients at each rigid node equal to 1?

7.2 什么是固端弯矩？如何计算约束力矩？为什么要将它反号才能进行分配？

7.2 What is the solid end bending moment? How to calculate the constraint moment? Why does it have to reverse its sign before distribution?

7.3 什么是传递力矩、传递系数？

7.3 What are distribution moment and distribution coefficient?

7.4 试述力矩分配法的基本运算步骤及每一步的物理意义。

7.4 Try to explain the basic concepts of moment distribution method and the physical meaning of each step.

7.5 为什么力矩分配法的计算过程是收敛的？

7.5 Why is the calculation process of moment distribution method convergent?

7.6 多节点的力矩分配中，每次是否只能放松一个节点，可以同时放松多个节点吗？

7.6 In moment distribution of multi-node, can only one node be relaxed at a time, and can multiple nodes be relaxed at the same time?

7.7 力矩分配法只适用于无节点线位移的结构。当这类结构发生已知支座移动时，节点是有线位移的，是否还可以用力矩分配法计算？

7.7 The moment distribution method is only suitable for structures without nodal linear displacement. When this kind of structure has support movement, the node has linear displacement. Can the moment distribution method be used to calculate this condition?

7.8 无剪力分配法的适用条件是什么？为什么称为无剪力分配？

7.8 What are the applicable conditions for the distribution method with no shear force? Why is it called distribution with no shear force?

习　题

Exercises

7.1 试用力矩分配法计算图 7.16 所示结构，并作 M 图。各杆 EI 为常数（注明者除外）。

7.1 Calculate the structures in Figure 7.16 by moment distribution method and draw M diagram. EI of each rod is constant (except for specified conditions).

7.2 试用力矩分配法计算图 7.17 所示

7.2 Calculate the continuous beams in

连续梁，并作M图。各杆EI为常数（注明者除外）。

Figure 7.17 by moment distribution method and draw M diagram. EI of each rod is constant (except for specified conditions).

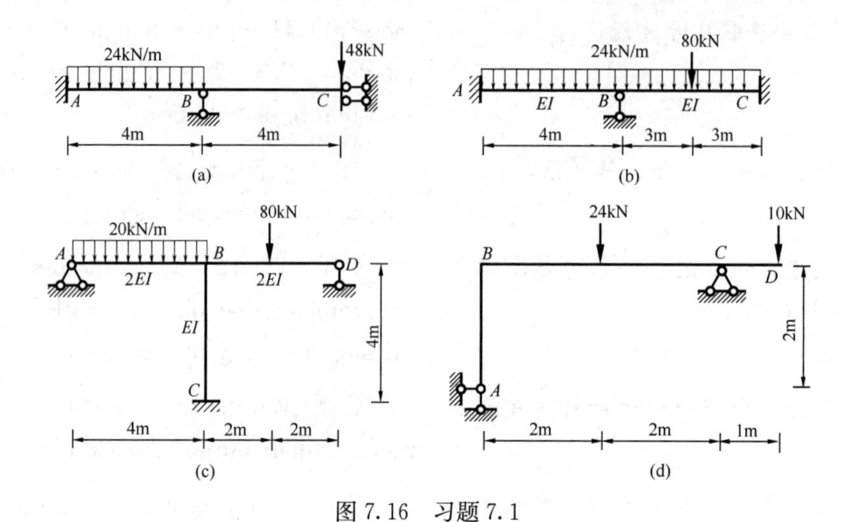

图 7.16　习题 7.1

Figure 7.16　Exercise 7.1

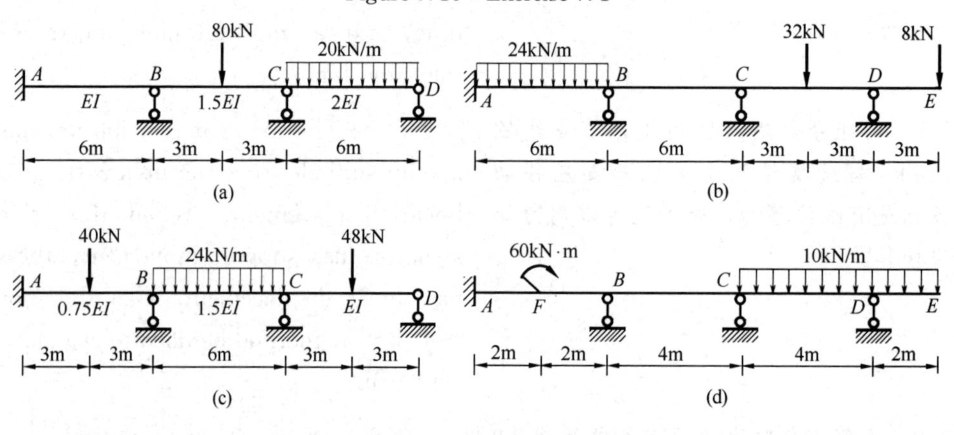

图 7.17　习题 7.2

Figure 7.17　Exercise 7.2

7.3　试用力矩分配法计算图7.18所示刚架，并作M图。各杆EI为常数(注明者除外)。

7.3　Calculate the rigid frames in Figure 7.18 by moment distribution method and draw M diagram. EI of each rod is constant (except for specified conditions).

7.4　试利用对称性按力矩分配法计算图7.19所示刚架，并作M图。各杆EI为常数。

7.4　Use symmetry to calculate the rigid frames in Figure 7.19 by moment distribution method and draw M diagram. EI of each rod is constant.

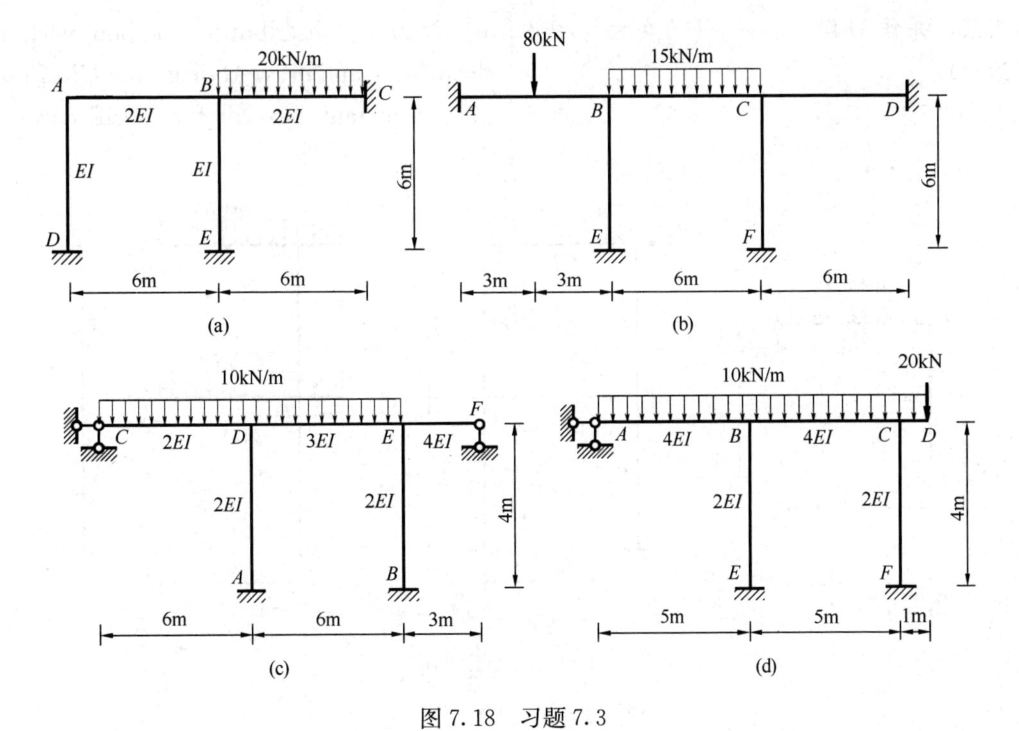

图 7.18 习题 7.3

Figure 7.18 Exercise 7.3

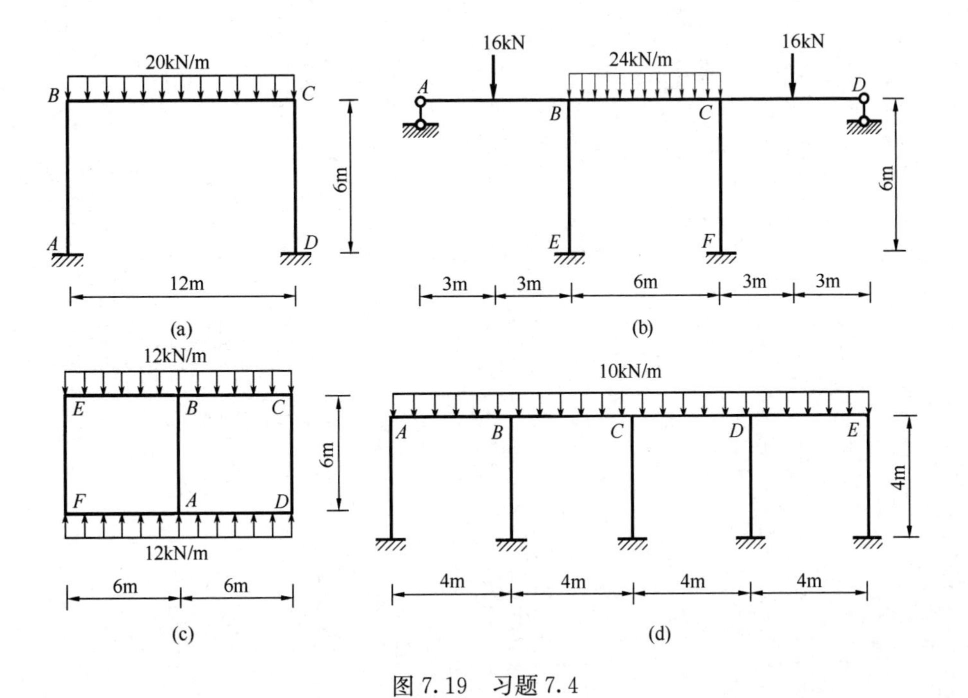

图 7.19 习题 7.4

Figure 7.19 Exercise 7.4

7.5　试用无剪力分配法计算图7.20所示刚架，并作M图。各杆EI为常数（注明者除外）。

7.5　Calculate the rigid frames in Figure 7.20 by distribution method with no shear force and draw M diagram. EI of each rod is constant (except for specified conditions).

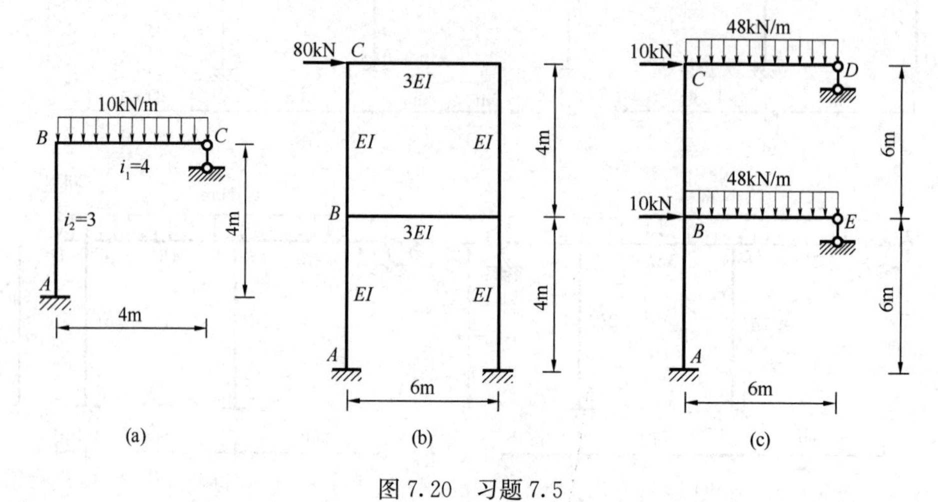

图7.20　习题7.5

Figure 7.20　Exercise 7.5

第8章 Chapter 8

影响线及其应用 Influence Line and Its Application

要点

- 影响线的概念
- 静力法作影响线
- 机动法作影响线
- 影响线的应用

Keys

- The concept of influence line
- Static method for influence line
- Kinematic method for influence line
- The application of influence line

8.1 移动荷载和影响线的概念

8.1 The Concept of Moving Load and Influence Line

移动荷载：作用位置按一定规律变化的荷载（但大小不随时间而改变），如图 8.1 所示。

Moving load: The load whose action position changes according to a certain rule (but the value of load does not change with time), as shown in Figure 8.1.

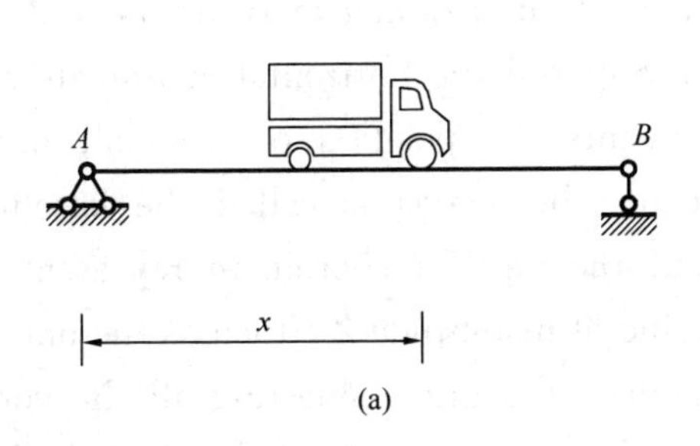

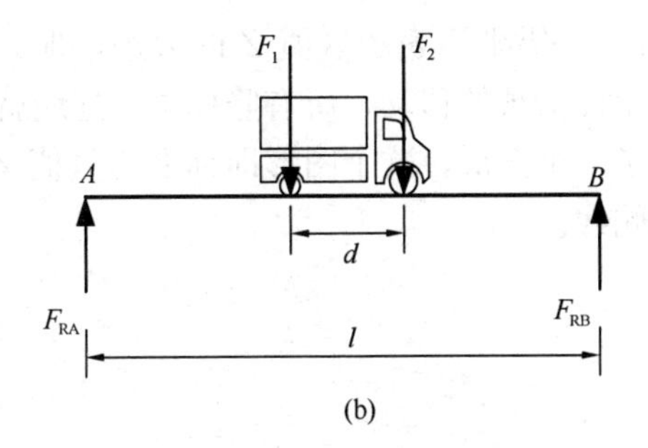

图 8.1 移动荷载

Figure 8.1 Moving load

响应：移动荷载作用下，结构的支反力、内力、位移等统称为结构对外界作用的响应，其值称为**响应量值**，简称**量值**，用 Z

Response: Under the action of moving load, the support reaction force, internal force, and displacement of the structure are

表示。

collectively referred to as the responses of the structure to the external action, and its value is called the **response value**, referred to as the **volume value**, expressed by Z.

一般来说，移动荷载作用下，响应量值可能会随之变化，使量值 Z 取极值（局部最大或最小值）时，荷载对应的位置称为**最不利位置**。

In general, under the action of moving load, the response magnitude may change accordingly, when the magnitude Z takes extreme values (local maximum or minimum), the location corresponding to the load is called the **most unfavorable location.**

由于移动荷载种类很多，可以是单个或多个，不可能逐一研究，因此，选择某一种最简单的荷载（通常是一个竖向单位力 $F=1$）沿结构移动，计算某一量值 Z 的响应，然后利用叠加原理可进一步研究各种移动荷载对该量值 Z 的影响。

Since there are many types of moving loads, which can be single or multiple, it is impossible to study them one by one. Therefore, the simplest load (usually a vertical unit force $F=1$) is selected to move along the structure, and the response of a certain magnitude Z is calculated, and then the effect of various moving loads on this magnitude Z can be further studied by using the superposition principle.

如图 8.2 所示简支梁，当单位移动荷载分别作用在①②③位置时，B 端支座反力 F_{RB}的数值大小分别为 1，1/2，0。如果以横坐标 x 表示单位移动荷载的位置（x 轴称为基线），纵坐标表示量值 Z 的大小，那么随着单位荷载的移动，所有竖标顶点连线就得到了一个图形，这个图形即称之为量值 Z 的影响线。

For example, in the simply supported beam shown in Figure 8.2, when the unit moving load is applied at positions①, ②, and ③, respectively, the values of the reaction force F_{RB} at end B are 1, 1/2, and 0. Suppose the horizontal coordinate x represents the position of the unit moving load (the x-axis is called the baseline), and the vertical coordinate represents the value of magnitude Z, then as the unit load moves, the line connecting all the vertical scale vertices gives a graph called the influence line of the magnitude Z.

影响线：当一个指向不变的单位集中荷载沿结构移动时，表示某一量值变化规律的图形，称为该量值的影响线。对应的函数 $Z=Z(x)$ 称为**影响线方程**。如图 8.3 所

Influence line: When a unit concentrated load with invariant orientation moves along with the structure, the graph that indicates change law of a certain value is

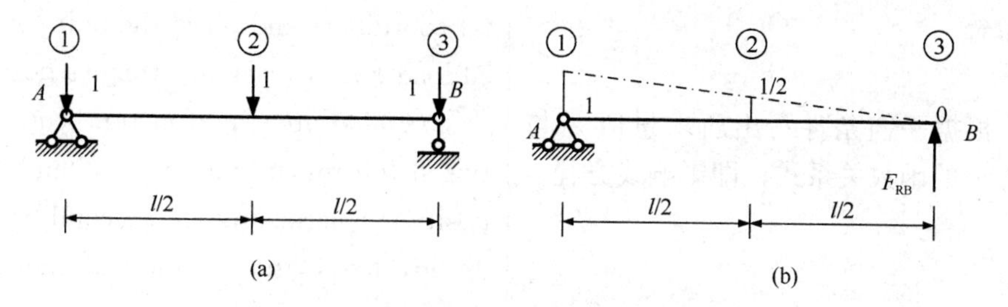

图 8.2 移动的单位荷载及对应的量值

Figure 8.2 Unit moving load and the corresponding volume value

示，影响线有正负，正值绘于基线上方，负值绘于基线下方，并注明正负。

called the influence line of the value. The corresponding function $Z=Z(x)$ is called the **influence line equation.** As shown in Figure 8.3, the influence line has positive or negative values, positive values are plotted above the baseline, and negative values are plotted below the baseline, with the indication of positive and negative signs.

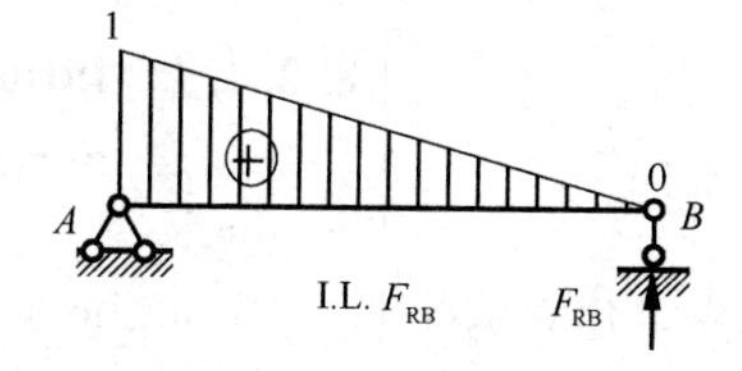

图 8.3 影响线的定义

Figure 8.3 Definition of the influence line

8.2 静力法作静定梁的影响线

8.2 Static Method for Influence Line

静力法是根据静力平衡条件，以单位移动荷载的作用点位置 x 为自变量，求出结构某响应量值 Z 的影响线方程，从而绘出该量值 Z 的影响线。其步骤可分为：

The static method is based on the static equilibrium conditions, with the position of the action point x of the unit moving load as the independent variable, to find out the influence line equation of the response value Z of the structure and draw the influence line of the value Z. The steps can be divided into following steps:

（1）以荷载的移动范围作为横坐标建立坐标系，并将单位集中荷载 $F_P=1$ 放在任

(1) Establish a coordinate system with the moving range of the load as the horizon-

意的 x 位置。

tal coordinate and place the unit concentrated load $F_P=1$ at an arbitrary x position.

(2) 根据平衡条件写出所求量值 Z 与荷载位置 x 的函数关系式，即影响线方程。

(2) Write the function equation between the magnitude value Z and the load position x according to the equilibrium condition, i. e. , the equation of the influence line.

(3) 以单位荷载的移动范围作为基线（横坐标），以量值 Z 的值作为纵坐标，根据影响线方程绘出该量值 Z 的影响线。

(3) Use the range of movement of the unit load as the baseline (horizontal coordinate) and the value of the magnitude Z as the vertical coordinate, plot the influence line of the magnitude Z according to the influence line equation.

8.2.1 简支梁的影响线

8.2.1 Influence Line of Simply Supported Beam

8.2.1.1 支座反力影响线

8.2.1.1 Influence Line of Support Reaction Force

对图 8.4 所示简支梁，首先，建立坐标系，以 A 为坐标原点，然后，由整体力矩平衡条件 $\sum M_B=0$ 得到影响线方程：$F_{RA}\cdot l-1\cdot(l-x)=0$，即：

For the simply supported beam shown in Figure 8.4, first, the coordinate system is established with point A as the coordinate origin, then, the influence line equation $F_{RA}\cdot l-1\cdot(l-x)=0$ is obtained from the overall moment balance condition $\sum M_B=0$, i. e. :

$$F_{RA}=\frac{l-x}{l}\quad(0\leqslant x\leqslant l)$$

可见，支座反力的影响线是线性函数，同理可得 B 支座反力的影响线方程：

It can be seen that the influence line of the support reaction force is a linear function, and similarly, the influence line equation of the support reaction force of support B can be obtained as follows:

$$F_{RB}=\frac{x}{l}\quad(0\leqslant x\leqslant l)$$

由于直线可以由两点确定，所以由 A、B 两点的位置很容易得到影响线，如图 8.4

Since two points can determine the line, the influence line is easily obtained

所示。

from the positions of points A and B, as shown in Figure 8.4.

几点说明：

Illustrations:

1. 影响线竖标 y_k 的含义是：当单位荷载作用在 K 位置时，量值 Z 的大小。

1. The meaning of the influence line's vertical coordinate y_k is the value of magnitude Z when the unit load acts on the position of point K.

2. 力的影响线量纲为一。

2. The influence line scale of the force is one.

3. 弯矩的影响线量纲为长度。

3. The influence line scale of the bending moment is the length.

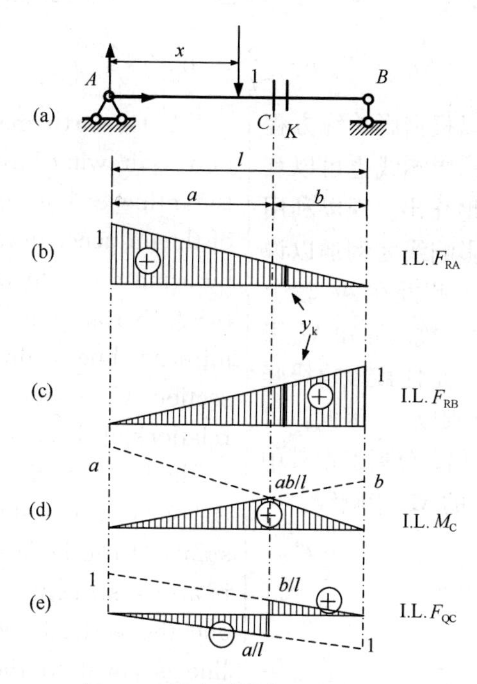

图 8.4　简支梁支座反力影响线

Figure 8.4　Influence line of support reaction force of simply supported beam

8.2.1.2　弯矩影响线

8.2.1.2　Influence Line of the Bending Moment

作弯矩影响线时，必须指定截面位置。现拟作指定截面 C 的弯矩 M_C 影响线，仍取上述坐标系。当 $F_P=1$ 作用在截面 C 以左

When making the bending moment influence line, the section position must be specified. It is proposed to make the influ-

或以右时，弯矩 M_C 的影响线方程具有不同的表达式，应分别考虑。

ence line of bending moment M_C of the specified section C, and still take the above coordinate system. When $F_P=1$ acts in left side or right side of section C, the influence line equation of bending moment M_C has different expressions that should be considered separately.

当 $F_P=1$ 在截面 C 左侧（AC 梁段）移动时，为计算方便，取截面 C 的右边为隔离体，由 $\Sigma M_C=0$ 可得：

When $F_P=1$ moves in the left side of section C (AC beam segment), for the convenience of calculation, the right side of section C is taken as the isolated body, by $\Sigma M_C=0$, it can be obtained:

$$M_C = F_{RB}b = \frac{x}{l}, \quad 0 \leqslant x \leqslant a$$

值得一提的是，除可以根据影响线方程绘制影响线外，有些量值的影响线也可以利用已知量值的影响线方便地作出。现以利用 F_{RB}影响线绘制 AC 段 M_C影响线为例加以说明。根据 M_C与 F_{RB}的关系，以及在 AC 段上 F_{RB}影响线为一直线，b 为常数，可知 M_C影响线在 AC 段上也为直线，而且其影响线竖标等于 F_{RB}影响线相应竖标乘以 b。因此，可把 F_{RB}影响线的竖标扩大 b 倍，然后保留其中的 AC 段，即可得到 AC 段的 M_C 影响线。

It is worth mentioning that, in addition to drawing influence lines according to the influence line equation, some quantities of the influence lines can also be easily made by using the influence lines of known quantities. Now, it is illustrated by using F_{RB} influence line to draw M_C influence lines of section AC as an example. According to the relationship between M_C and F_{RB}, as well as the fact that the F_{RB} influence line on the AC segment is a straight line and b is a constant, it can be seen that M_C influence line is also a straight line on the AC segment, and the vertical coordinate of its influence line is equal to the corresponding vertical coordinate of F_{RB} multiplied by b. Therefore, the vertical coordinate of F_{RB} influence line can be enlarged by b times, and then the AC segment can be retained to obtain M_C influence line of AC segment.

当 $F_P=1$ 在截面 C 右侧（CB 梁段）移动时，取截面 C 的左边为隔离体，可得：

When $F_P=1$ moves on the right side of section C (CB beam segment), the left side of section C is taken as the isolated body, therefore:

$$M_C = F_{RA}a，a \leqslant x \leqslant l$$

可知，M_C影响线在 CB 段上也为一直线，而且其影响线竖标等于 F_{RA}影响线相应竖标乘以 a。

It can be seen that M_C influence line is also a straight line on CB segment, and its vertical coordinate of influence line is equal to the corresponding vertical coordinate of F_{RA} influence line multiplied by a.

综上所述，可把 F_{RA}影响线的竖标扩大 a 倍，然后保留其中的 CB 段；把 F_{RB}影响线的竖标扩大 b 倍，保留其中的 AC 段，就得到 M_C的影响线（见图 8.4d），其中 C 点的竖标为 ab/l。从图 8.4（d）可以看出，M_C的影响线分成 AC 和 CB 两段，每一段都是直线，形成一个三角形。当 $F_P=1$ 作用在 C 点时，弯矩 M_C为极大值；当 $F_P=1$ 由 C 点向梁的两端移动时，弯矩 M_C逐渐减小至 0。

To sum up, the vertical coordinate of F_{RA} influence line can be expanded a times, and then the CB segment can be retained; Expand the vertical coordinate of F_{RB} influence line by b times, and keep the AC segment, the influence line of M_C can be obtained (Figure 8.4d), where the vertical coordinate of point C is ab/l. It can be seen from Figure 8.4 (d) that the influence line of M_C is divided into AC and CB segments, and each segment is a straight line, forming a triangle. When $F_P=1$ acts on point C, the bending moment M_C is the maximum value; When $F_P=1$ moves from point C to both ends of the beam, the bending moment M_C gradually decreases to 0.

8.2.1.3 剪力影响线

8.2.1.3 Shear Force Influence Line

作剪力影响线时，也必须先指定截面位置。现拟作指定截面 C 的剪力 F_{QC}影响线，当 $F_P=1$ 作用在截面 C 以左或以右时，剪力 F_{QC}的影响线方程也具有不同的表达式，仍应分段考虑。

To make shear force influence line, section location must also be specified. Now make the shear force F_{QC} influence line of specified section C, when $F_P=1$ acts at left or right sides of section C, the shear force F_{QC} influence line equation also has different expressions, which should still be considered in segments.

当 $\Sigma F_y = 0$　可得：

When　$\Sigma F_y = 0$:

$$F_{QC} = -F_{RB}，\quad 0 \leqslant x \leqslant a:$$

当 $F_P=1$ 在截面 C 右侧（CB 梁段）移动时，取截面 C 的左边为隔离体，可得：

When $F_P=1$ moves on the right side of section C (CB beam segment), the left side of section C is taken as the isolated body:

$$F_{QC}=F_{RA},\quad a\leqslant x\leqslant l$$

因此，F_{QC}影响线在AC段部分可将F_{RB}影响线画在基线的下方并保留C以左部分得到；在CB段部分可由F_{RA}影响线保留C以右部分得到，如图 8.4（e）所示。由图 8.4（e）可知，F_{QC}影响线由两段相互平行的直线组成，其竖标在C处有一突变。当$F_P=1$作用在C截面稍左处时，$F_{QC}=-a/l$；当$F_P=1$作用在C截面稍右处时，$F_{QC}=b/l$；当$F_P=1$越过C点由其左侧移到右侧时，截面C的剪力将发生突变，突变量等于 1。当$F_P=1$正好作用在C点时，F_{QC}的影响线是没有意义的。

Therefore, the F_{QC} influence line in AC segment can be obtained by drawing F_{RB} influence line below the baseline and keeping the left part of section C; The CB segment can be obtained by keeping the right part of section C of F_{RA} influence line, as shown in Figure 8.4 (e). It can be seen from Figure 8.4 (e) that F_{QC} influence line is composed of two mutually parallel straight lines, of which the vertical coordinate has a sudden change at point C. When $F_P=1$ acts on the slightly left of section C, $F_{QC}=-a/l$; when $F_P=1$ acts on the slightly right of section C, $F_{QC}=b/l$; when $F_P=1$ moves from the left to the right across point C, the shear force of section C will suddenly change, and the abrupt variable is equal to 1. When $F_P=1$ just acts on point C, the influence line of F_{QC} is meaningless.

8.2.2 伸臂梁的影响线

8.2.2 Influence Line of the Cantilever Beam

如图 8.5 所示，以A点为坐标原点，水平轴为横坐标x，将$F_P=1$放在x处。支座反力以向上为正，截面弯矩以使梁下侧纤维受拉为正，截面剪力以使隔离体有顺时针转动趋势者为正，先作F_{RA}、F_{RB}的影响线。由平衡条件可求得两支座反力分别为（$-l_1\leqslant x\leqslant l+l_2$）：

As in Figure 8.5, point A is the coordinate origin, and the horizontal axis is the horizontal coordinate x where $F_P=1$ is acted. The support reaction force is positive in the upward direction, the section bending moment is positive to make the lower side fiber of the beam under tension, and the section shear force is positive to make the isolated body with a trend of clockwise rotation. The influence lines of F_{RA} and F_{RB} are made first. The two support reaction forces can be obtained from the equilibrium condition, respectively ($-l_1\leqslant x\leqslant l+l_2$):

$$F_{RA}=\frac{l-x}{l}$$

$$F_{RB}=\frac{x}{l}$$

F_{RA}、F_{RB}的影响线方程与简支梁的相同，但要注意的是，这两个影响线方程对$F_P=1$在梁全长范围内都是适用的，F_{RA}、F_{RB}的影响线图形分别如图 8.5（b）、(c）所示，在梁全长范围内均为一条直线，并且在AB段内的影响线与相应简支梁的影响线完全相同，而伸臂部分的影响线可由AB段内部分延长得到。

The influence line equations of F_{RA}、F_{RB} are the same with those of the simply supported beam, but it should be noted that both influence line equations are applicable for $F_P=1$ over the full length of the beam. The influence line graphs of F_{RA}、F_{RB} are shown in Figure 8.5 (b) and Figure 8.5 (c), respectively, which are a straight line over the full length of the beam, and the influence line within the AB segment is exactly the same as that of the corresponding simply supported beam, while the influence line of the cantilever segment can be obtained from the extension in AB segment.

M_C的影响线见图 8.5（d），可以看出，伸臂梁跨内截面内力M_C的影响线也可由相应简支梁对应截面的弯矩影响线向伸臂部分延伸得到。

The influence line of M_C is shown in Figure 8.5 (d), it can be seen that the influence line of internal force M_C in the cantilever segment can also be obtained by extending the influence line of bending moment in the corresponding section of the corresponding simply supported beam to the extended-arm segment.

由上述内容可以看出，简支梁支座反力和内力影响线是最基本的影响线，由简支梁影响线向两端伸臂部分延伸，即可得到伸臂梁的支座反力和支座间截面内力的影响线。

From the above, we can see that the influence lines of support reaction force and internal force of simply supported beam are the most basic influence lines. By extending the influence lines of the simply supported beam to both ends of the cantilever parts, we can get the influence lines of support reaction force and internal force of sections between supports of cantilever beam.

伸臂段某一截面的弯矩和剪力影响线见图 8.5（f）、（g）。可以看出，只有当单位移动荷载作用于该截面以外的伸臂段上时，影响线竖标才有非零值。这个特点取决于静

The bending moment and shear force influence lines for a section of the extended arm are shown in Figures 8.5 (f) and (g). It can be seen that the vertical coordinate of

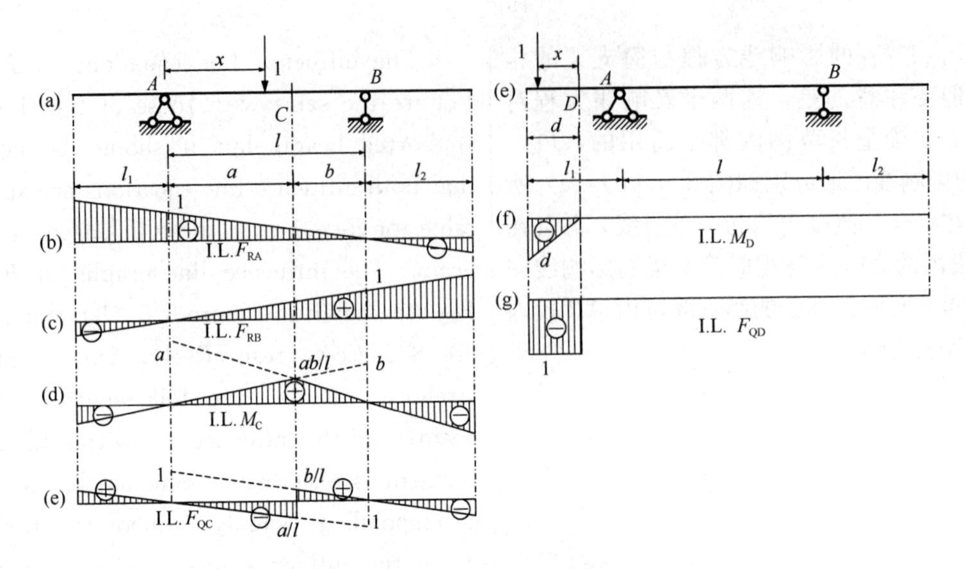

图 8.5 伸臂梁影响线

Figure 8.5 Influence line of the cantilever beam

定结构特性。

the influence line has a non-zero value only when the unit moving load acts on the extended arm segment. This feature depends on the statically determinate structure properties.

8.2.3 影响线和内力图的比较

8.2.3 Comparison of Influence Lines and Internal Force Diagrams

影响线与内力图是完全不同的两个概念，其含义有着本质不同，其比较如表 8.1 所示。

Influence lines and internal force diagrams are two completely different concepts with fundamentally different meanings, which are compared in Table 8.1.

内力影响线和内力图对比 **表 8.1**

Comparison of influence lines and internal force diagrams **Table 8.1**

	内力影响线 Influence lines of internal force	内力图 Internal force diagrams
荷载 Load	单位移动荷载 Unit moving load	实际荷载 Actual load
荷载位置 Load position	变化的 Changing	固定的 Fixed

续表
Continued

	内力影响线 Influence lines of internal force	内力图 Internal force diagrams
横坐标含义 Meaning of horizontal coordinate	单位移动荷载的位置 Location of the unit moving load	竖标所在截面的位置 The position of the section where the vertical coordinate is located
竖标意义 Meaning of vertical coordinate	指定量值 Z 的大小 The value of the specified magnitude Z	所在截面内力大小 The value of the internal force in the section
图形范围 Graphical range	移动荷载的移动范围 Range of movement of the moving load	整个结构 Whole structure
基线 Baseline	移动荷载作用点的连线 Connections of moving load acting points	杆轴线 Rod axis
作图的一般规定 General provisions of the drawing	正号在基线上方，标符号 The positive sign is above the baseline, marked with the sign	弯矩图不标符号，画在受拉侧，轴力、剪力图标符号 The bending moment diagram is not marked with a sign, which is drawn on the tensile side, while axis and shear forces have signs
内力的量纲 The magnitude of internal force	弯矩的影响线：长度 L 剪力、轴力、反力影响线：1 Influence line of bending moment: length L Influence lines of shear force、axis force and reaction force: 1	弯矩：L^2MT^{-2} 力：LMT^{-2} Bending moment: L^2MT^{-2} Force: LMT^{-2}

8.3 间接荷载作用下的影响线

8.3 Influence Line under Indirect Load

直接荷载：荷载直接作用于梁上。

Direct load: The load acts directly on the beam.

间接荷载：实际工程中有些结构或构件承受的荷载是由其他构件传递过来的，对主梁来说，只在节点处受到集中力的作用，这种荷载称为间接荷载，又叫节点荷载，如图8.6所示。

Indirect load: In actual engineering, some structures or members are subjected to loads that are transferred from other members, and for main beams, they are subjected to concentrated forces only at the nodes and such loads are called indirect loads, also called nodal loads, as shown in Figure 8.6.

下面以图8.7（a）中C截面弯矩为例，说明间接荷载作用下C截面弯矩影响线的做法。

The following is an example of the bending moment of section C in Figure 8.7 (a) to illustrate the practice of the influence line of the bending moment of section C under indirect load.

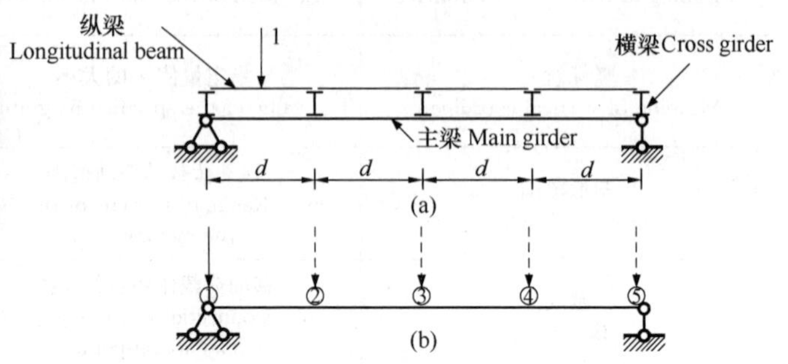

图8.6 间接荷载示意图

Figure 8.6 Schematic diagram of indirect load

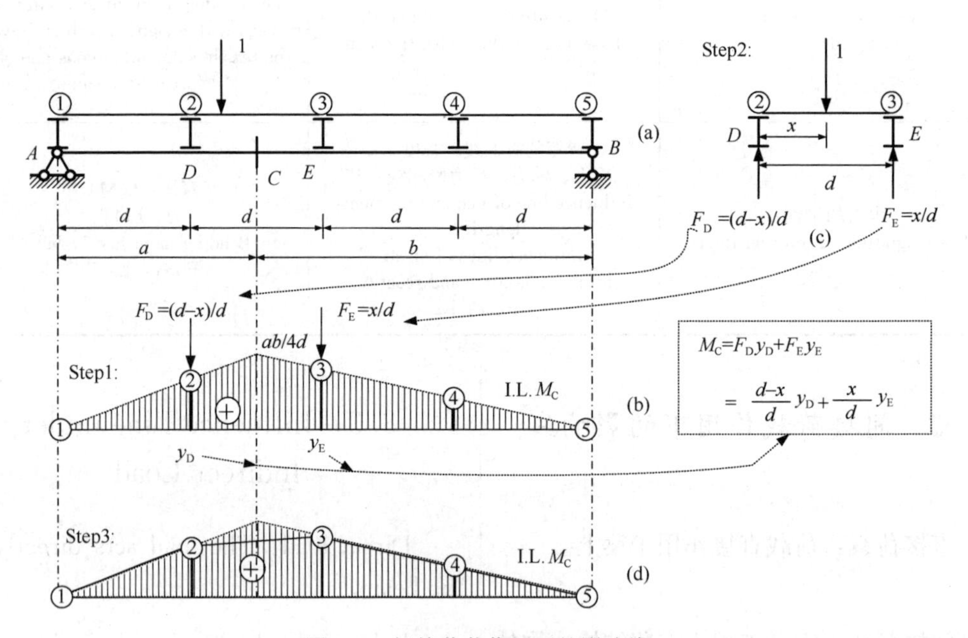

图8.7 间接荷载作用下的影响线

Figure 8.7 Influence line under indirect load

首先，当移动荷载作用在节点上时，其影响线竖标和直接荷载作用是一样的（图8.7b）。

First, when the moving load acts on the nodes, the vertical coordinates of influence line are same with the direct load acting (Figure 8.7b).

第二，考虑移动荷载作用在节点间，如图8.7（c）所示，此时可将其转化为两个

Second, consider the moving load acting between the nodes, as in Figure 8.7

节点反力作用于主梁（F_D、F_E），由影响线概念可知，此时造成的M_C可以写成：

(c), which can be transformed into two nodal reaction forces acting on the main beam (F_D, F_E), and according to the influence line concept, M_C can be written as:

$$M_C = \frac{d-x}{d}y_D + \frac{x}{d}y_E$$

这是一次函数，即一条直线，而这条直线的上 D、E 两点又已经确定，因此，将②③对应的 y_D、y_E 直接用直线相连即可，如图 8.7（d）所示。

This is a linear function, i.e., a straight line, and the points D and E on this line have been determined, so y_D and y_E corresponding to ② and ③ can be directly connected, as shown in Figure 8.7 (d).

综上所述，可以得到这样的结论：

In summary, the following conclusions can be drawn:

（1）节点荷载作用与直接荷载作用下，影响线在节点处的竖标是相同的。

(1) The vertical coordinates of the influence line at the node under the nodal load are the same with that under direct load.

（2）在节间荷载作用下，相邻节点间的影响线为连接节点处竖标的一段直线。

(2) Under the load between nodes, the influence line between adjacent nodes is a straight line connecting the vertical coordinate at the nodes.

容易证明，当 $F_P=1$ 在其他节点间时，上述结论依然成立。

It is easy to prove that the above conclusion still holds when $F_P=1$ is between other nodes.

因此可得，作节点荷载作用下主梁影响线的一般步骤为：

Therefore, it can be obtained that the general steps for making the influence line of the main beam under the nodal load are:

（1）首先作出直接荷载作用下所求量值的影响线；

(1) First, make the influence line for the requested quantity under direct load;

（2）然后用直线连接相邻节点的竖标顶点。

(2) Then connect the vertices of the vertical coordinates of the adjacent nodes with straight lines.

由图 8.7（d）可见，二者在大部分节点间的直线段是重合的，只是在截面 C 所在的节点间内，虚线所示的两段直线被修正为一段了。因此，作主梁影响线的步骤也可以说成是：首先作出直接荷载作用下所求量

As shown in Figure 8.7 (d), the straight lines of the two load conditions coincide at most of the nodes, except that the two straight lines by the dotted line are corrected to one in the segment where section

值的影响线；然后将节点间不是一段直线的部分修正为一段直线。

C is located. Therefore, the procedure of making the influence line of the main beam can also be said as follows: Firstly, the influence line of the requested quantity under direct load is made; then, the part between adjacent nodes that is not a straight line is corrected to a straight line.

8.4 静定桁架轴力的影响线

8.4 Influence Line of Axial Force of Statically Determinate Truss

理想情况下，静定桁架的荷载都是间接荷载/节点荷载。为便于绘图，图 8.8（a）和图 8.8（b）含义是一致的。

Ideally, the loads on the statically determinate trusses are all indirect loads/nodal loads. For drawing purposes, the meanings of Figures 8.8(a) and 8.8(b) are the same.

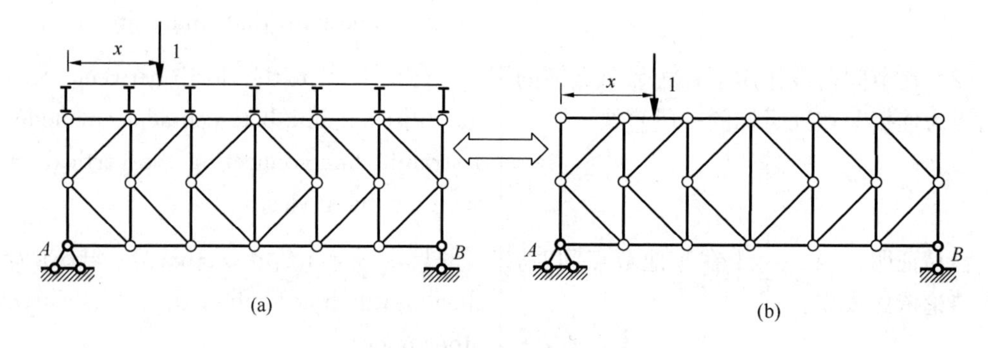

图 8.8 桁架的荷载示意图

Figure 8.8 Load schematic diagram of truss

桁架结构影响线的做法如下：

The steps of the influence line of truss are as follows:

(1) 将单位集中荷载 $F_P=1$ 依次放置于各节点上，计算所求量值的大小，即为该量值在各相应节点处的影响线竖标；

(1) Place the unit concentrated load $F_P=1$ on each node in turn and calculate the magnitude of the requested quantity, which is the vertical coordinate of the influence line for that quantity at each corresponding node;

(2) 再将相邻节点处的竖标连以直线即为该量值的影响线。

(2) Then connect the vertical coordinates at the adjacent nodes with a straight line, which is the influence line of the quantity.

当节点较多时，逐点求值再作影响线很

When there are many nodes, it is in-

不方便，可以先求影响线方程再作影响线。

convenient to find the value point by point and then make the influence line, so we can find the influence line equation first and then make the influence line.

对于单跨梁式静定桁架，其支座反力的计算与相应单跨梁相同，二者的支座反力影响线也完全一样。

For single-span beam-type statically determinate truss, the calculation of support reaction forces is the same as the corresponding single-span beam, and the support reaction force influence lines are the same as well.

计算桁架内力的方法通常有节点法和截面法，用静力法作桁架内力影响线时，同样用这些方法，只不过所作用的荷载为一个单位移动荷载。与作梁的影响线方法一样，一般是首先写出 $F_P=1$ 在不同部分移动时所求杆件内力的影响线方程，然后根据方程作出影响线；另外，也可以利用已知量值的影响线绘制其他量值的影响线。

The methods for calculating the internal force of the truss are usually the nodal method and section method, which are also used to make the influence line of internal force of the truss by static method, but the difference is that the load is a unit moving load. As with the influence line method for beams, the influence line equation for the internal force of the rod is usually written first when $F_P=1$ is moved in different segments, and then the influence line is made according to the equation; in addition, the influence line for other quantities can be drawn using the influence line for known quantities.

值得一提的是，桁架结构承受荷载分上承式和下承式两种情形，所谓上承式即荷载沿桁架的上弦移动；下承式则荷载沿桁架的下弦移动。有时这两种情况下所作出的影响线是不相同的。

It is worth mentioning that the truss subjected to load is divided into two cases: upper bearing and lower bearing, the so-called upper bearing means the load moves along the upper chord of the truss; while the lower bearing moves along the lower chord of the truss. Sometimes the influence lines made in these two cases are different.

【例 8.1】以图 8.9（a）所示上承式桁架为例，求①号杆轴力影响线。

［**Example 8.1**］Take the upper bearing truss shown in Figure 8.9 (a) as an example, find out the axial force influence line of the rod ①.

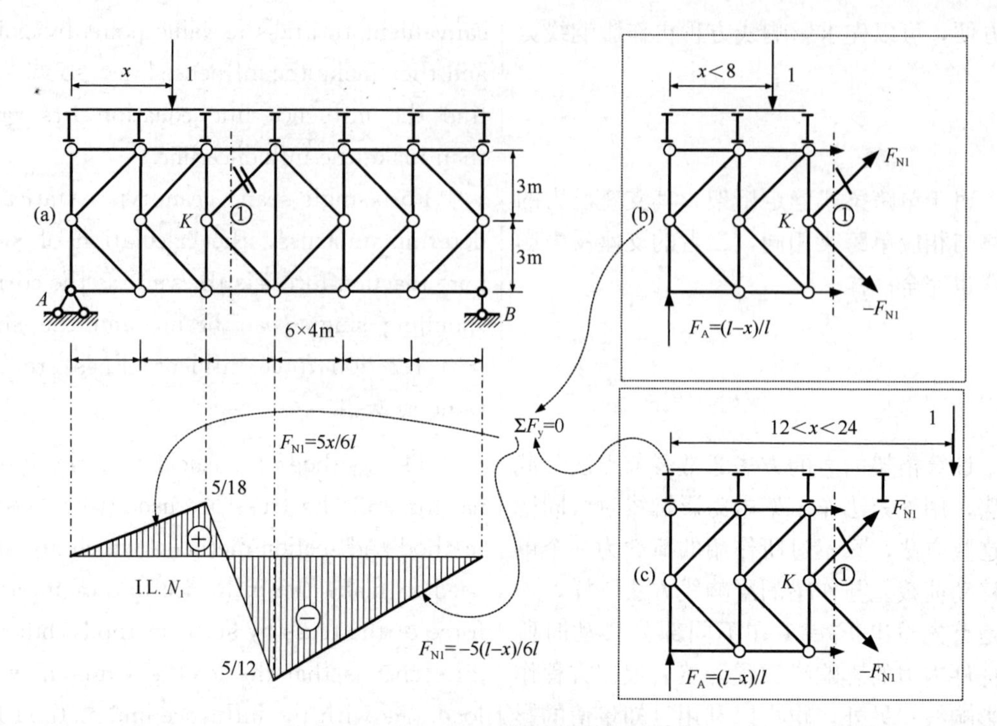

图 8.9 节点法求解桁架轴力的影响线

Figure 8.9 The nodal method for solving the axial force influence line of the truss

【解】本题关键点在于 K 节点，其所连两斜杆的轴力大小相等、方向相反，取隔离体如图 8.9（b）所示，然后根据竖向合力为零可以求出 F_{N1}。

[Solution] The key point of this example lies in the node K, of which the axial force of two diagonal rods connected are equal in value and opposite in direction, take the isolated body as shown in Figure 8.9 (b), then F_{N1} can be obtained according to the vertical combined force being zero.

在此思路下，分为三段，一是单位移动荷载在 1～2 节，二是单位荷载在 4～6 节，在求出 F_{N1} 表达式之后，剩下第 3 节只可以用左右直线连接即可得到最终影响线。

In this idea, the structure can be divided into three segments, one is that the unit moving load acts in 1-2 segments, the second is in 4-6 segments, after obtaining the expression of F_{N1}, the left third segment can only connect the left and right nodes and the influence line can finally be obtained.

上述求解过程中，既用到了截面法也用到了节点法，此外有的桁架（如平行弦桁架）也可以用力矩法求解，并建立和等跨简

During the above solving process, section method as well as nodal method is used, in addition, some trusses (such as

支梁的弯矩/剪力影响线的比例关系，在实际计算中可酌情采用。

parallel chord trusses) can also be solved by the moment method and establish the proportional relationship with the moment/shear force influence line of the equal span simply supported beam, which can be used as appropriate in the actual calculation.

8.5 机动法

8.5 Kinematic Method

8.5.1 机动法原理

8.5.1 Principle of Kinematic Method

作静定结构支座反力或内力量值影响线时，除可采用静力法外，还可以采用机动法。机动法作影响线的理论依据是虚功原理的虚位移原理，即：

To make influence lines of support reaction force or internal force of statically determinate structure, except that the static method can be used, the kinematic method can also be used. The theoretical basis for the influence line of the kinematic method is the virtual displacement principle of the virtual work principle, that is:

刚体体系在力系作用下处于平衡 ⇔ 在任何微小虚位移中，力系所做虚功总和为零。

Rigid body system is in equilibrium under the action of the force system ↔ In any small virtual displacement, the sum of virtual work done by the force system is zero.

下面以图 8.10 为例进行说明。

Figure 8.10 is an example for illustration.

从图 8.10 可以看出，虚位移图即影响线，图形在基线上方时标正号，反之标负号。

As shown in Figure 8.10, the virtual displacement graph is the influence line, which is marked with a positive sign when the graph is above the baseline and with a negative sign in opposite situation.

由上述可知，欲作某一量值 Z（反力或内力）的影响线，只需将与 Z 相应的约束解除后使体系沿 Z 的正方向产生单位位移，由此得到的荷载 $F_P=1$ 作用点的竖向位移图就是 Z 的影响线，这种作影响线的方法称之为**机动法**。总结起来，机动法作某一量

As can be seen from the above, to make the influence line of a certain magnitude Z (reaction force or internal force), only the corresponding constraints with Z are removed so that the system along the positive direction of Z occurs unit displace-

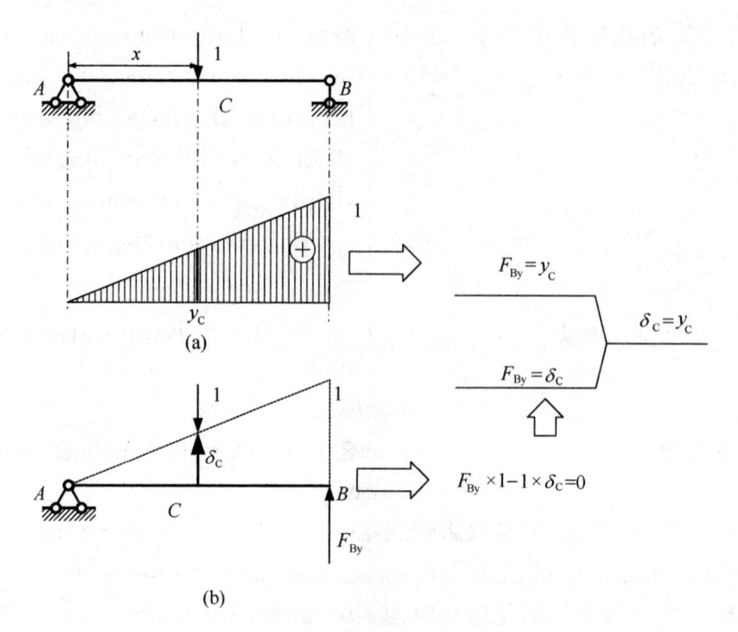

图 8.10　机动法原理

（a）I. L. F_{By}；（b）虚位移

Figure 8.10　Principle of kinematic method

（a）I. L. F_{By}；（b）Virtual displacement

值 Z（反力或内力）影响线的步骤如下：

ment, the resulting vertical displacement of the acting point of load $F_P=1$ is influence line of Z, this method of making the influence line is called the **kinematic method.** In summary, the steps of the kinematic method for making the influence line of a certain magnitude Z (reaction force or internal force) are as follows:

（1）撤去与量值 Z 相应的约束，代以未知力 Z；

(1) Replace the constraints corresponding to the magnitude Z by the unknown force Z;

（2）使体系沿 Z 的正方向发生单位位移，作出荷载 $F_P=1$ 作用点的竖向位移图，即为量值 Z 的影响线；

(2) The system occurs unit displacement along the positive direction of Z, make the vertical displacement graph of the action point of the load $F_P=1$, this is the influence line for magnitude Z;

（3）基线以上的图形，影响线竖标取正号；基线以下的图形，影响线竖标取负号。

(3) For the graph above the baseline, the vertical coordinate of the influence line is positive; below the baseline, it is nega-

tive.

8.5.2 用机动法作静定梁的影响线

8.5.2 Influence Line of Statically Determinate Beam by Kinematic Method

利用机动法也可进行直接荷载/间接荷载作用下主梁影响线的绘制，下面分别进行举例。

The drawing of the influence line of the main beam under direct load/indirect load can also be carried out using the kinematic method, respectively, as the following examples.

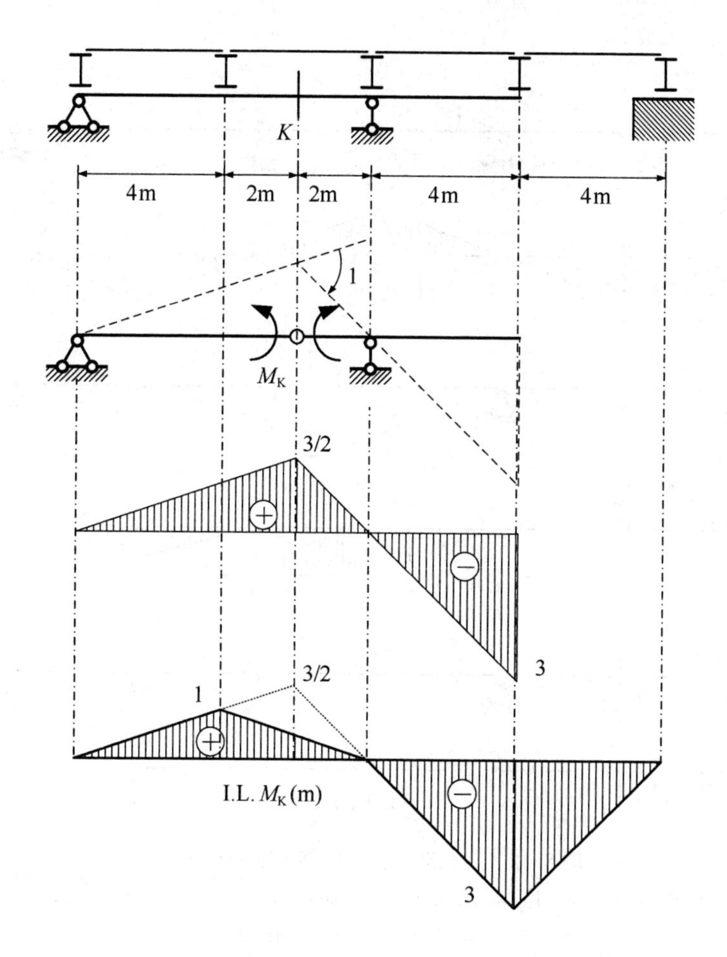

图 8.11 机动法作间接作用下的简支梁影响线

Figure 8.11 Kinematic method for the influence line of simply supported beam under indirect load

从图 8.11 和图 8.12 可以看出，机动法作静定梁的影响线快捷简单，但要注意的

From Figures 8.11 and 8.12, it can be seen that the kinematic method is fast and

是，机动法对含斜杆刚架以及桁架结构求影响线时应慎重，因其几何关系较为复杂。

simple for the influence line of the statically determinate beam, but it should be noted that the kinematic method should be careful with the influence line of the rigid frame containing diagonal rods and the truss structure because the geometric relationship is more complicated.

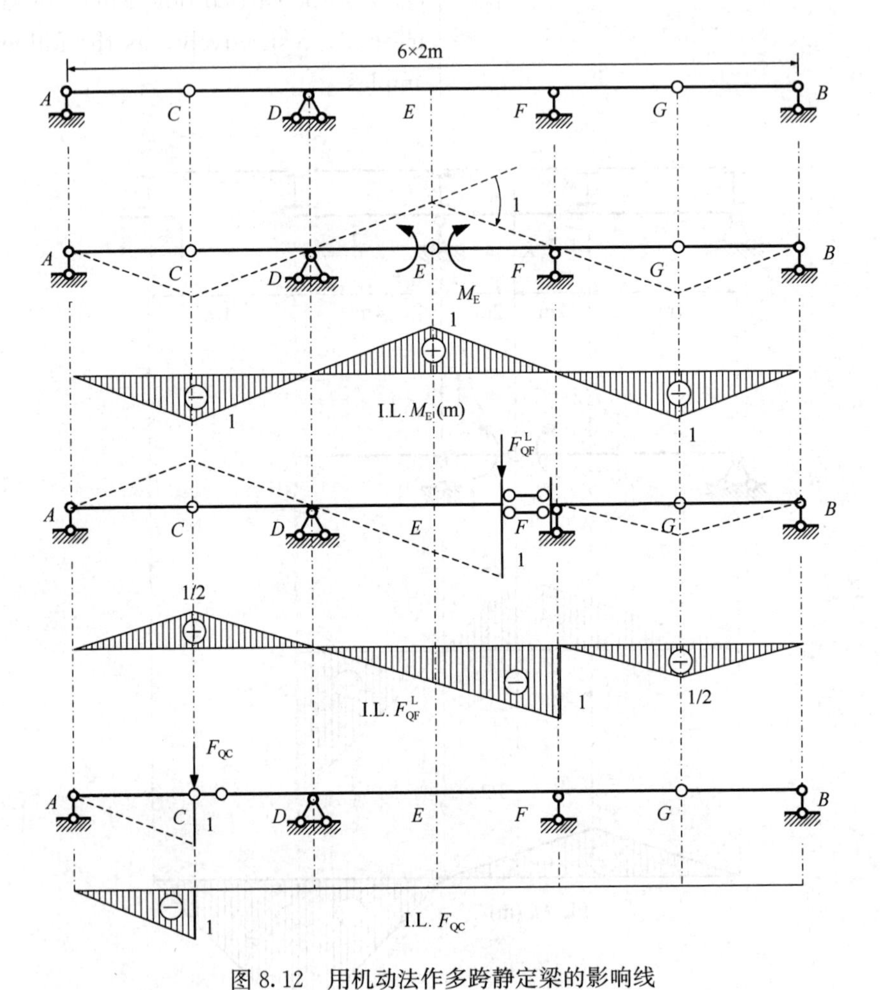

图 8.12　用机动法作多跨静定梁的影响线

Figure 8.12　Influence line of multi-span statically determinate beam by kinematic method

8.6　组合结构的影响线

8.6　Influence Line of the Composite Structure

对组合结构的影响线而言，依然可以用

For the influence line of the composite

机动法进行求解，然而，在某些刚架中，由于几何构成的复杂性，虚位移图中的几何量值不一定容易确定。

structure, kinematic method can also be used, however, in some rigid frames, the geometric quantities in the virtual displacement diagram are not always easy to determine due to the complexity of the geometric composition.

而考虑到静定结构的量值影响线都是直线或折线图形，因此可以将机动法和静力法联合使用，这样既简单又准确。

And considering that the influence lines of the static structure are all straight lines or folded lines, the kinematic method and the static method can be used jointly, which is simple and accurate.

图 8.13（a）中组合结构 M_K 的影响线就是由机动法所得。为了说明机动法和静力法的联合使用，参考图 8.13（b），其步骤可分为：

The influence line of M_K of the composite structure in Figure 8.13 (a) is obtained by kinematic method. To illustrate the combination of the kinematic method and the static method, the steps can be divided into the following steps referred to Figure 8.13 (b):

（1）使用机动法先确定直线/折线轮廓；

(1) Use the kinematic method first to determine the linear/folded line contours;

（2）使用静力法，直接将单位移动荷载放置在某几个特殊节点，计算出量值 Z，即为该处影响线的竖标，然后依次连线即可。

(2) Use the static method, place the unit moving load directly at some special nodes, calculate the magnitude Z, which is the vertical coordinate of the influence line, and then connect the lines in turn.

如图 8.13（b）中，将单位移动荷载放置于 C 处，可求得 M_K 大小，又因为 D 处影响线竖标为零，所以即可得到最终影响线，其结果和图 8.13（a）一致。

As shown in Figure 8.13 (b), the unit moving load is placed at point C, the value of M_K can be found, and because the vertical coordinate of the influence line at point D is zero, the final influence line can be obtained, which is consistent with Figure 8.13 (a).

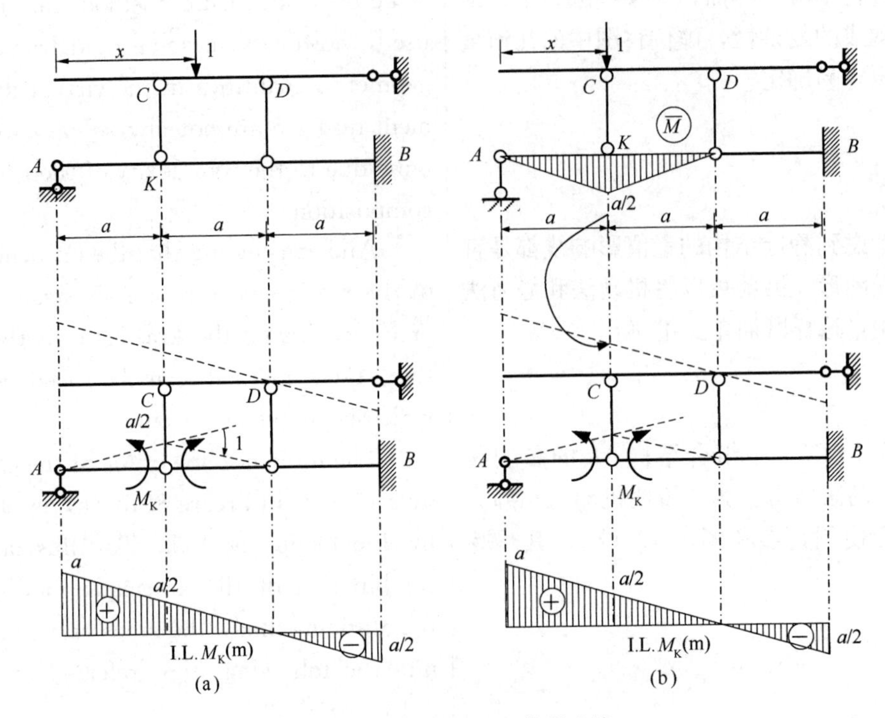

图 8.13　组合结构的量值影响线

Figure 8.13　Influence line of the composite structure

8.7　影响线的应用

绘制某量值的影响线，可以起到以下作用：

（1）研究实际移动荷载在结构上移动时对某一量值的影响规律；

（2）利用影响线求实际荷载作用于结构某一位置（即固定荷载）所产生的该量值的总量，称为影响量；

（3）利用某一量值的影响线确定移动荷载的最不利位置，并找出该量值的最大值，以作为结构设计的依据。

8.7　Application of Influence Line

Plotting the influence line of a quantity can serve the following purposes:

(1) To study the law of the influence on a quantity when an actual moving load moves on the structure;

(2) Use the influence line to find the total amount of a quantity produced by the actual load acting on a location of the structure (i. e. , a fixed load), and the total amount is called influence quantity;

(3) Use the influence line of a quantity to determine the most unfavorable position of the moving load and find the maximum value of the quantity to be used as the basis

for structural design.

由于影响线是研究单位集中移动荷载 $F_P=1$ 作用于结构上时对某一量值的影响规律，因此实际移动荷载作用下该量值的影响规律可以直接从影响线图形上观察出来。下面主要讨论影响线的后两个应用。

Since the influence line studies the law of influence on a quantity when a unit concentrated moving load $F_P=1$ acts on the structure, the law of influence on this quantity under the actual moving load can be observed directly from the influence line graph. The latter two applications of the influence line are discussed below.

8.7.1 利用影响线求量值

8.7.1 Using the Influence Line to Find the Value of the Magnitude

8.7.1.1 集中荷载作用

8.7.1.1 Under the Action of Concentrated Load

在已知某量值 Z 影响线的情况下，利用影响线求实际固定集中荷载 F_P 作用下该量值的大小，只需首先确定该固定集中荷载作用位置处的影响线竖标 y，然后利用叠加原理，则影响量为：

In the case of a known magnitude Z influence line, for the purpose to the use of the influence line to find the value of the magnitude under the actual fixed concentrated load F_P, first determine the vertical coordinate y of the influence line at the position of the fixed concentrated load, and then use the principle of superposition, the influence quantity is:

$$Z = F_P \times y \tag{8.1}$$

若结构承受一系列固定集中荷载 F_{P1}，F_{P2}，…，F_{Pn}，各荷载作用位置处某量值 Z 影响线的竖标分别为 y_1，y_2，…，y_n，则根据叠加原理，这一系列固定集中荷载引起的该量值影响量为：

If the structure is subjected to a series of fixed concentrated loads F_{P1}, F_{P2}, …, F_{Pn}, the vertical coordinate of the influence line of a magnitude Z at the position of each load is y_1, y_2, …, y_n, respectively, then according to the principle of superposition, the influence quantity of this magnitude caused by this series of fixed concentrated loads is:

$$Z = F_{P1} \times y_1 + \cdots + F_{Pi} \times y_i + \cdots + F_{Pn} \times y_n = \sum_{i=1}^{n} F_{Pi} \times y_i \tag{8.2}$$

式（8.1）、式（8.2）中正负号的规定

The provisions for positive and nega-

如下：F_{Pi}方向与作影响线时单位集中荷载$F_P=1$方向一致时为正，一般向下为正；y_i在坐标轴上方取正。

tive signs in Equations (8.1) and (8.2) are: The direction of F_{Pi} is consistent with the unit concentration load $F_P = 1$ when making the influence line, generally it is positive downward; it is positive when y_i is above the axis.

如果多个荷载作用在某一段直线影响线上，则有：

If multiple loads act on a certain segment of the linear influence line:

$$Z = \sum_{i=1}^{n} F_{Pi} \times y_i = \sum_{i=1}^{n} F_{Pi} x_i \tan\alpha \tag{8.3a}$$

又根据合力矩定理，有：

According to the resultant moment theorem, we have:

$$\Sigma F_{Pi} x_i = F_R \overline{x} \tag{8.3b}$$

联立式（8.3a）和式（8.3b），可得：

Combining Equations (8.3a) and (8.3b), it is obtained that:

$$Z = F_R \overline{x} \tan\alpha = F_R \overline{y} \tag{8.3c}$$

式（8.3c）表明，同一段直线影响线上多个荷载可以用它们的合力代替，如图8.14所示。

Equation (8.3c) shows that multiple loads can be replaced by their combined loads on the same straight influence line, as shown in Figure 8.14.

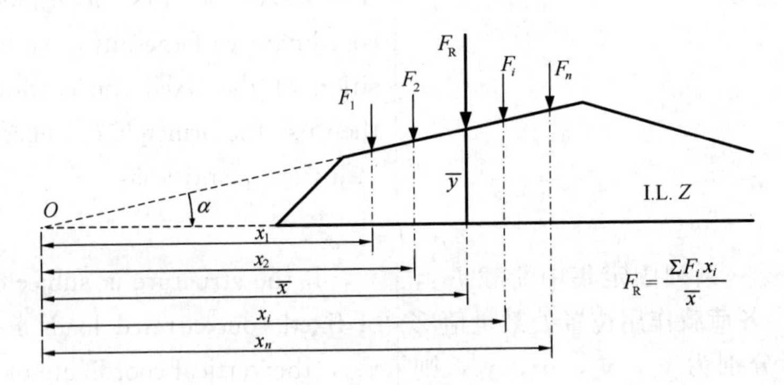

图8.14 合力矩定理

Figure 8.14 Resultant moment theorem

8.7.1.2 分布荷载作用

8.7.1.2 Under the Action of Distributed Load

如图8.15（a）所示，如果沿分布荷载长度方向分成许多无穷小微段，则每一段dx上的荷载$q_x dx$都可看作一个集中荷载，

As in Figure 8.15 (a), if many infinitesimal micro-ends are divided along the distributed load's direction, the load $q_x dx$ on

所以在 a，b 区间分布荷载所产生的量值 Z 为：

each segment dx can be regarded as a concentrated load, so the magnitude Z resulting from the distributed load in the interval a, b is:

$$Z = \int_a^b q_x y dx \tag{8.4}$$

如果 $q_x = q$ 为常数，即为均布荷载，式（8.4）可以简化为：

If $q_x = q$ is a constant, i.e., a uniform load, Equation (8.4) can be simplified as:

$$Z = \int_a^b q_x y dx = q\int_a^b y dx = qA_0 \tag{8.5}$$

其中，A_0 表示影响线图形在受荷段 AB 上的面积。应用式（8.5）时，注意 A_0 应为正负面积的代数和，其中，影响线基线上方部分为正面积，下方部分为负面积。上式表明，均布荷载作用下量值 Z 等于荷载集度 q 乘以受载段的影响线的图形面积。

Where A_0 is the area of the loaded segment AB of the influence line graph. When applying Equation (8.5), note that A_0 should be the algebraic sum of the positive and negative areas, where the part above the baseline of the influence line is the positive area and the part below is the negative area. The above equation shows that the magnitude Z under uniform load is equal to the load density q multiplied by the area of the loaded section of the influence line.

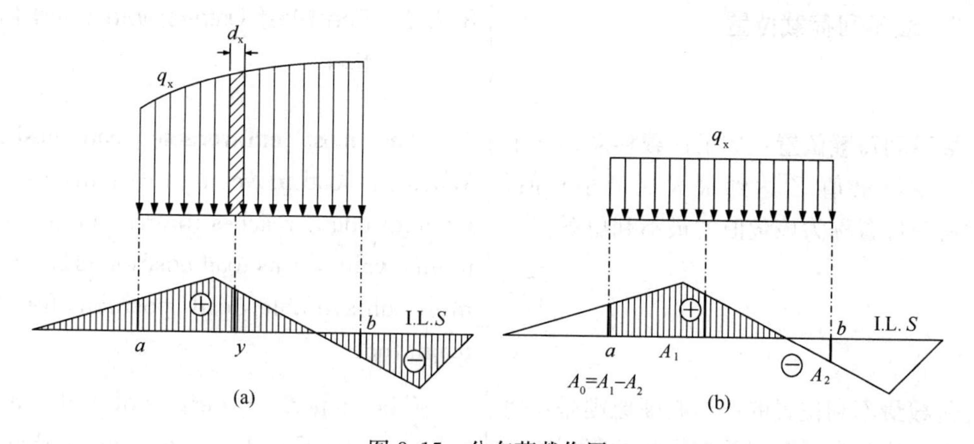

图 8.15 分布荷载作用

Figure 8.15 Under the action of distributed load

【例 8.2】 利用影响线求图 8.16 中 K 截面弯矩。

[Example 8.2] Use the influence line to find the bending moment of section K in Figure 8.16.

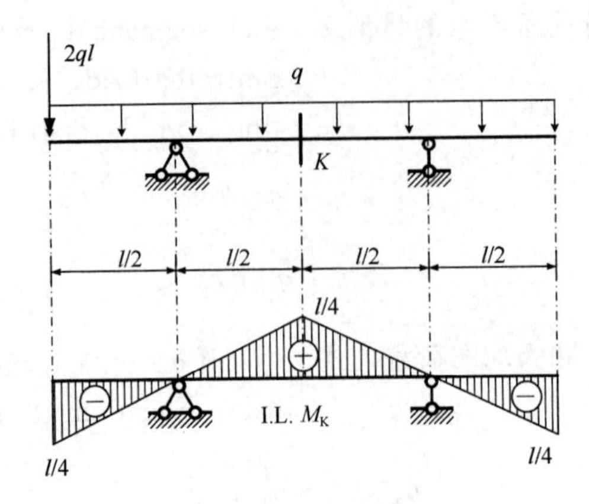

图 8.16　用影响线求量值

Figure 8.16　Using the influence line to find the magnitude

【解】（1）用机动法作出 M_K 影响线；

［**Solution**］（1）Make M_K influence line by kinematic method;

$$(2)\ M_K = 2ql \times (-l/4) + q \times \left(\frac{1}{2} \cdot \frac{l}{2} \cdot \frac{l}{4} \cdot 2 - \frac{1}{2} \cdot \frac{l}{2} \cdot \frac{l}{4} \cdot 2\right) = -ql^2/2$$

此题中由于影响线上下面积相等，故 $qA_0=0$，可快速计算。

In this example, because the area above the baseline is equal to the area below the baseline, $qA_0=0$, which can be calculated quickly.

8.7.2　最不利荷载位置

8.7.2　The Most Unfavorable Load Position

最不利荷载位置：如果荷载移动到某个位置，使某量值 Z 达到最大（最小）值，则此荷载位置称为该量值的最不利位置。

The most unfavorable load position: When the load moves to a certain position, a magnitude Z reaches its maximum (minimum) value, this load position is called the most unfavorable load position for that magnitude.

荷载最不利位置的一般判断原则是：将数量大、排列密的荷载放在影响线竖标较大的部位。

The general principle of judging the most unfavorable load position is that the load with a large quantity and dense arrangement is placed in the part with a larger vertical coordinate of the influence line.

8.7.2.1 单个移动集中荷载

如果移动荷载是单个集中荷载 F_P，则通过观察易知：最不利荷载位置是这个集中荷载作用在影响线的竖标最大处，即：将 F_P置于影响线的正最大竖标（$+y_{max}$）处产生 Z_{max}，将 F_P置于影响线的负最小竖标($-y_{max}$)处产生 Z_{min}。

8.7.2.1 Single Moving Concentrated Load

If the moving load is a single concentrated load F_P, then it is easy to know by observation: The most unfavorable load location is this concentrated load acting at the maximum vertical coordinate of the influence line, i. e. , F_P at the positive maximum vertical coordinate（$+y_{max}$）of the influence line produces Z_{max}, F_P at the negative minimum vertical coordinate（$-y_{max}$）of the influence line produces Z_{min}.

8.7.2.2 可任意布置的均布荷载

若移动荷载是均布荷载，且可以按任意方式分布，则其最不利荷载位置（图 8.17）是在影响线正号部分布满荷载产生 Z_{max}，在影响线负号部分布满荷载产生 Z_{min}。

8.7.2.2 Uniformly Distributed Load that Can Be Arbitrarily Arranged

If the moving load is an evenly distributed load and can be distributed anywhere, the most unfavorable load position (Figure 8.17) is in the positive sign part of the influence line with full distributed loads that produces Z_{max} and the negative sign part of the influence line with full distributed loads that produces Z_{min}.

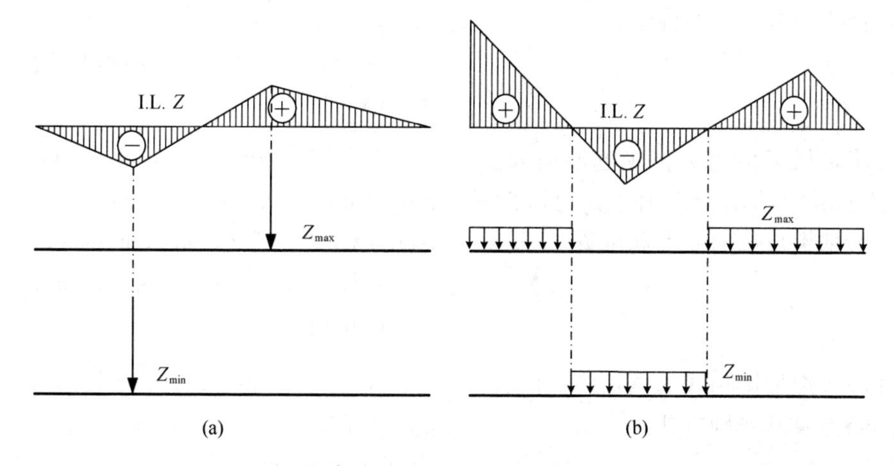

图 8.17 最不利荷载布置

(a) 单个移动集中荷载；(b) 均布荷载

Figure 8.17 The most unfavorable load position

(a) Single moving concentrated load; (b) Uniformly distributed load

8.7.2.3 行列荷载

行列荷载：间距不变的一组移动集中荷载，如火车、汽车的轮压荷载等。行列荷载的最不利位置很难直观判断。

在移动荷载作用下，量值 Z 的大小随着荷载位置（记为 x）的变化而变化，也就是说 Z 是 x 的函数，记为 $Z=f(x)$。要确定量值 Z 的最不利荷载位置，实际上就是要找出使 Z 取最大（小）值的 x。

若要找出使 Z 取最值的 x，可以首先找出使 Z 取极值的 x，然后进行比较。因此，确定行列荷载作用下最不利荷载位置通常分成两步进行：

第一步，求出使量值 Z 达到极值的荷载位置，称之为荷载的**临界位置**；

第二步，从 Z 的极大值中选出最大值，从 Z 的极小值中选出最小值，其对应的荷载临界位置即为荷载的最不利位置。

下面以多边形影响线为例，说明荷载临界位置的特点及其判别方法。

如图 8.18（a）所示，可以把各直线段

8.7.2.3 Row Load

Row load: It is a set of moving concentrated loads with constant spacing, such as the wheel pressure load of trains and automobiles, etc. The most unfavorable load position of the row load is difficult to determine intuitively.

Under a moving load, the value of magnitude Z varies with the location of the load (denoted as x), i. e., Z is a function of x, denoted as $Z=f(x)$. To determine the most unfavorable load position of magnitude Z is to find x that makes the maximum (minimum) value of Z.

Try to find x first that makes Z take extreme values and then compare these extreme values to find the maximum or minimum value of Z. Therefore, the determination of the most unfavorable load position under the action of the row load is usually divided into two steps:

In the first step, find out the location of the load that makes Z take an extreme value, which is called **the critical position of the load;**

In the second step, the maximum and minimum values are selected from the extreme values of Z, the corresponding critical position is the most unfavorable position of the load.

The following is an example of a polygon influence line to illustrate the characteristics of the critical position of the load and its discriminatory method.

As in Figure 8.18 (a), a combined

的集中荷载取合力，则由影响线定义，量值 Z 可表示为：

force is taken by the sum of the concentrated loads of each straight line segment, then by the definition of the influence line, the magnitude Z can be expressed as:

$$Z = F_{R1}y_1 + \cdots + F_{Ri}y_i + \cdots + F_{Rn}y_n = \sum_{i=1}^{n} F_{Ri}y_i \tag{8.6a}$$

当荷载向右移动 Δx 时（图 8.18b），y_i 的增量为 $\Delta y_i = \Delta x \tan\alpha_i$，则：

When the load moves Δx to the right (Figure 8.18b), the increment of y_i is $\Delta y_i = \Delta x \tan \alpha_i$, then:

$$\Delta Z = \Delta x \sum_{i=1}^{n} F_{Ri} \tan\alpha_i \tag{8.6b}$$

$$\frac{\Delta Z}{\Delta x} = \sum_{i=1}^{n} F_{Ri} \tan\alpha_i \tag{8.6c}$$

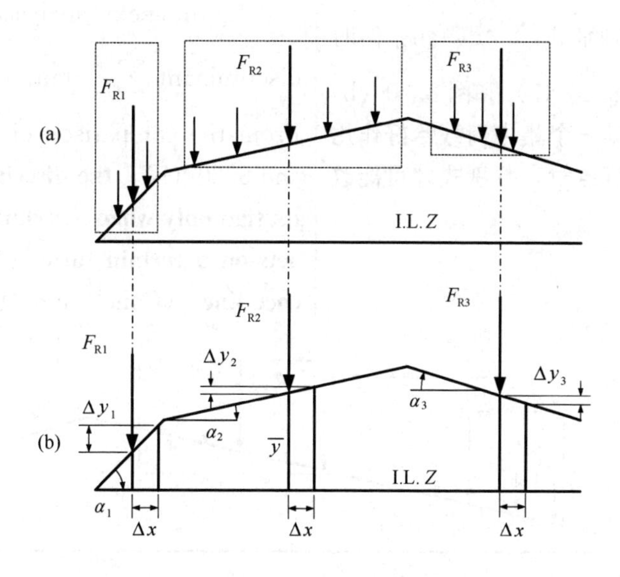

图 8.18 行列荷载的分段合力

Figure 8.18 Segmented combined forces of row load

从图 8.19 中可以看出，**如果 Z 取极值（无论极大值极小值）：则荷载左右发生微小移动，$\sum_{i=1}^{n} F_{Ri} \tan\alpha_i$ 必须改变正负号。**（结论 1）

As can be seen from Figure 8.19, **if Z takes extreme values (regardless of the extreme big or small values): a slight shift occurs to the left and right of the load, $\sum_{i=1}^{n} F_{Ri} \tan\alpha_i$ must change its signs.** (Conclusion 1)

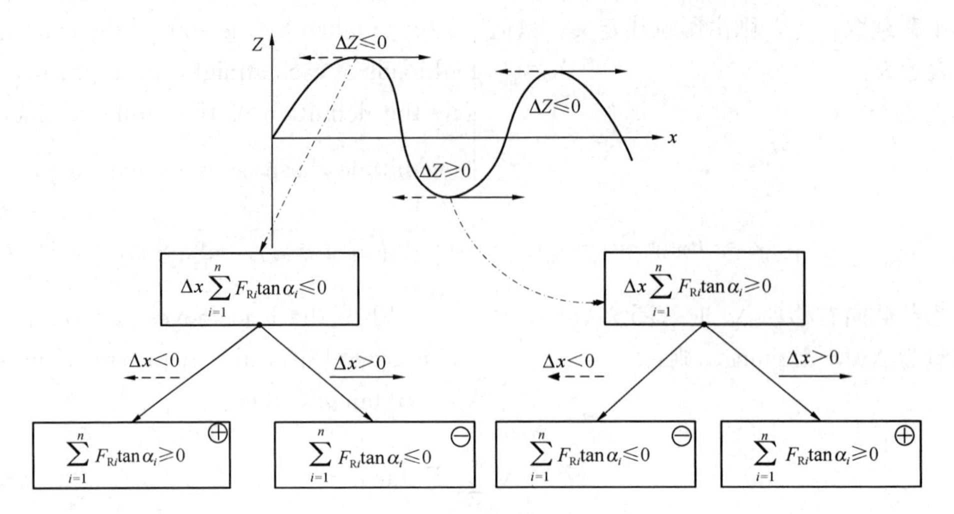

图 8.19 极值判断逻辑图

Figure 8.19 Logic diagram of extreme value judgment

接下来分析，判别式 $\sum_{i=1}^{n}F_{Ri}\tan\alpha_i$ 何时会改变符号。从图 8.20（a）和图 8.20（b）比较来看，**只有当某一个集中荷载恰好作用于影响线的某一转折点时，判别式才可能改变符号。**（结论 2）

In the next analysis, when does the discriminant $\sum_{i=1}^{n}F_{Ri}\tan\alpha_i$ change its sign? From the comparison of Figures 8.20（a）and 8.20（b），**the discriminant may change its sign only when a certain concentrated load acts on a certain turning point of the influence line.**（Conclusion 2）

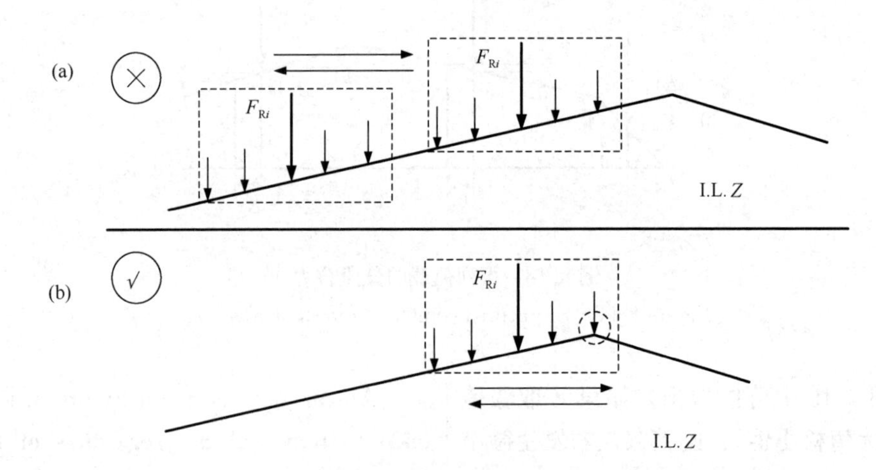

图 8.20 判别式的符号改变

Figure 8.20 Change in sign of discriminant

（1）当然，并非所有集中荷载位于转折点时都能使判别式变号。

（1）Of course, not all concentrated loads located at the turning point can cause the discriminant to change its sign.

（2）能使判别式变号的集中荷载称为**临界荷载 F_{Pcr}**，此时的荷载位置称为**临界位置**。

（2）The concentrated load that can make the discriminant change its sign is called the **critical load F_{Pcr}**，and the load position is called the **critical position.**

综上所述，确定荷载最不利位置的步骤为：

In summary，the steps to determine the most unfavorable load position are：

（1）将估计有可能成为临界荷载的集中力 F_{Pi}依次置于影响线竖标较大的顶点上。

（1）Place the concentrated forces F_{Pi}，which is estimated to be possible to be the critical loads，on the vertices of the larger vertical coordinate of the influence line in order.

（2）令荷载稍向左、向右移动，分别求 $\sum_{i=1}^{n} F_{Ri}\tan\alpha_i$ 的数值。若 $\sum_{i=1}^{n} F_{Ri}\tan\alpha_i$ 变号（或由零变为非零），则说明此荷载为临界荷载 F_{Pi}，此荷载位置为荷载临界位置；如果 $\sum_{i=1}^{n} F_{Ri}\tan\alpha_i$ 不变号，则此荷载位置不是荷载临界位置。

（2）Make the load move slightly to the left and right，and find the values of $\sum_{i=1}^{n} F_{Ri}\tan\alpha_i$，respectively. If $\sum_{i=1}^{n} F_{Ri}\tan\alpha_i$ change signs（or change values from zero to non-zero），then this load is the critical load F_{Pi}，this load position is the critical load position；if $\sum_{i=1}^{n} F_{Ri}\tan\alpha_i$ do not change signs，then this load position is not the critical load position.

（3）对每个临界位置可求出一个 Z 的极值，然后从中选取最大值或最小值，该最大值或最小值所对应的临界位置就是荷载的最不利位置。

（3）For each critical position，an extreme value of Z can be found，and then the maximum or minimum value is selected from them，and the critical position corresponding to the maximum or minimum value is the most unfavorable load position.

特殊地，如果影响线形状为三角形时，由图 8.21 可知其临界荷载的判别方法。

Specially，if the influence line shape is triangular，the critical load discriminatory method can be known from Figure 8.21.

【例 8.3】图 8.22（a）中的梁承受一组间距不变的集中力作用，其中 $F_{P1}=F_{P2}=8\text{kN}$，$F_{P3}=F_{P4}=10\text{kN}$。试确定在行列荷载自左向右移动时 E 截面正弯矩的最不利荷载位置和 M_E的最大值。

［**Example 8.3**］The beam in Figure 8.22（a）is subjected to a set of concentrated forces with constant spacing，where $F_{P1}=F_{P2}=8\text{kN}$ and $F_{P3}=F_{P4}=10\text{kN}$. Try to determine the most unfavorable load position and the maximum value of M_E for the positive bending moment of section E

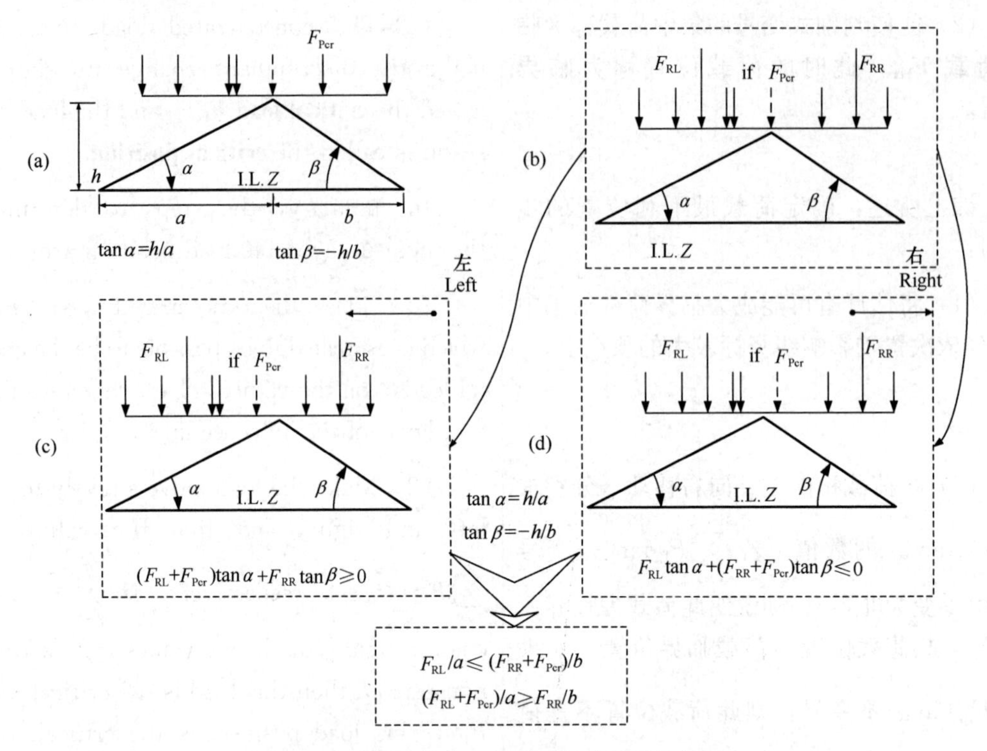

图 8.21　影响线为三角形时的临界荷载判别

Figure 8.21　The discrimination of critical load when the influence line is triangular

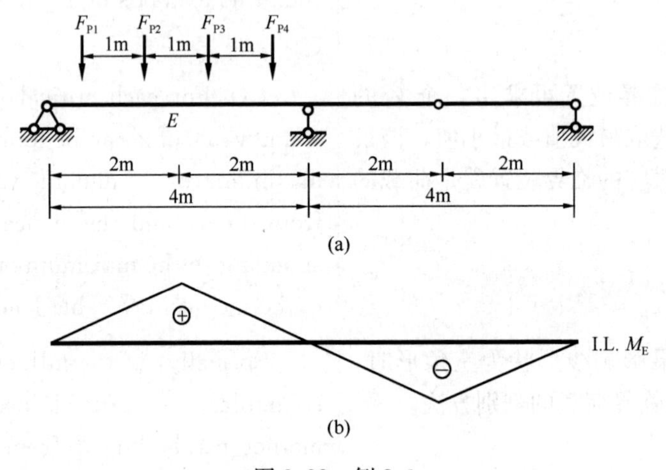

图 8.22　例 8.3

Figure 8.22　Example 8.3

when the row load is moved from left to right.

【解】首先作出 M_E 的影响线，如图 8.22（b）所示，从左至右三段直线倾角的

[**Solution**] First make the influence line of M_E, as shown in Figure 8.22 (b),

正切分别为 $\tan\alpha_1 = 0.5$，$\tan\alpha_2 = -0.5$，$\tan\alpha_3 = 0.5$。

the tangent of the inclination of the three straight lines from left to right are $\tan\alpha_1 = 0.5$, $\tan\alpha_2 = -0.5$, $\tan\alpha_3 = 0.5$.

下面检查 F_{P1}、F_{P2}、F_{P3}、F_{P4} 是否是临界荷载。

Then check whether F_{P1}、F_{P2}、F_{P3}、F_{P4} are critical loads.

1. F_{P1} 作用于 E 截面

1. F_{P1} acts on section E

荷载右移：

Load moves right:

$$F_{R1} = 0, F_{R2} = 36\text{kN}, F_{R3} = 0,\ \sum F_{Ri}\tan\alpha_i = 0 + 36 \times (-0.5) < 0$$

荷载左移：

Load moves left:

$$F_{R1} = 8\text{kN}, F_{R2} = 28\text{kN}, F_{R3} = 0, \sum F_{Ri}\tan\alpha_i = 8 \times 0.5 + 28 \times (-0.5) + 0 < 0$$

$\sum_{i=1}^{n} F_{Ri}\tan\alpha_i$ 未改变符号，所以 F_{P1} 不是临界荷载。

$\sum_{i=1}^{n} F_{Ri}\tan\alpha_i$ do not change signs, so F_{P1} is not the critical load.

2. F_{P2} 作用于 E 截面

2. F_{P2} acts on section E

荷载右移：

Load moves right:

$$F_{R1} = 8\text{kN}, F_{R2} = 28\text{kN}, F_{R3} = 0,\ \sum F_{Ri}\tan\alpha_i = 8 \times 0.5 + 28 \times (-0.5) < 0$$

荷载左移：

Load moves left:

$$F_{R1} = 16\text{kN}, F_{R2} = 20\text{kN}, F_{R3} = 0, \sum F_{Ri}\tan\alpha_i = 16 \times 0.5 + 20 \times (-0.5) < 0$$

$\sum_{i=1}^{n} F_{Ri}\tan\alpha_i$ 未改变符号，所以 F_{P2} 不是临界荷载。

$\sum_{i=1}^{n} F_{Ri}\tan\alpha_i$ do not change signs, so F_{P2} is not the critical load.

3. F_{P3} 作用于 E 截面

3. F_{P3} acts on section E

荷载右移：

Load moves right:

$$F_{R1} = 16\text{kN}, F_{R2} = 20\text{kN}, F_{R3} = 0,\ \sum F_{Ri}\tan\alpha_i = 16 \times 0.5 + 20 \times (-0.5) < 0$$

荷载左移：

Load moves left:

$$F_{R1} = 18\text{kN}, F_{R2} = 10\text{kN}, F_{R3} = 0, \sum F_{Ri}\tan\alpha_i = 18 \times 0.5 + 10 \times (-0.5) > 0$$

$\sum_{i=1}^{n} F_{Ri}\tan\alpha_i$ 改变符号，所以 F_{P3} 是临界荷载。

$\sum_{i=1}^{n} F_{Ri}\tan\alpha_i$ change signs, so F_{P3} is the critical load.

4. F_{P4} 作用于 E 截面

4. F_{P4} acts on section E

荷载右移：

Load moves right:

$$F_{R1} = 18\text{kN}, F_{R2} = 10\text{kN}, F_{R3} = 0,\ \sum F_{Ri}\tan\alpha_i = 18 \times 0.5 + 10 \times (-0.5) > 0$$

荷载左移：

Load moves left:

$$F_{R1}=20\text{kN}, F_{R2}=0\text{kN}, F_{R3}=0, \sum F_{Ri}\tan\alpha_i=20\times 0.5>0$$

$\sum_{i=1}^{n} F_{Ri}\tan\alpha_i$ 未改变符号，所以 F_{P4} 不是临界荷载。

$\sum_{i=1}^{n} F_{Ri}\tan\alpha_i$ do not change signs, so F_{P4} is not the critical load.

由于仅有一个临界荷载 F_{P3}，因此 F_{P3} 位于 E 截面，即为 E 截面正弯矩的最不利荷载位置，M_E的最大值为：

Since there is only one critical load F_{P3}, so F_{P3} located at section E is the most unfavorable load position of section E for the positive bending moment, the maximum value of M_E is:

$$M_{E\max}=\sum F_{Pi}y_i=8\times 0+8\times 0.5+10\times 1+10\times 0.5=19\text{kN}\cdot\text{m}$$

8.7.2.4 定长均布荷载

8.7.2.4 Fixed-length Uniform Load

对于定长均布荷载，可以由 $\frac{dZ}{dx}=\sum_{i=1}^{n} F_{Ri}\tan\alpha_i=0$ 来确定临界位置。

For a fixed-length uniform load, the critical position can be determined by $\frac{dZ}{dx}=\sum_{i=1}^{n} F_{Ri}\tan\alpha_i=0$.

特殊地，某量值影响线为三角形（图8.23），则其荷载最不利位置可推导如下：

Specially, the influence line of certain magnitude is triangular (Figure 8.23), then the most unfavorable load position can be deduced as follows:

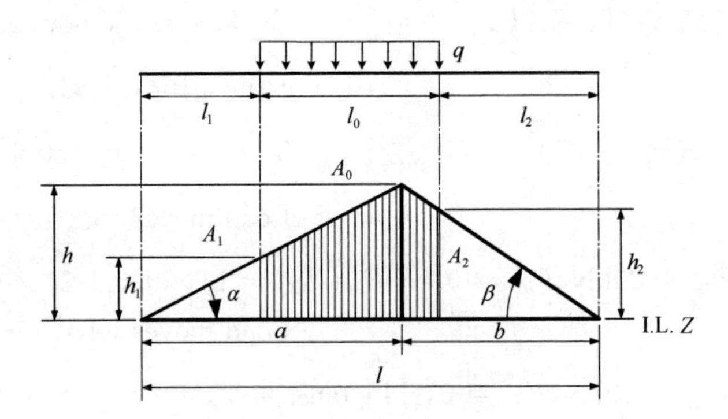

图 8.23 定长均布荷载

Figure 8.23 Fixed-length uniform load

$$Z=qA=q(A_0-A_1-A_2)=q\frac{1}{2}lh-q\frac{1}{2}l_1^2\tan\alpha-q\frac{1}{2}l_2^2\tan\beta$$

$$=\frac{q}{2}[lh-l_1^2\tan\alpha-(l-l_0-l_1)^2\tan\beta]$$

当荷载 q 移动时，量纲 Z 与 l_1 呈二次函数关系，取极值条件，有：

When the load q moves, the dimension of magnitude Z is quadratically related to

l_1, and taking the extreme value condition, we have:

$$\frac{dZ}{dl_1} = q[-l_1\tan\alpha + (l - l_0 - l_1)\tan\beta] = q(-l_1\tan\alpha + l_2\tan\beta) = 0$$

$$l_1\tan\alpha = l_2\tan\beta$$

$$h_1 = h_2$$

其位置如图 8.24 所示。

Its location is shown in Figure 8.24.

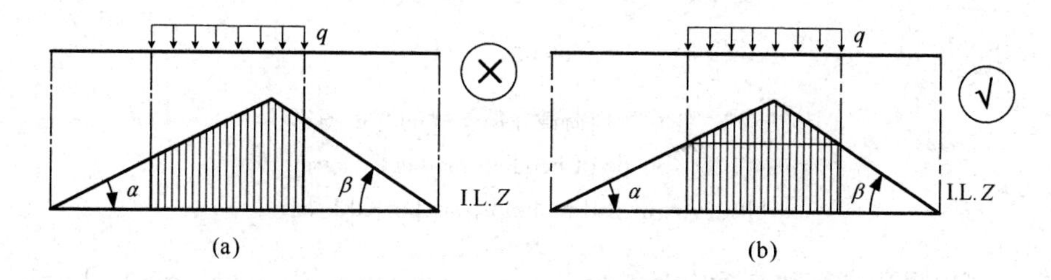

图 8.24 定长均布荷载在三角形影响线的最不利位置

Figure 8.24 Fixed-length uniform load at the most unfavorable load position of the triangular influence line

8.8 简支梁的绝对最大弯矩和内力包络图

8.8 Absolute Maximum Bending Moment and Internal Force Envelope of Simply Supported Beam

8.8.1 简支梁的绝对最大弯矩

8.8.1 Absolute Maximum Bending Moment of Simply Supported Beam

前文所述方法可以确定任一截面的最大弯矩，但在各截面弯矩中，有最大的，称为**绝对最大弯矩**，随之而来有两个问题：

The method described in the previous section can determine the maximum bending moment of any section, but in each section's bending moment, the largest is called the **absolute maximum bending moment**, followed by two questions:

1. 绝对最大弯矩发生在哪一个截面？

1. In which section does the absolute maximum bending moment occur?

2. 此截面发生绝对最大弯矩时荷载如何分布？

2. How is the load distributed when the absolute maximum bending moment occurs in this section?

下面以行列荷载为例进行讨论。

The following discussion is based on the example of row load.

首先，如图 8.25 所示，绝对最大弯矩一定在某个荷载作用处。

First, as shown in Figure 8.25, the absolute maximum bending moment must be at certain load action.

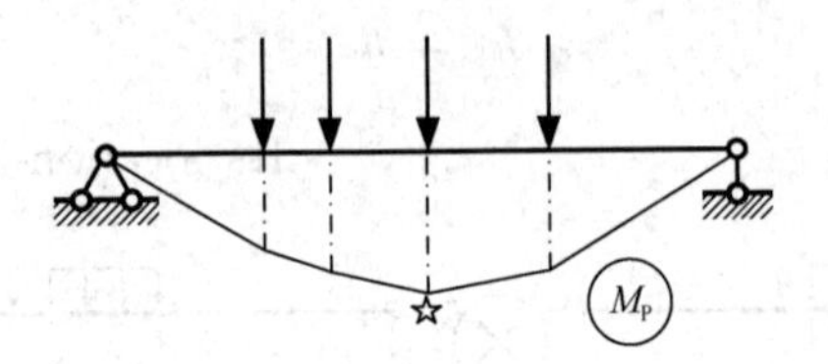

图 8.25 任意行列荷载在简支梁的弯矩图形状

Figure 8.25 Shape of bending moment diagram for arbitrary row load in simply supported beam

所以，可以采用试算法来确定是哪个集中力作用下的截面产生绝对最大弯矩。

Therefore, a trial method can determine which concentrated force acts on the section to produce the absolute maximum bending moment.

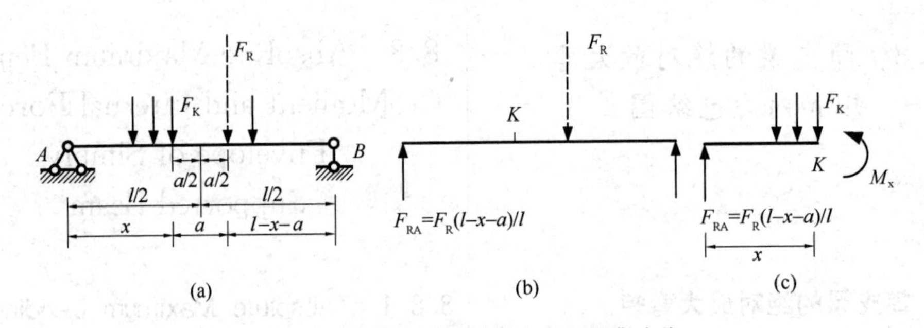

图 8.26 试算产生绝对最大弯矩的集中力

Figure 8.26 Calculation of the concentrated force generating the absolute maximum bending moment

如图 8.26（a）所示，试算 F_K 是否产生绝对最大弯矩，取隔离体如图 8.26（c）所示，有：

As shown in Figure 8.26 (a), try to calculate whether F_K produces the absolute maximum bending moment, and take the isolator as shown in Figure 8.26 (c):

$$M_x = F_{RA}x - M_L$$

其中，M_L 为 F_K 左侧力对 K 截面的弯矩，从图中可知，这是一个定值，和 x 无关，因此：

Where M_L is the bending moment from the left force of F_K to section K, it can be seen from the figure that this is a constant value and is irrelevant with x, therefore:

$$M_x = \frac{F_R}{l}(l-x-a)x - M_L$$

取极值条件：

Take extreme condition:

$$\frac{dM_x}{dx} = \frac{F_R}{l}(l-2x-a) = 0$$

$$x = \frac{l-a}{2} \tag{8.7}$$

式（8.7）表明，把F_K和合力对称分布于跨中两侧时，F_K之下的截面弯矩达到最大值（图 8.26a）。

Equation (8.7) shows that when F_K and resultant force are symmetrically distributed on both sides of the midspan, the section bending moment under F_K reaches the maximum value (Figure 8.26a).

该最大值为：

The maximum value is:

$$M_{max} = \frac{F_R}{l}\left(\frac{l}{2} - \frac{a}{2}\right)^2 - M_L \tag{8.8}$$

需要注意的是：

It should be noted that:

（1）应只考虑作用于梁上的集中力，已移除梁的不考虑；

(1) Only the concentrated forces acting on the beam should be considered, and those that have removed from the beam are not considered;

（2）如果合力F_R在F_K左侧，则a要取负值。

(2) If the combined force F_R is to the left of F_K, a is to be taken as a negative value.

利用上述结论，可将各个集中荷载作用点截面的最大弯矩求出来，再将它们加以比较从而得到绝对最大弯矩。但是，当梁上集中荷载数目较多时，这种方法是相当烦琐的，最好能先估计发生绝对最大弯矩的临界荷载。大量的计算实例表明，简支梁的绝对最大弯矩总是发生于梁的中点附近，因此使梁的中点截面产生最大弯矩的临界荷载，就是发生绝对最大弯矩的临界荷载。经验表明，这样的设想通常情况下都是正确的。

Using the conclusions drawn above, the maximum bending moment of the section at each point of concentrated load can be found out, and then they can be compared to obtain the absolute maximum bending moment. However, when the number of concentrated loads on the beam is large, this method is quite tedious, and it is better to estimate the critical loads where the absolute maximum moment occurs first. Numerous calculation examples show that the absolute maximum moment of a simply supported beam always occurs near the midpoint of the beam, so the critical load that causes the maximum bending moment in the midpoint section of the beam is the critical load where the absolute maximum bending moment occurs. Experience has shown that such a scenario is usually correct.

综上所述，计算简支梁绝对最大弯矩的步骤如下：

In summary, the steps for calculating the absolute maximum bending moment of a simply supported beam are as follows:

(1) 确定使梁中点截面产生最大弯矩的临界荷载 F_K。

(1) Determine the critical load F_K that causes the maximum bending moment at the midpoint section of the beam.

(2) 假设梁上荷载的个数并求其合力 F_R（大小及位置）。

(2) Assume the number of loads on the beam and find out their combined force F_R (magnitude and location).

(3) 移动荷载组使 F_K 与全跨荷载的合力 F_R之间的距离被梁的中点平分，此时应注意检查梁上荷载是否与求合力时相符，若不相符（即有荷载离开梁上或有新的荷载作用到梁上），则应重新计算合力，再行安排直至相符。

(3) Move the load group so that the distance between F_K and the combined force F_R of the full span is equally divided by the midpoint of the beam, at which point it should be taken to check whether the loads on the beam match the combined force sought, and if it does not (i. e., a load leaves the beam or a new load is applied on the beam), the combined force should be recalculated and rearranged until it matches.

(4) 计算 F_K 作用点处截面的弯矩，通常即为绝对最大弯矩 M_{max}。

(4) Calculate the bending moment of the section at the point F_K acts, which is usually the absolute maximum bending moment M_{max}.

【例 8.4】 求图 8.27（a）所示吊车梁在两台同吨位吊车作用下的绝对最大弯矩，并将其与跨中截面的最大弯矩进行比较。其中，$F_{P1} = F_{P2} = F_{P3} = F_{P4} = 290kN$。

[Example 8.4] Find the absolute maximum bending moment of the crane beam shown in Figure 8.27 (a) under the action of two cranes of the same tonnage and compare it with the maximum bending moment of the mid span. $F_{P1} = F_{P2} = F_{P3} = F_{P4} = 290kN$.

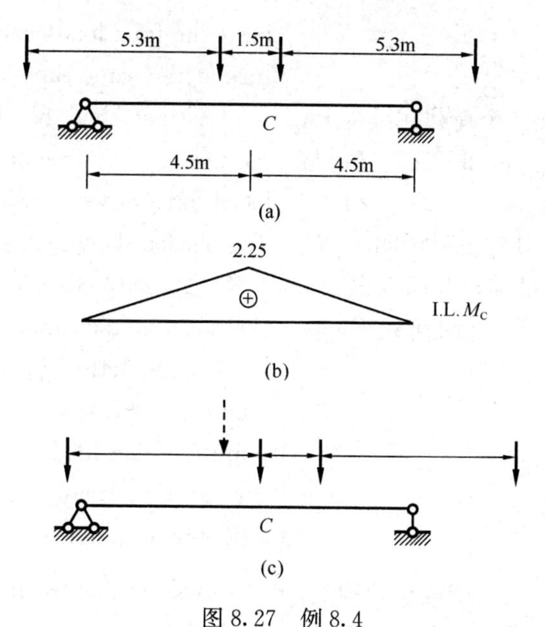

图 8.27 例 8.4
Figure 8.27 Example 8.4

【解】(1) 求 M_{Cmax}

[Solution] (1) Solve M_{Cmax}

绘出跨中截面弯矩 M_C 的影响线(图 8.27b),通过三角形影响线临界位置判别式,很容易知道只有 F_{P2} 和 F_{P3} 才可能为临界荷载。

Draw the influence line of bending moment M_C of mid span (Figure 8.27b), it is easy to know that only F_{P2} and F_{P3} can be the critical loads through the discriminant of the critical position of the triangle influence line.

将 F_{P2} 置于跨中 C 点时,有:

When F_{P2} is placed at point C of the mid span:

$$M_C = 290 \times (2.25 + 1.5) = 1087.5\text{kN} \cdot \text{m}$$

由对称性可知,当将 F_{P3} 置于 C 点时,仍有:

By symmetry, when F_{P3} is placed at point C, we still have:

$$M_C = 1087.5\text{kN} \cdot \text{m}$$

(2) 求 M_{max}

(2) Solve M_{max}

由上述计算可知,绝对最大弯矩将发生在荷载 F_{P2} 或 F_{P3} 对应的截面上。

From the above calculation, it is clear that the absolute maximum bending moment will occur at the corresponding section of the load F_{P2} or F_{P3}.

先考虑 F_{P2} 为临界荷载求 M_{max}。根据荷载间距可知,四个荷载不可能同时在结

First, consider F_{P2} as the critical load to obtain M_{max}. According to the load spac-

构上。

ing, the four loads cannot be on the structure at the same time.

假设 F_{P1} 、F_{P2} 、F_{P3} 三个荷载在结构上，此时合力 $F_R=870kN$（图 8.27c）。F_R 与 F_{P2} 之间的距离 $a=(290\times1.5-290\times5.3)/870=-1.267m$，其中负号说明 F_R 的作用点在 F_{P2} 的左边。可知，此时荷载 F_{P2} 位于 C 点右侧 0.633m 处，此时荷载 F_{P1} 在结构之外，与假设不符。

Assume that the three loads F_{P1} 、F_{P2} 、F_{P3} are on the structure, at which the combined force is $F_R=870kN$ (Figure 8.27c). The distance between F_R and F_{P2} is $a=(290\times1.5-290\times5.3)/870=-1.267m$, where the negative sign indicates that the point of F_R is to the left of F_{P2} . It is known that the load F_{P2} is located at 0.633m to the right of point C. The load F_{P1} is outside the structure at this time, which is not consistent with the assumption.

假设 F_{P2}、F_{P3}、F_{P4} 三个荷载在结构上，此时合力 $F_R=870kN$（图略）。F_R 与 F_{P2} 之间的距离 $a=(290\times1.5+290\times5.3)/870=2.267m$，$F_R$ 的作用点在 F_{P2} 的右边。易知，此时 F_{P4} 在结构之外，亦与假设不符。

Assume that the three loads F_{P2}、F_{P3}、F_{P4} are on the structure, at which the combined force is $F_R=870kN$ (The figure is elided) . The distance between F_R and F_{P2} is $a=(290\times1.5+290\times5.3)/870=2.267m$, the point of F_R is to the right of F_{P2} . It is easy to see that F_{P4} is outside the structure, which is also inconsistent with the assumption.

将 F_{P2} 、F_{P3} 两个荷载放于结构上，此时合力 $F_R=580kN$。F_R与 F_{P2} 之间的距离 $a=0.75m$，F_R 的作用点在 F_{P2} 的右边。此时结构实际分布与假设相符。由式（8.8）可求出：

F_{P2}、F_{P3} are placed on the structure, at this time the combined force is $F_R=580kN$. The distance between F_R and F_{P2} is $a=0.75m$, the point of F_R is to the right of F_{P2} . The actual distribution of the structure is consistent with the assumption. It can be derived from Equation (8.8):

$$M_{max}^{l}=\frac{580\times(9-0.75)^2}{4\times9}-0=1096.56kN\cdot m$$

考虑 F_{P3} 为临界荷载求 M_{max}。根据对称性可知，荷载 F_{P2} 和 F_{P3} 作用截面处的最大弯矩相等，因此梁的绝对最大弯矩为 $M_{max}=1096.56kN\cdot m$。

Consider F_{P3} as the critical load to find M_{max}. According to the symmetry, the maximum bending moments at the acting points of F_{P2} and F_{P3} are equal, so the absolute maximum bending moment of the beam is $M_{max}=1096.56kN\cdot m$.

从上述计算可以看出，行列荷载作用下

From the above calculation, it can be

该简支梁的绝对最大弯矩产生的位置不在梁的跨中，而在距离跨中 0.375m 的截面上。而其与跨中截面最大弯矩的差值为：

seen that the absolute maximum bending moment of the simply supported beam under row load is not at the midspan of the beam, but at the section 0.375m away from the mid span. The difference between it and the maximum bending moment of mid span is:

$$M_{max} - M_{Cmax} = 1096.56 - 1087.5 = 9.06\text{kN} \cdot \text{m}$$

可以看出，若以跨中截面最大弯矩近似代替简支梁的绝对最大弯矩，误差不大。

It can be seen that if the absolute maximum bending moment of the simply supported beam is approximately replaced by the maximum bending moment of the mid span, the error is small.

8.8.2 简支梁的内力包络图

8.8.2 Envelope of Internal Force of Simply Supported Beam

1. 在荷载（包括恒荷载和活荷载）作用下，结构任一截面的某一内力 Z 都有一个最大值 Z_{max} 和一个最小值 Z_{min}。

1. Under the action of loads (including constant and live loads), a certain internal force Z in any section of the structure has a maximum value Z_{max} and a minimum value Z_{min}.

2. 结构任一截面某一内力的 Z_{max} 和 Z_{min} 都是随移动（活）荷载位置而变化的。

2. Z_{max} and Z_{min} of a certain internal force in any section of the structure are varied with the position of the moving (live) loads.

3. 同一结构上不同截面的 Z_{max} 和 Z_{min} 并不相同，即 Z_{max} 和 Z_{min} 是截面位置 x 的函数。

3. Z_{max} and Z_{min} for different sections on the same structure are not the same, i.e., Z_{max} and Z_{min} are functions of the section position x.

4. 将各截面产生的某一内力 Z 的最大值 Z_{max} 和最小值 Z_{min} 分别连成一条光滑的曲线，称该图形为**内力 Z 的包络图。**

4. Connect respectively the maximum internal force value Z_{max} and the minimum value Z_{min} of a certain internal force Z generated in each section into smooth curves, and the graph is called **the envelope of the internal force *Z*.**

5. 在结构上各截面的最大内力中的最大值称为**绝对最大内力**，该绝对最大内力不难从内力包络图上找出。

5. The largest value among maximum internal forces in each section on the structure is called the **absolute maximum internal force**, which is not difficult to find out from the envelope of internal force.

下面先以简支梁在单个移动集中荷载F_P作用下的内力包络图为例说明其做法。

The following is an example of the envelope of internal force of a simply supported beam under the action of a single moving concentrated load F_P.

图 8.28（a）所示简支梁受到单个移动集中荷载F_P作用，某截面C的弯矩和剪力影响线如图 8.28（b）和图 8.28（c）所示。由影响线的形状可知：

The simply supported beam shown in Figure 8.28（a）is subjected to a single moving concentrated load F_P and the bending moment and shear force influence lines for section C are shown in Figures 8.28（b）and 8.28（c）. From the shape of the influence lines，it can be seen that：

当F_P恰好作用于C点时，M_C达到最大值：

When F_P acts exactly on point C，M_C reaches its maximum value：

$$M_{Cmax} = F_P \cdot \frac{ab}{l}$$

当F_P恰好作用于支座处时，M_C达到最小值：

When F_P acts exactly at the support，M_C reaches its minimum value：

$$M_{Cmin} = 0$$

当F_P恰好作用于C点左侧时，F_{QC}达到最小值：

When F_P acts exactly to the left of point C，F_{QC} reaches its minimum value：

$$F_{QCmin} = -F_P \cdot \frac{a}{l}$$

当F_P恰好作用于C点右侧时，F_{QC}达到最大值：

When F_P acts exactly to the right of point C，F_{QC} reaches its maximum value：

$$F_{QCmax} = F_P \cdot \frac{b}{l}$$

当荷载 F_P 在梁上移动时，只要逐个算出荷载作用点的弯矩最大、最小值和剪力最大、最小值，即可分别得到弯矩包络图和剪力包络图。为方便起见，将梁 10 等分，逐个算出每个等分截面处的弯矩、剪力的最大值或最小值。计算结果如表 8.2 所示。

When the load F_P moves on the beam, the maximum and minimum values of bending moment and maximum and minimum values of shear force at the acting point of load are calculated one by one to obtain the bending moment envelope and shear force envelope, respectively. For convenience, the beam is divided into 10 equal parts, and the maximum and minimum values of bending moment and shear force at each part of the section are calculated one by one. The calculation results are shown in Table 8.2.

弯矩、剪力计算表 **表 8.2**

Calculation table of bending moment and shear force **Table 8.2**

截面 Section	a	b	最大弯矩 Maximum bending moment $F_P \cdot ab/l$	最大剪力 Maximum shear force $F_P \cdot b/l$	最小剪力 Minimum shear force $-F_P \cdot a/l$
0	0	l	0	F_P	0
1	$0.1\,l$	$0.9\,l$	$0.09F_Pl$	$0.9\,F_P\,l$	$-0.1\,F_P\,l$
2	$0.2\,l$	$0.8\,l$	$0.16\,F_P\,l$	$0.8\,F_P\,l$	$-0.2\,F_P\,l$
3	$0.3l$	$0.7\,l$	$0.21\,F_P\,l$	$0.7\,F_P\,l$	$-0.3\,F_P\,l$
4	$0.4l$	$0.6\,l$	$0.24\,F_P\,l$	$0.6\,F_P\,l$	$-0.4\,F_P\,l$
5	$0.5l$	$0.5\,l$	$0.25\,F_P\,l$	$0.5\,F_P\,l$	$-0.5\,F_P\,l$
6	$0.6l$	$0.4l$	$0.24\,F_P\,l$	$0.4\,F_P\,l$	$-0.6\,F_P\,l$
7	$0.7l$	$0.3\,l$	$0.21\,F_P\,l$	$0.3\,F_P\,l$	$-0.7\,F_P\,l$
8	$0.8\,l$	$0.2\,l$	$0.16\,F_P\,l$	$0.2\,F_P\,l$	$-0.8\,F_P\,l$
9	$0.9\,l$	$0.1\,l$	$0.09\,F_P\,l$	$0.1\,F_P\,l$	$-0.9\,F_P\,l$
10	l	0	0	0	F_P

根据逐点算出的最大弯矩值而连成的图形即为弯矩包络图，如图 8.28（c）所示；根据逐点算出的最大（小）剪力值而连成的图形即为剪力包络图，如图 8.28（e）所示。

The point-by-point calculations of the maximum bending moment values are connected to form the bending moment envelope, as shown in Figure 8.28 (c); the point-by-point calculations of the maximum (minimum) shear values are connected to form the shear force envelope, as shown in Figure 8.28 (e).

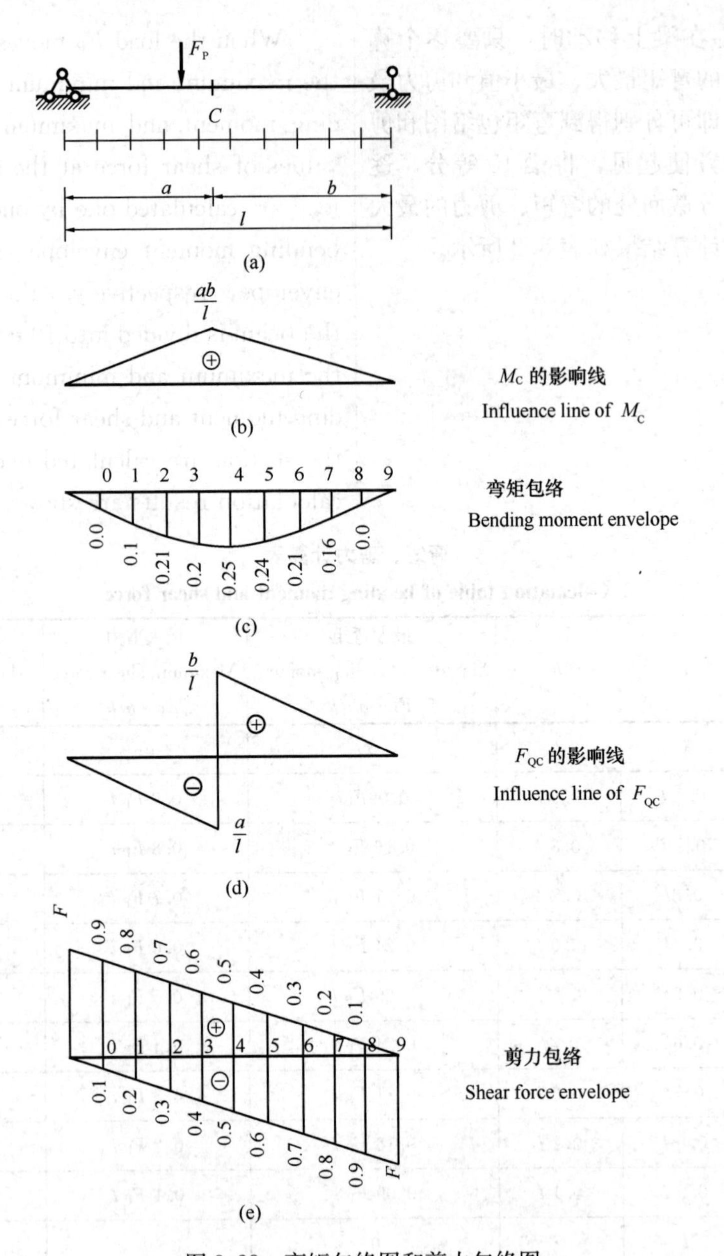

图 8.28　弯矩包络图和剪力包络图

Figure 8.28　The bending moment envelope and the shear force envelope

8.9　超静定结构影响线概述

8.9　Overview of Influence Lines of Statically Indeterminate Structure

作超静定结构反力、内力影响线也有两种方法，一种是静力法，即按力法先求出量值的影响线方程，然后根据影响线方程直接

There are also two methods to make influence lines of reaction force and internal force in statically indeterminate structure,

作出该量值的影响线；另一种方法是机动法，即利用位移图来作影响线。

one is the static method, that is, to find out the influence line equation of magnitude first according to the force method, and then directly to make the influence line by equation; another method is the kinematic method, that is to use displacement diagrams to make the influence lines.

8.9.1 静力法

8.9.1 Static Method

下面以图 8.29 为例，对静力法作超静定结构影响线进行说明。

The following is an example of the static method for statically indeterminate structure influence lines in Figure 8.29.

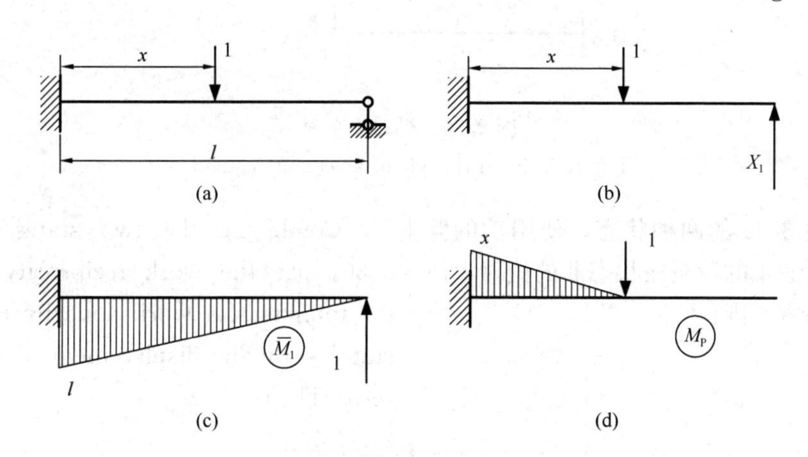

图 8.29 单位移动荷载作用于超静定结构

Figure 8.29 Unit moving load acting on a statically indeterminate structure

首先，采用力法，建立基本体系，写出力法的典型方程：

First, use the force method, establish the basic system and write the typical equation of the force method:

$$\delta_{11}X_1+\delta_{1P}=0$$

求出自由项和系数：

Find the free terms and coefficients:

$$\delta_{11}=\frac{l^3}{3EI}\quad \delta_{1P}=-\frac{x^2(3l-x)}{6EI}$$

进而解力法方程可得：

Further, solve the force method equation:

$$X_1=-\frac{\delta_{1P}}{\delta_{11}}=\frac{x^2(3l-x)}{2l^3}$$

这就是 X_1 的影响线方程，这是一条关

This is the influence line equation of

于 x 的三次函数曲线。

X_1，which is a cubic function curve of x.

8.9.2　机动法

8.9.2　Kinematic Method

仍以图 8.29 中结构举例，以图 8.30（a）中 B 支座反力为研究对象。

Still taking the structure in Figure 8.29 as an example，the support reaction force of support B in Figure 8.30（a）is studied.

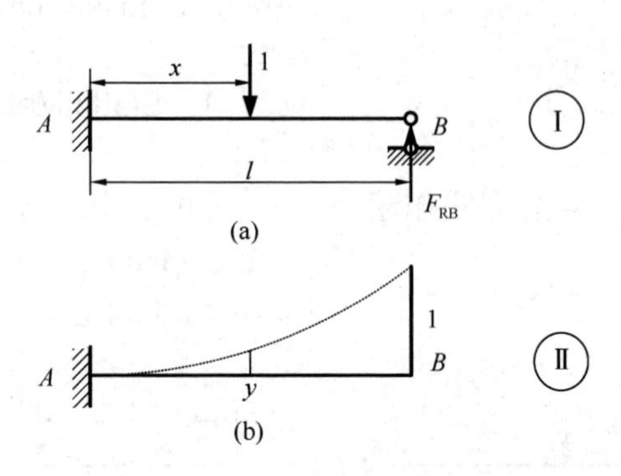

图 8.30　机动法原理

Figure 8.30　Principle of kinematic method

结合图 8.30 的两种状态，使用功的互等定理：状态Ⅰ的荷载在状态Ⅱ的位移上所做总虚功为零。则：

Combining the two states of Figure 8.30，use the work reciprocity theorem：the total virtual work done by the load in state Ⅰ on the displacement of state Ⅱ is zero. Then：

$$-1\times y+F_{RB}\cdot 1=0$$

$$F_{RB}=y$$

这说明撤去对应约束后的梁的变形图即为量值的影响线。

This shows that the deformation diagram of the beam after the withdrawal of the corresponding constraint is the influence line of the magnitude.

注：

1. 撤除一个多余约束后梁的超静定次数降低了一次，但并没有变为机动体系，上述方法称为“机动法”并不十分妥帖，因此有时被称为“挠曲线比拟法”。

Note：

1. The degree of indeterminacy of beam is reduced once after the removal of a redundant restraint，but it does not change into a kinematic system，the above method is called the “kinematic method” is not very appropriate，so it is sometimes called “deflection curve comparison method”.

2. 因静力法过程复杂，故一般使用机动法作超静定结构的影响线。

2. Because of the complicated process of the static method, the kinematic method is generally used for the influence line of the statically indeterminate structure.

【例 8.5】用机动法大致绘出图 8.31（a）所示连续梁中 F_{QCR}的影响线形状。

[**Example 8.5**] Use the kinematic method to roughly plot the shape of the influence line of F_{QCR} in the continuous beam shown in Figure 8.31 (a).

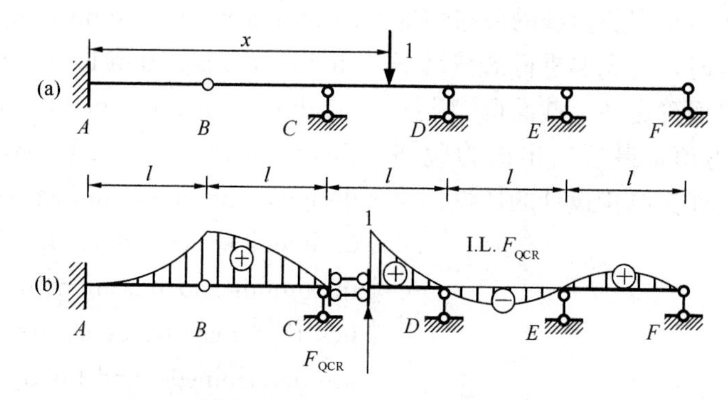

图 8.31 超静定梁的影响线

Figure 8.31 Influence line of statically indeterminate beam

【解】去掉与 F_{QCR}相应的约束，然后使之沿 F_{QCR}的正方向发生单位相对位移 1，则梁的变形曲线就是 F_{QCR}的影响线轮廓，其中截面 C 两侧曲线的相对位移为 1，如图 8.31（b）所示。

[**Solution**] Remove the constraint corresponding to F_{QCR}, and then make the unit relative displacement 1 along the positive direction of F_{QCR}, then the deformation curve of the beam is the influence line profile of F_{QCR}, where the relative displacement of the curve on both sides of section C is 1, as shown in Figure 8.31 (b).

由上例可知，用机动法描绘影响线的大致形状是很方便的。在活荷载为可移动荷载的大多数情况下，只要有了影响线的大致形状，就足以确定活荷载的最不利布置。在上例中，对 F_{QCR}而言，活荷载在第二、第三、第五跨为正剪力最不利布置。

As can be seen from the above example, it is convenient to depict the approximate shape of the influence line by the kinematic method. In most cases where the live load is moveable, the approximate shape of the influence line is sufficient to determine the most unfavorable live load arrangement. In the above example, for F_{QCR}, the live loads in the second, third and fifth spans are the most unfavorable arrangement for the positive shear force.

8.10 连续梁的内力包络图

8.10 Internal Force Envelope of Continuous Beam

连续梁受到恒载和活载的共同作用，对某一截面来说，恒载产生的弯矩是固定不变的，而活载产生的弯矩则随活载的分布情况不同而改变。设计时，首先要进行活载最不利分布的判断，进而确定各个截面可能产生的最大和最小内力值，最后绘出内力包络图，选出最大内力作为结构设计的依据。

Continuous beams are subjected to actions of constant load and live load. For a certain section, the bending moment generated by the constant load is fixed, while the bending moment generated by live load varies with the live load distribution. In the design, the most unfavorable distribution of live load is judged first, and then the maximum and minimum internal force values that may be generated in each section are determined, and finally the envelope of internal force is drawn and the maximum internal force is selected as the basis for structural design.

8.10.1 活荷载最不利分布的确定

8.10.1 Determination of the Most Unfavorable Distribution of Live Load

活荷载引起的内力随活荷载分布的不同而改变，因此求连续梁各截面最大内力的主要问题是确定活荷载的影响。连续梁受到的多为可动均布荷载情况，只要知道影响线的轮廓，就可确定活荷载的最不利分布。例如图8.32所示。

The internal forces due to live load vary with the distribution of live loads, so the main problem in finding the maximum internal forces in each section of a continuous beam is to determine the influence of live loads. Continuous beams are mostly subjected to movable uniform loads, and the most unfavorable distribution of live loads can be determined by knowing the contours of the influence lines. An example is shown in Figure 8.32.

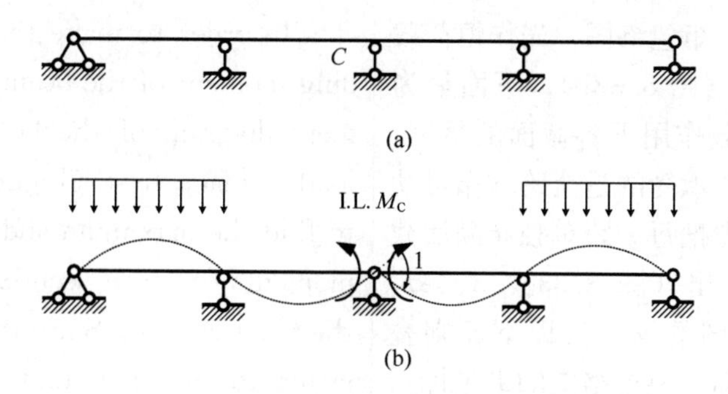

图 8.32 活荷载的最不利分布举例

Figure 8.32 Example of the most unfavorable distribution of live load

8.10.2 求解内力包络图

当确定了活荷载的最不利分布后，即可求出活荷载作用下某一截面的最大、最小内力值，然后叠加上恒载作用下产生的内力，就可以得到恒载和活载共同作用下该截面的最大内力和最小内力。将梁上各截面的最大内力和最小内力用图形表示出来，即为连续梁的内力包络图。连续梁的内力包络图通常包括弯矩包络图和剪力包络图。

图 8.33（a）为一三跨连续梁，该梁承受 g=20kN/m 的均布恒荷载和 p=30kN/m 的全跨均布可动活荷载，下面讨论如何作其弯矩包络图。其中，“全跨均布可动活荷载”是指对某一跨而言，要么不受活荷载作用，要么全跨布满活荷载，不考虑跨内局部受活荷载的情况。

8.10.2 Solve for Internal Force Envelope

When the most unfavorable distribution of live load is determined, the maximum and minimum internal forces of a section under live load can be found, and then the internal forces generated under constant load can be superimposed to obtain the maximum and minimum internal forces of the section under the combined action of constant load and live load. The maximum internal force and minimum internal force of each section of the beam are graphically represented as the envelope of internal force of the continuous beam. The envelope of internal force of a continuous beam usually includes the envelope of bending moment and the envelope of shear force.

Figure 8.33 (a) shows a three-span continuous beam subjected to a uniform constant load of g=20kN/m and a full-span uniformly movable live load of p=30kN/m, the envelope of bending moment is discussed below. The "full-span uniformly movable live load" means that either no live load is acted on a certain span or the whole span is covered with live load without considering local live load within the span.

为了作出梁的弯矩包络图，先作恒荷载作用下梁的弯矩图（图 8.33b）。下面最为关键的是求出活荷载作用下各截面的最大、最小弯矩。由于结构承受的是全跨均布可动活荷载，因此，只需把每一跨单独布满活载时的弯矩图逐一作出（图 8.33c、d、e），然后同一截面弯矩图的纵标按以下原则叠加，就可得到该截面弯矩包络图的纵坐标：

In order to make the envelope of bending moment of the beam，the bending moment diagram of the beam under constant load is made first（Figure 8.33b）. Then to find the maximum and minimum bending moments for each section under live load is the most critical. Since the structure is subjected to full-span uniformly movable live load，it is only necessary to make the moment diagrams for each span individually when the live load is applied（Figures 8.33c，d and e），then the vertical coordinates of the bending moment diagrams of the same section are superimposed according to the following principles to obtain the vertical coordinates of the bending moment envelope of the section：

最大弯矩值＝恒荷载下的弯矩值＋Σ活荷载下的正弯矩值

Maximum bending moment＝bending moment under constant load＋Σ positive bending moments under live loads

最小弯矩值＝恒荷载下的弯矩值＋Σ活荷载下的负弯矩值

Minimum bending moment＝bending moment under constant load＋Σ negative bending moments under live loads

将各跨分为 4 等份，在每一等分点截面处将各活荷载弯矩图和恒荷载弯矩图中数值按上述原则叠加在一起，即可得到各截面的最大和最小弯矩值。如 B 支座截面处，由图 8.33（b）和图 8.33（e）叠加，可得最大弯矩为（－32＋8）＝－24kN·m；由图 8.33（b）和图 8.33（c）、图 8.33（d）叠加，可得最小弯矩为（－32－32－24）＝－88kN·m。最后，将每个截面弯矩的最大值或最小值的点连成光滑曲线，即为弯矩包络图（图 8.33f）。

The span is divided into 4 equal parts，and the values in the live load bending moment diagram and constant load bending moment diagram are superimposed at each equal diversion point section according to the above principle to obtain the maximum and minimum bending moment values of each section. For example at the section of support B，the maximum bending moment is（－32＋8）＝－24kN·m by superimposing Figure 8.33（b）and Figure 8.33（e）；the minimum bending moment is（－32－32－24）＝－88kN·m by superimposing Figures 8.33（b），8.33（c）and 8.33（d）. Finally，each section's maximum or minimum moment values are connected into a smooth curve, which is the bending moment envelope (Figure 8.33f).

按照同样的方法亦可作出连续梁的剪力包络图，不再赘述。

The shear force envelope of the continuous beam can also be made in the same way, which will not be described here.

综上所述，受到恒载和可动均布荷载作用的连续梁，其内力包络图可按以下步骤进行：

In summary, the envelope of internal force of a continuous beam subjected to constant and movable uniform loads can be carried out as follows:

（1）绘制恒载作用下的某一内力图。

(1) Plot the internal force diagram under constant load.

（2）依次在每一跨上单独布满活载，并绘制与步骤（1）相应的内力图。

(2) Place live load on each span individually in turn and draw the internal force diagram corresponding to step (1) .

（3）将各跨若干等分，对每一等分点处截面，将步骤（1）和步骤（2）中得到的各内力图中竖标按下述原则叠加在一起，即可分别得到截面该内力的最大值和最小值：

(3) Divide each span into several equal parts, and for the section at each equal diversion point, according to the following principles, superimpose the vertical coordinates in each internal force diagram from steps (1) and (2) to obtain the maximum and minimum values of that internal force of the section, respectively:

最大值＝恒荷载下的内力值＋Σ活荷载下的正内力值

Maximum bending moment = bending moment under constant load $+\Sigma$ positive internal forces under live loads

最小值＝恒荷载下的内力值＋Σ活荷载下的负内力值

Minimum bending moment = bending moment under constant load $+\Sigma$ negative internal forces under live loads

（4）将步骤（3）所得到的最大（小）内力值分别用光滑的曲线连接起来，即为内力包络图。

(4) The maximum (minimum) internal force values obtained from step (3) are connected to form smooth curves, respectively, which is the internal force envelope.

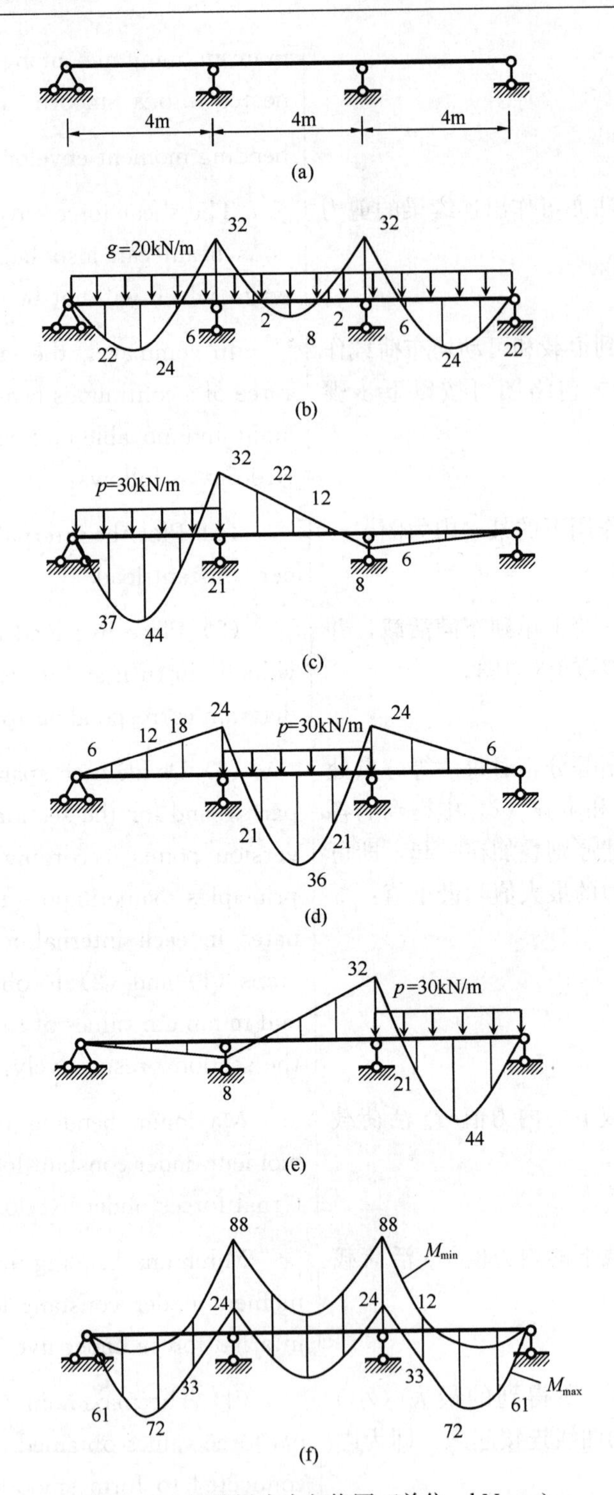

图 8.33　连续梁内力包络图（单位：kN・m）

Figure 8.33　The internal force envelope of continuous beam（Unit：kN・m）

思 考 题

8.1 何谓影响线？影响线横坐标和纵坐标的物理意义是什么？

8.2 影响线与内力图有何不同？

8.3 为什么静定结构内力、反力的影响线一定是由直线段组成的图形？

8.4 如何作简支梁在单位移动力偶作用下的反力和内力影响线？

8.5 何谓间接荷载？如何作间接荷载下的影响线？

8.6 机动法作影响线的原理是什么？

8.7 某截面剪力影响线是否一定有突变？为何剪力影响线左、右两支一定平行？

8.8 桁架影响线为何要分上弦承载和下弦承载？两种承载方式下影响线是否相同？

8.9 如何作刚架、拱的内力、反力影响线？

8.10 如何求静定结构位移影响线？

8.11 利用影响线求固定荷载作用下内

Questions

8.1 What is the influence line? What are the physical meanings of the horizontal coordinate and vertical ordinate of the influence line?

8.2 What is the difference between the influence line and the internal force diagram?

8.3 Why are the internal force influence line and reaction force influence line of statically determinate structure composed of straight line segments?

8.4 How to make the influence lines of reaction and internal force of simply supported beam under the action of a unit moving couple?

8.5 What is indirect load? How to make influence line under indirect load?

8.6 What is the principle of kinematic method to make influence line?

8.7 Is there a sudden change in the shear force influence line of a section? Why are the left and right branches of the shear force influence line parallel?

8.8 Why should the truss influence line be divided into upper chord bearing and lower chord bearing? Are the influence lines the same under the two bearing modes?

8.9 How to make the influence lines of internal force and reaction force of rigid frame and arch?

8.10 How to find the displacement influence line of statically determinate structure?

8.11 What is the premise of using in-

力的前提是什么？

fluence line to calculate internal force under fixed load?

8.12　若给定荷载中含有集中力偶，该如何利用影响线求指定内力？

8.12　If a given load contains a concentrated couple, how to use the influence line to calculate the specified internal force?

8.13　什么是移动荷载的临界位置？它与荷载的最不利位置有什么关系？

8.13　What is the critical position of a moving load? What is the relationship between it and the most unfavorable load position?

8.14　何谓内力包络图？它与内力图、影响线有何区别？绝对最大弯矩所对应的荷载对于其他截面是否一定是临界荷载？

8.14　What is the internal force envelope diagram? What is the difference among it, internal force diagram and influence line? Is the load corresponding to the absolute maximum bending moment the critical load for other sections?

8.15　试述用机动法作超静定结构内力、反力影响线的原理与步骤，它与静定结构的机动法作影响线有何异同？

8.15　Try to describe the principle and steps of using the kinematic method as the influence line of internal force and reaction force of statically indeterminate structures. What are the similarities and differences between it and the kinematic method to make the influence line of statically determinate structures?

习　题

Exercises

8.1　试用静力法作图8.34所示悬臂梁的反力F_{yA}、M_A和C截面内力M_C、F_{QC}的影响线。

8.1　Try to use static method to make the influence lines of reaction forces F_{yA}, M_A and internal forces M_C、F_{QC} of section C of cantilever beam in Figure 8.34.

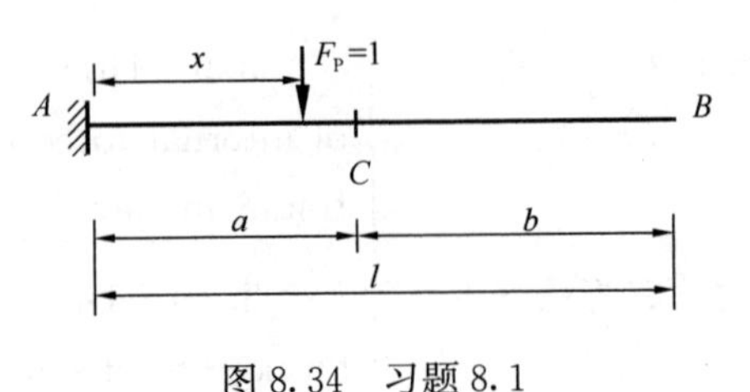

图8.34　习题8.1

Figure 8.34　Exercise 8.1

8.2　试作图 8.35 所示斜梁 F_{yA}和 C 截面内力 M_C、F_{QC}、F_{NC}的影响线。

8.2　Try to make influence lines of F_{yA} and internal forces M_C、F_{QC}、F_{NC} of section C of the inclined beam in Figure 8.35 .

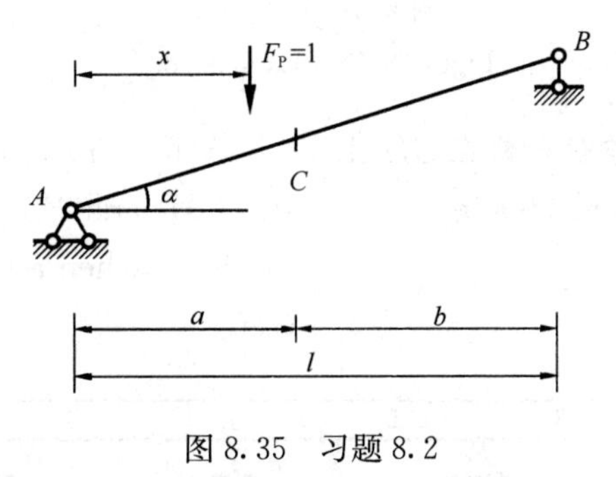

图 8.35　习题 8.2

Figure 8.35　Exercise 8.2

8.3　试用静力法作图 8.36 所示梁中截面 C 剪力 F_{QC}和弯矩 M_C，截面 B 左侧截面剪力 F_{QB}^L 、右侧截面 F_{QB}^R 及弯矩 M_B的影响线。

8.3　Try to use static method to make influence lines of shear force F_{QC} and bending moment M_C of section C, shear force F_{QB}^L to the left side of section B and F_{QB}^R to the right side of section B and bending moment M_B of section B in the beam in Figure 8.36.

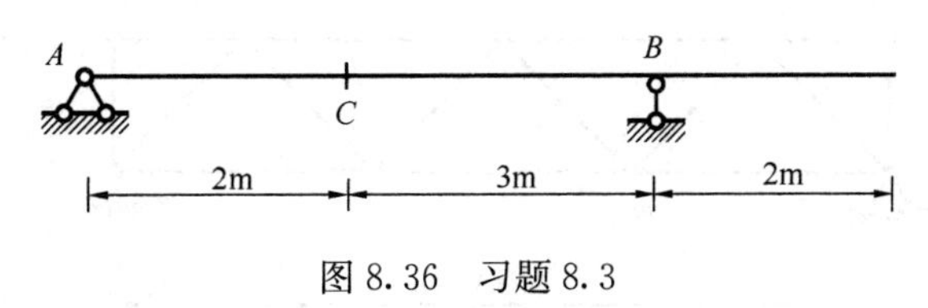

图 8.36　习题 8.3

Figure 8.36　Exercise 8.3

8.4　试用机动法作图 8.37 指定量值的 F_{RB}、F_{RD}、M_B、M_D、F_{QB}^R 、F_{QC}的影响线。

8.4　Try to use kinematic method to make the influence lines of F_{RB}、F_{RD}、M_B、M_D、F_{QB}^R 、F_{QC} in Figure 8.37.

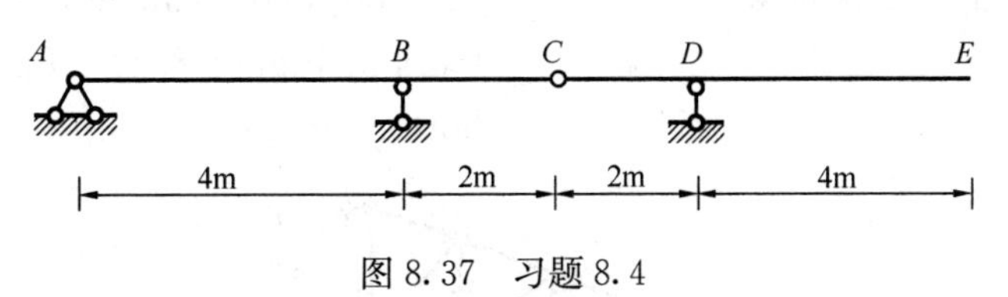

图 8.37　习题 8.4

Figure 8.37　Exercise 8.4

8.5　试作图 8.38 中指定量值 M_A、F_{RB}、F_{QE}、M_B、F_{QC}的影响线。

8.5　Try to make influence lines of M_A、F_{RB}、F_{QE}、M_B、F_{QC} in Figure 8.38.

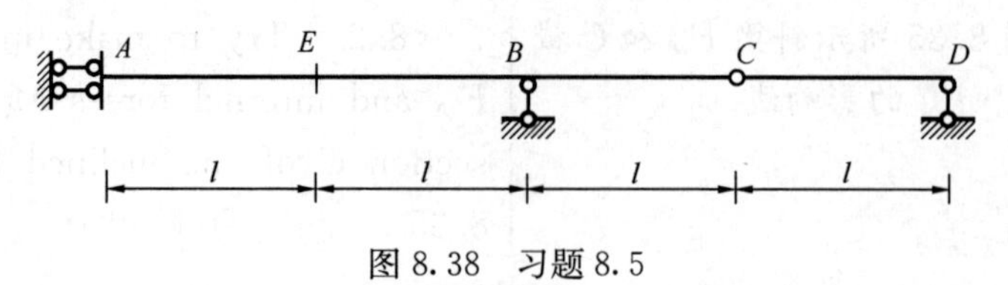

图 8.38　习题 8.5

Figure 8.38　Exercise 8.5

8.6　试作图 8.39 中单位荷载在 GH 上移动，主梁 F_{yA}、M_E、F_{QE}的影响线。

8.6　Try to make the influence lines of F_{yA}、M_E and F_{QE} of the main beam in Figure 8.39 when a unit moving load moves on GH.

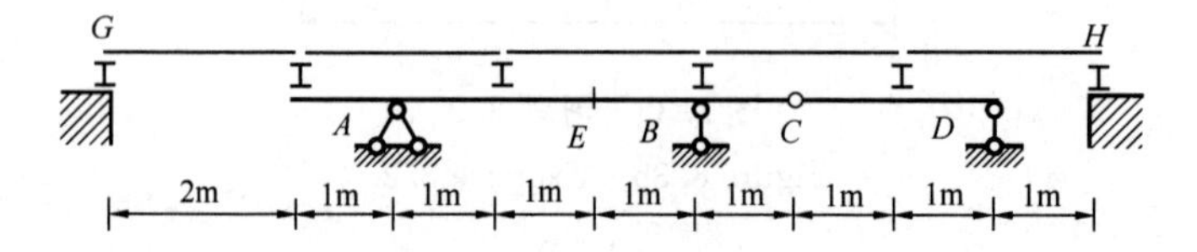

图 8.39　习题 8.6

Figure 8.39　Exercise 8.6

8.7　作图 8.40 所示桁架中指定杆件轴力的影响线，单位荷载分别在上弦或下弦移动。

8.7　Draw the influence line of the specified members' axial forces in the truss in Figure 8.40, and the unit load moves at the top chord or bottom chord respectively.

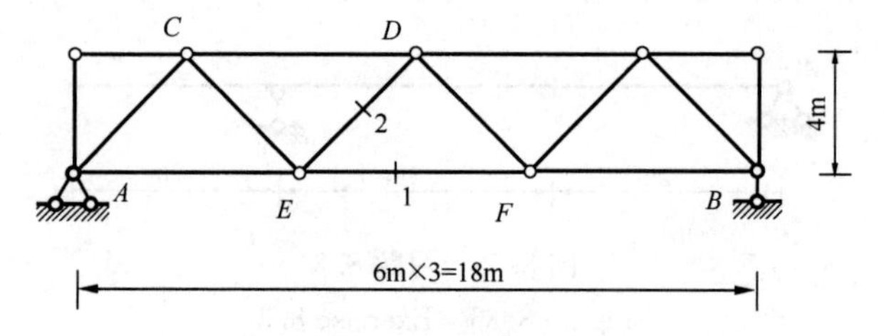

图 8.40　习题 8.7

Figure 8.40　Exercise 8.7

8.8　作图 8.41 所示桁架中指定杆件轴力的影响线，$F_P=1$ 在下弦移动。

8.8　Try to make the influence line of the specified members' axial force in the truss in Figure 8.41, $F_P=1$ moves at the bottom chord.

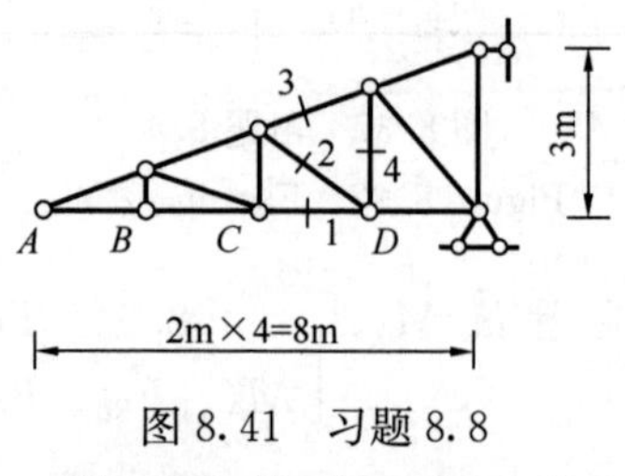

图 8.41　习题 8.8

Figure 8.41　Exercise 8.8

8.9　单位荷载 $F_P=1$ 在图 8.42 所示刚架的横梁 CD 上移动时，试求 K 截面的弯矩 M_K 和剪力 F_{QK} 的影响线。

8.9　When the unit load $F_P=1$ moves on the cross beam CD of the rigid frame shown in Figure 8.42, try to find the influence lines of the bending moment M_K and the shear force F_{QK} of section K.

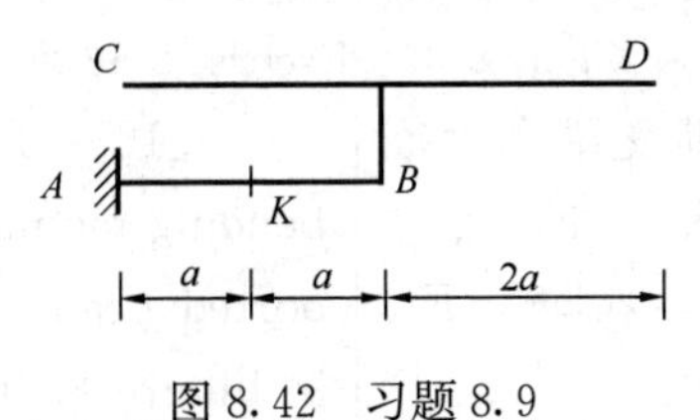

图 8.42　习题 8.9

Figure 8.42　Exercise 8.9

8.10　试作图 8.43 所示组合结构 F_{N1}、F_{N2}、M_D、F_{QD}^L、F_{QD}^R 的影响线，$F_P=1$ 在 AB 上移动。

8.10　Draw the influence lines of F_{N1}、F_{N2}、M_D、F_{QD}^L、F_{QD}^R of the composite structure in Figure 8.43, $F_P=1$ moves on AB.

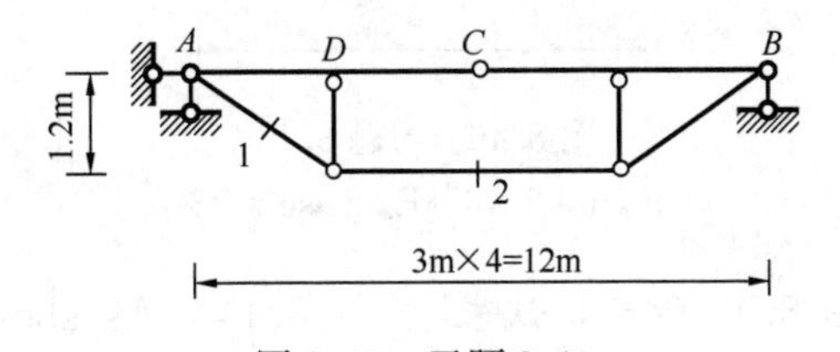

图 8.43　习题 8.10

Figure 8.43　Exercise 8.10

8.11　利用影响线计算图 8.44 所示梁的 M_C 和 F_{QC}。

8.11　Try to use influence line to calculate diagrams of M_C and F_{QC} of the beams in Figure 8.44.

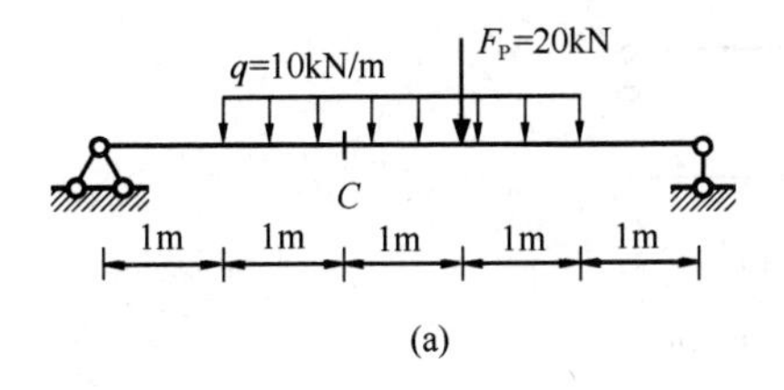

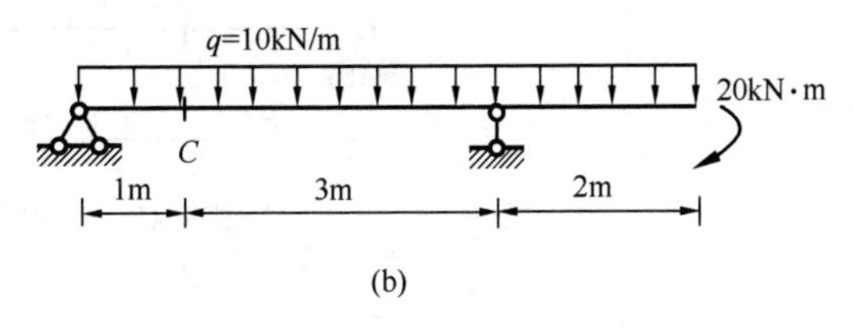

图 8.44　习题 8.11

Figure 8.44　Exercise 8.11

8.12　作图 8.45 所示梁 F_{yA}、M_A 的影响线，并利用影响线计算图中荷载作用下的 F_{yA}、M_A 的影响线。

8.12　Draw the influence lines of F_{yA}、M_A in Figure 8.45, and use the influence line to calculate the influence lines of F_{yA}、M_A under the illustrated loads.

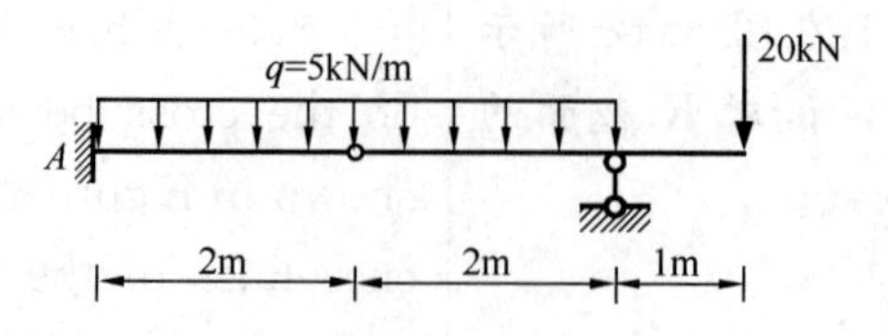

图 8.45　习题 8.12

Figure 8.45　Exercise 8.12

8.13　试求图 8.46 所示简支梁在所给移动荷载作用下截面 C 的最大弯矩，其中 $F_{P1}=40\text{kN}$，$F_{P2}=60\text{kN}$，$F_{P3}=20\text{kN}$，$F_{P4}=30\text{kN}$。

8.13　Try to calculate the maximum bending moment of section C of simply supported beam under the given moving loads in Figure 8.46，$F_{P1}=40\text{kN}$，$F_{P2}=60\text{kN}$，$F_{P3}=20\text{kN}$，$F_{P4}=30\text{kN}$.

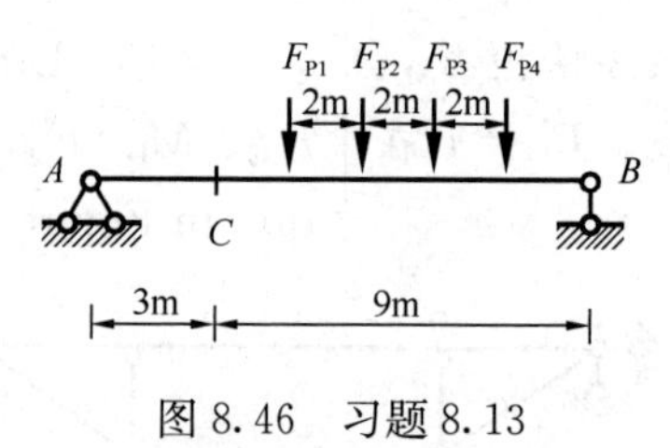

图 8.46　习题 8.13

Figure 8.46　Exercise 8.13

8.14　两台吊车如图 8.47 所示，试求吊车梁的 M_C、F_{QC} 的荷载最不利位置，并计算其最大值（和最小值）。

8.14　As shown in Figure 8.47，try to find the most unfavorable load position of M_C，F_{QC} of crane beam，and calculate the maximum value（and minimum value）.

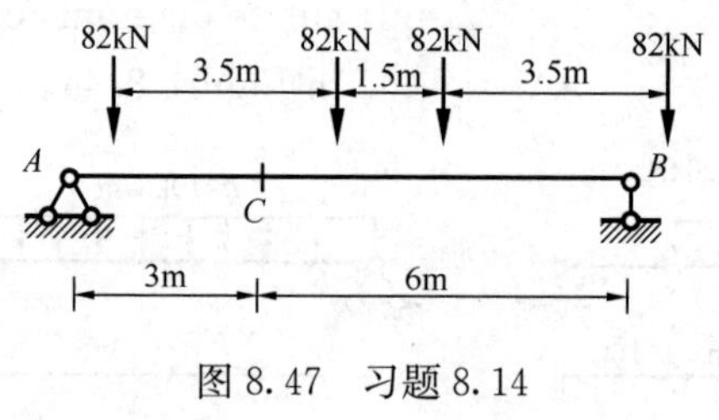

图 8.47　习题 8.14

Figure 8.47　Exercise 8.14

8.15　图 8.48 所示两台吊车在简支梁上移动，试求截面 C 弯矩 M_C 的最大值及剪力 F_{QC} 的最大值、最小值。$F_{P1}=F_{P2}=435\text{kN}$，$F_{P3}=F_{P4}=295\text{kN}$。

8.15　As shown in Figure 8.48，two cranes move on a simply supported beam，try to find the maximum value of bending moment M_C of section C and the maximum and minimum values of shear force F_{QC}. $F_{P1}=F_{P2}=435\text{kN}$，$F_{P3}=F_{P4}=295\text{kN}$.

8.16　如图 8.49 所示，两台吊车在梁上移动，试求支座 B 反力的最大值。$F_{P1}=$

8.16　As shown in Figure 8.49，two cranes move on the beam，try to find the

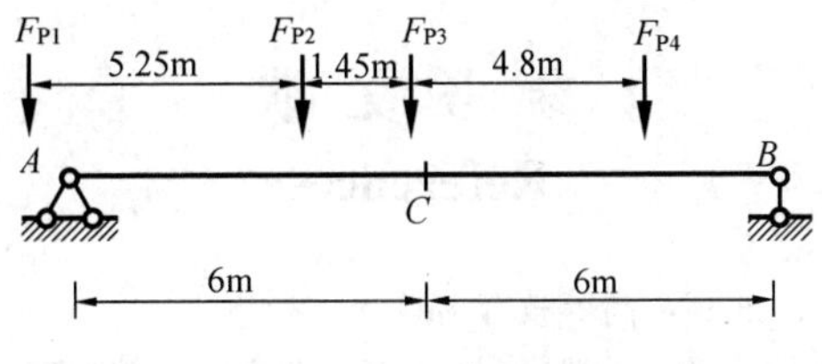

图 8.48 习题 8.15

Figure 8.48 Exercise 8.15

$F_{P2}=F_{P3}=F_{P4}=105kN$。

maximum value of the reaction force of support B. $F_{P1}=F_{P2}=F_{P3}=F_{P4}=105kN$.

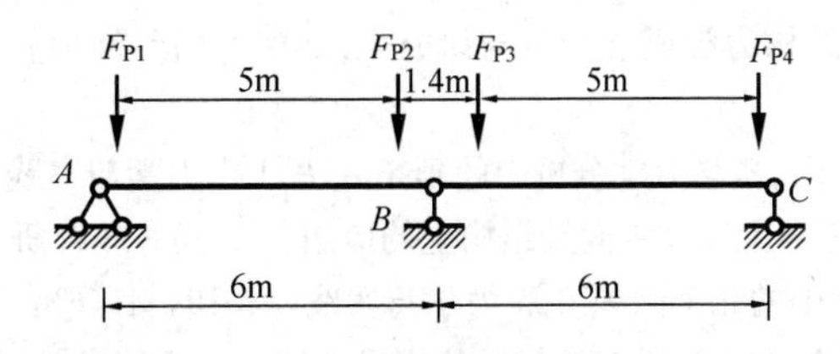

图 8.49 习题 8.16

Figure 8.49 Exercise 8.16

8.17 试求简支梁在图 8.50 所示荷载作用下的绝对最大弯矩。

8.17 Try to calculate the absolute maximum bending moment of simply supported beams under the load shown in Figure 8.50.

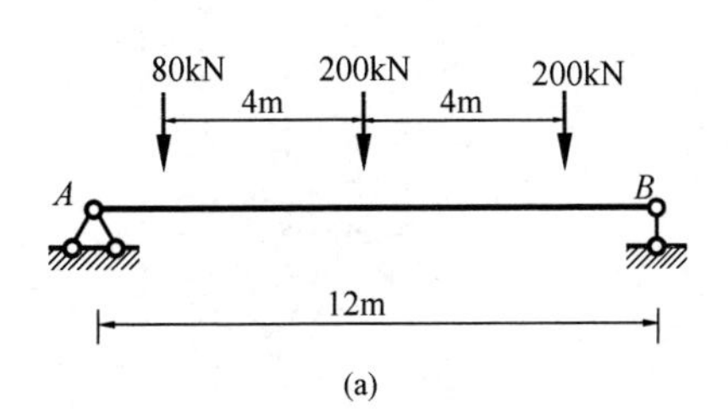

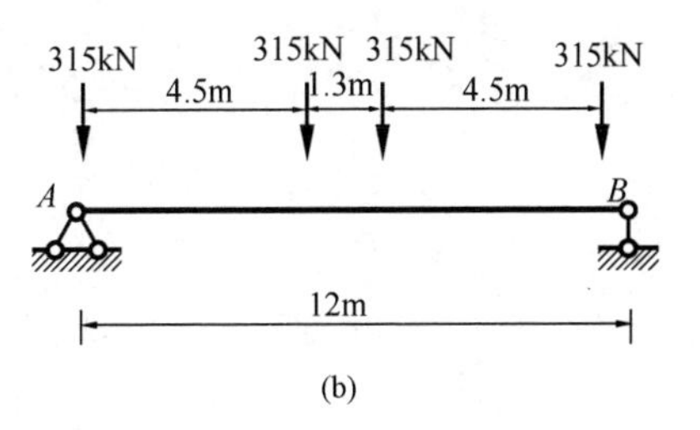

图 8.50 习题 8.17

Figure 8.50 Exercise 8.17

8.18 试绘制图 8.51 所示连续梁 F_{RB}、M_A、M_C、M_K、F_{QK}、F_{QB}^L 影响线的形状。

8.18 Draw the influence line shapes of F_{RB}, M_A, M_C, M_K, F_{QK}, F_{QB}^L of continuous beam shown in Figure 8.51.

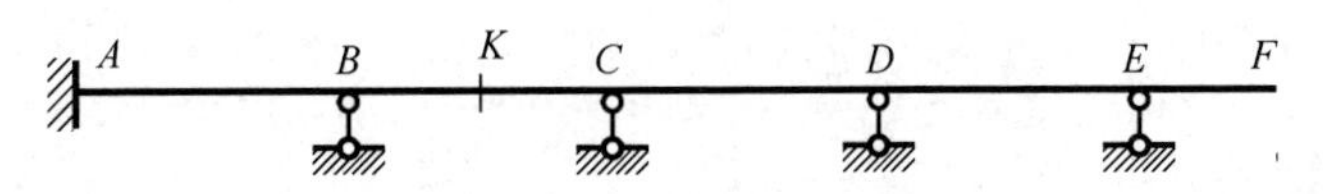

图 8.51 习题 8.18

Figure 8.51 Exercise 8.18

参考文献

References

[1] 单建．趣味结构力学[M]．北京：高等教育出版社，2008.

[2] 杨弟康，李家宝，等．结构力学：上册[M]．6版．北京：高等教育出版社，2016.

[3] 李廉锟，等．结构力学：上册[M]．6版．北京：高等教育出版社，2017.

[4] 包世华，等．结构力学教程[M]．武汉：武汉理工大学出版社，2017.

[5] 缪加玉．结构力学的若干问题[M]．成都：成都科技大学出版社，1993.

[6] 樊有景．结构力学[M]．郑州：郑州大学出版社，2021.

[7] 包世华．《结构力学》学习指导及题解大全[M]．武汉：武汉理工大学出版社，2003.

[8] 罗永坤，蔺安林，等．结构力学概念分析与研究生入学考试指导[M]．成都：西南交通大学出版社，2008

[9] 朱慈勉，郭志刚，张伟平．概念力学分析中的延拓方法[J]．力学与实践，2011，33(4)：61-63.

[10] 蒋庆．简易渐近法-力矩分配法在中间铰结构上的应用[J]．高等教育研究，1981，(4)：86-92.

[11] 田振国．结构力学教学中的几个问题[J]．力学与实践，2019，41(6)：728-732.

[12] 黄亮，陈国栋，王博．结构力学中三铰斜拱等代梁的探索与应用[J]．力学与实践，2015，37(4)：525-527.

[13] 宁宝宽，鲍文博，鲁丽华，等．M_p图与$\overline{M}_1$图选用不同基本结构的力法分析[J]．力学与实践，2009，31(6)：69-71.

[14] 张波．力法基本结构选取探讨[J]．长江大学学报(自科版)理工卷，2010，7(3)：644-645，648.

[15] 杨立军，邓志恒，陆守明，等．基于不同基本结构求解超静定问题的力法方程[J]．河南理工大学学报(自然科学版)，2012，31(5)：594-597.